Energy Efficient Building Construction in Florida

9th Edition
(with references to
the Florida Building Code, 6th Edition [2017])

Craig Miller

James Sullivan

Sherry Ahrentzen

Published by the University of Florida

Shimberg Center for Housing Studies,
M.E. Rinker, Sr. School of Building Construction,
College of Design, Construction and Planning

and

Program for Resource Efficient Communities
Institute of Food and Agricultural Sciences (IFAS)

Ninth Edition, printed January 2018

ISBN: 978-0-9852487-4-1

Brands mentioned are for educational purposes only
and are not directly or indirectly endorsed or recommended by
program developers or the University of Florida.

The Program for Resource Efficient Communities (PREC) is a unit within University of Florida IFAS Extension.

PREC integrates and applies the University of Florida's educational and analytical assets to promote the adoption of best design, construction, and management practices that measurably reduce energy and water consumption and environmental degradation in master planned residential community developments.

Cover photograph: Doug Finger / Gainesville Magazine

To order additional copies, visit www.buildgreen.ufl.edu or call (352) 392-5864.

www.shimberg.ufl.edu

www.buildgreen.ufl.edu

Acknowledgements

The original version of this publication was adapted from *A Builder's Guide to Energy Efficient Homes in Georgia* by the University of Florida's Program for Resource Efficient Communities[1]. This project was tremendously facilitated by the generosity of the Georgia Environmental Facilities Authority—Division of Energy Resources and Southface Energy Institute. We would like to thank Melinda Koken, a licensed Florida Building Contractor and former member of the Construction Industry Licensing Board, for encouraging us to undertake this project and for her insightful suggestions during its initial development.

On-going development and updating of *Energy Efficient Building Construction in Florida* is the result of a collaborative effort between the University of Florida Program for Resource Efficient Communities, Shimberg Center for Housing Studies and M.E. Rinker, Sr. School of Building Construction.

Over the years, many people have contributed to the updates of *Energy Efficient Building Construction in Florida.* Unless otherwise noted, the individuals listed below are with the University of Florida:

B. Haldeman – Academic Assistant, Program for Resource Efficient Communities

L. Jarrett, M.S. – Water Resources Engineer, Program for Resource Efficient Communities

H.S. Knowles III, PhD. – Research Associate, Program for Resource Efficient Communities

B. Larson, PhD – Extension Assistant Scientist, Soil and Water Sciences Department.

X. Lo, MS – Graduate Research Assistant, Shimberg Center for Housing Studies

J. Michael – Director of Housing Development, Neighborhood Housing & Development Corporation

W. Porter, PhD, PE – Senior Lecturer, Agricultural and Biological Engineering Department

K.C. Ruppert, EdD – Emeritus Extension Scientist, Program for Resource Efficient Communities

A. Stewart – Residential and Commercial Energy Rater, President at AZS Consulting, Inc.

C. Swanson – Program Assistant (Retired), Program for Resource Efficient Communities

N. Taylor, PhD – Research Associate, Program for Resource Efficient Communities

L. Wetherington, PhD – Retired Professor, M.E. Rinker, Sr. School of Building Construction

To offer comments or suggestions for improvement, please contact one of the authors:

Craig R. Miller, M.S.
Program for Resource Efficient Communities
University of Florida
PO Box 110940
Gainesville, FL 32611-0940
Phone: 352-392-5684
Fax: 352-392-9033
Email: craigmil@ufl.edu

James Sullivan, PhD
M.E. Rinker, Sr. School of Building Construction
University of Florida
PO Box 115703
Gainesville, FL 32611-5703
Phone: 352-392-5198
Email: sullj@ufl.edu

Sherry Ahrentzen, PhD
Shimberg Professor of Housing Studies
University of Florida
P.O. Box 115703
Gainesville, FL 32611-5703
Phone: 352-273-1229
Email: ahrentzen@ufl.edu

[1] The Program for Resource Efficient Communities (www.buildgreen.ufl.edu) is directed by Dr. Pierce Jones and operates through the Extension Service within the Institute of Food and Agricultural Sciences at the University of Florida.

Table of Contents

INTRODUCTION

When used together, the phrases *energy efficiency* and *building construction* are often assumed to mean greater costs with no incentives. But actually, energy efficient building construction practices can mean greater profits. *Energy Efficient Building Construction in Florida* is an adaptation for Florida conditions and code requirements of *A Builder's Guide to Energy Efficient Homes in Georgia.*

This is the ninth edition of this handbook (2018) printed since the Florida Legislature amended Chapter 553, Florida Statutes, Building Construction Standards, to create a single state building code. It has been updated to reflect the current Florida Building Code referred to throughout this publication as either the Florida Building Code or the FBC.

The following (listed alphabetically) are effective on December 31, 2017:

- *Florida Building Code Sixth Edition (2017), Accessibility*
- *Florida Building Code Sixth Edition (2017), Building*
- *Florida Building Code Sixth Edition (2017), Energy Conservation*
- *Florida Building Code Sixth Edition (2017), Existing Building*
- *Florida Building Code Sixth Edition (2017), Fuel Gas*
- *Florida Building Code Sixth Edition (2017), Mechanical*
- *Florida Building Code Sixth Edition (2017), Plumbing*
- *Florida Building Code Sixth Edition (2017), Residential*
- *Florida Building Code Sixth Edition (2017), Test Protocols for High Velocity Hurricane Zones*

These codes are primarily based on the 2015 International Building Codes. The Florida Building Code can be viewed online, or copies ordered, through http://www.floridabuilding.org

These codes are primarily based on the 2015 International Building Codes. The Florida Building Code can be viewed online, or copies ordered, through http://www.floridabuilding.org

Note: References to specific sections or volumes of the Florida Building Code in this publication appear in the following format in boldface: FBC, [Volume], [Section number and name]. For instance, a reference to Section R402.4, Air leakage (Mandatory)* in the volume *Florida Building Code Sixth Edition (2017)*, Energy Conservation, will appear as **FBC, Energy Conservation, Section R402.4, Air leakage (Mandatory).**

* *The "R" prefix in R402.4 refers to the Residential Provisions of the Energy Conservation code, as opposed to the "C" prefix, which refers to the Commerical Provisions.*

Greater profits can be realized around four key concepts: increased sales, improved energy efficiency, reduced system costs, and improved products.

- **Increased Sales** – Increase sales volume by offering more value for less total monthly cost. Typically energy efficient construction costs more to build, but will be less expensive for owners to operate, where monthly energy savings exceed any small increase in the monthly mortgage. Builders can profit by selling extra energy features, attracting more buyers looking for energy efficient buildings, and reducing expensive call-backs where buildings are more comfortable and well-built.
- **Improved Energy Efficiency** – Increase spending on critical features that provide a wide range of owner benefits in addition to impressively lower utility bills. In particular, builders can look for significant energy savings with high-efficiency building envelopes; heating, cooling and ventilation (HVAC) systems; lighting; and appliances.
- **Reduced System Costs** – Realize significant construction cost savings with an integrated approach to mechanical system design. Opportunities to reduce initial construction costs are based on more efficient building envelopes, which require smaller HVAC equipment and distribution systems.
- **Improved Products** – Use the benefits offered by energy efficient features to differentiate your buildings. In addition to significantly lower utility bills, you can look to offer improved comfort, quieter interiors, better indoor air quality, high quality construction, positioning for high resale value, and an environmentally friendly building.

With mortgage considerations and allowances available to purchasers of energy-efficient buildings, *Energy Efficient Building Construction in Florida* makes both cents and sense.

1
Step-by-Step Energy Efficient Construction

This quick reference guide shows the key elements of energy efficient construction. These features save money, improve indoor air quality, enhance comfort, prevent moisture problems, and increase the long term durability of the building.

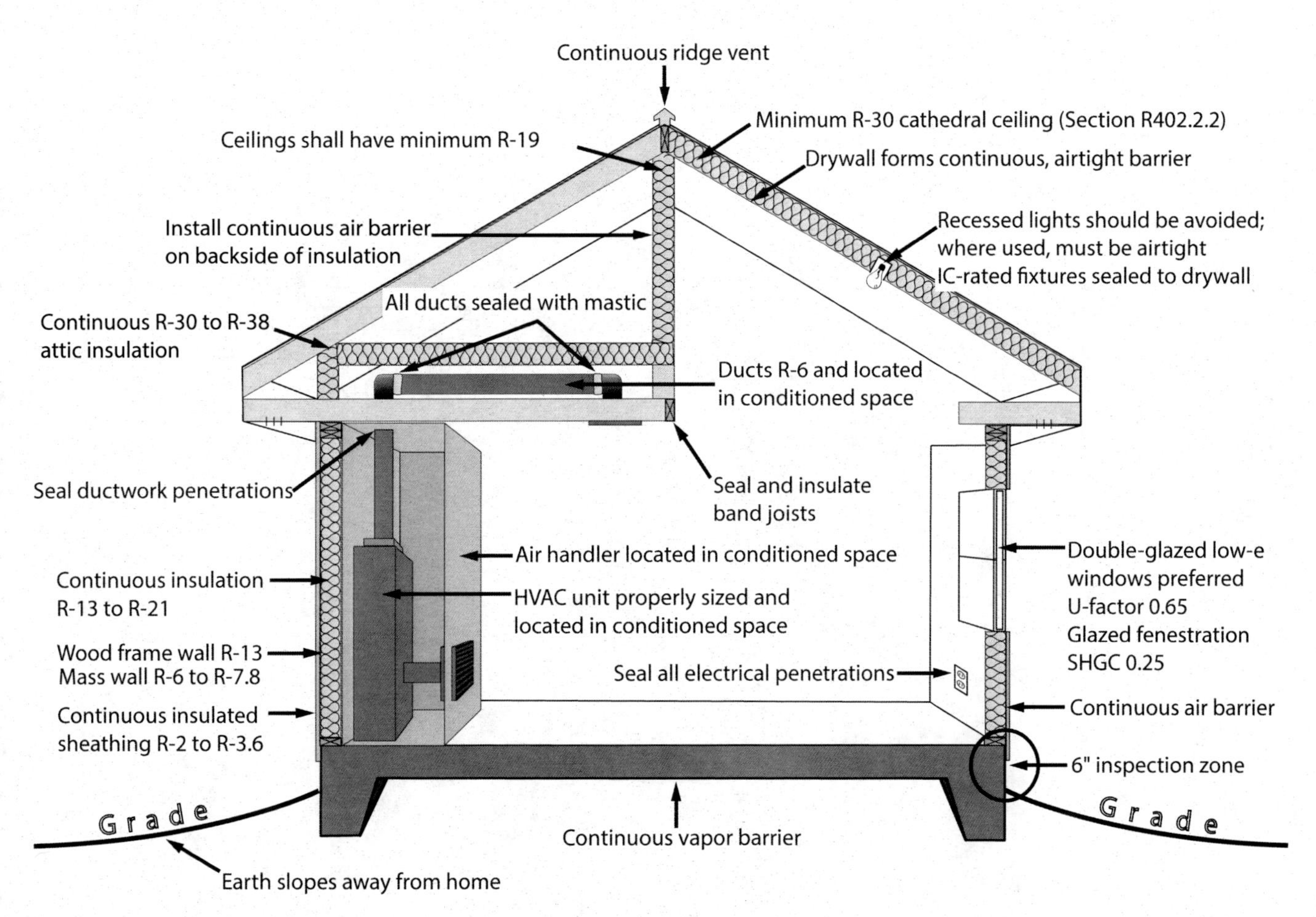

Figure 1-1 Building section

Energy Efficient Buildings: The Key Features

1. *Carefully consider the placement of the building on the property to provide maximum shade in the summer and sunlight in the winter (Figure 1-2).*

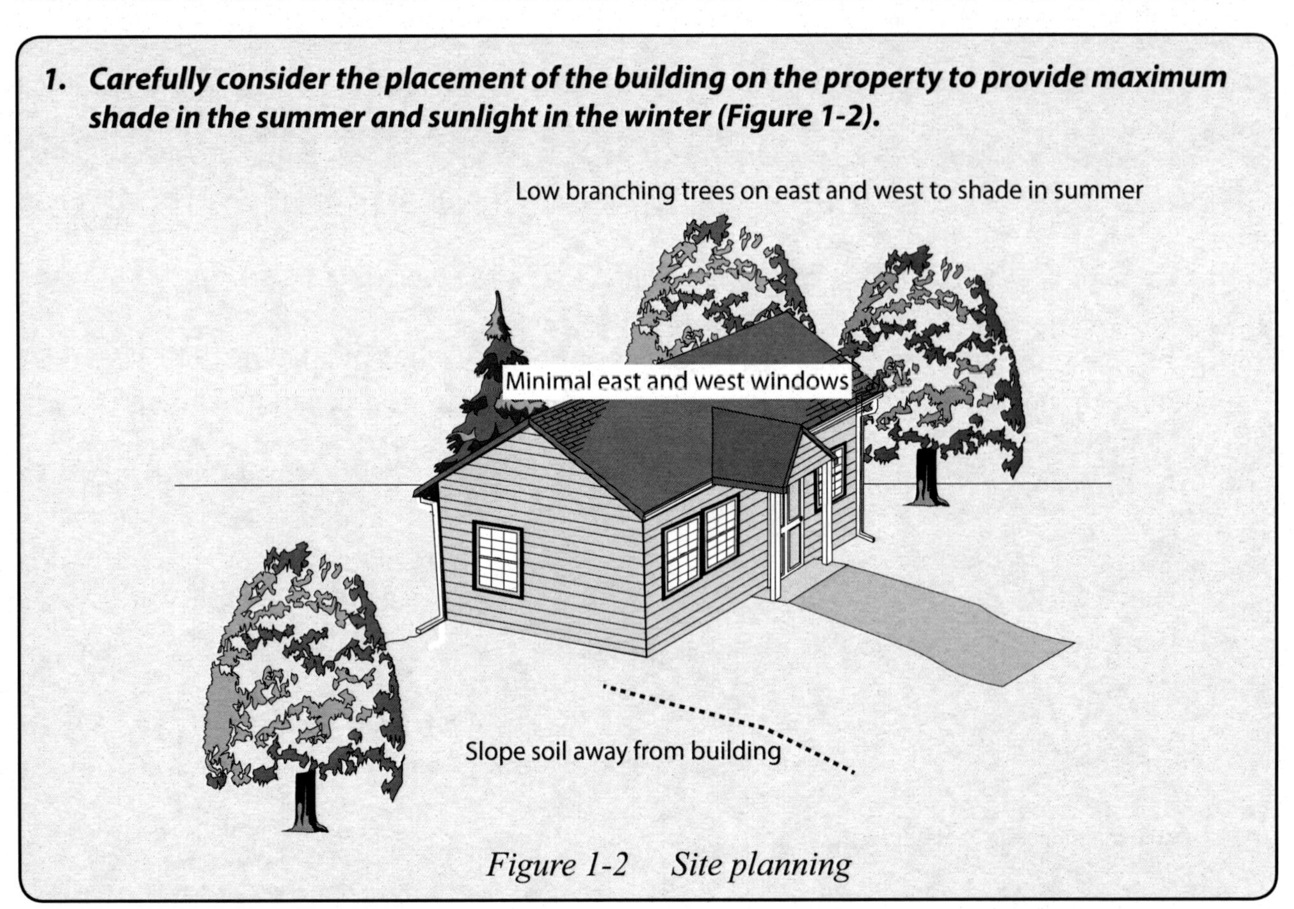

Figure 1-2 Site planning

2. *Air barrier system*

For slab-on-grade construction, design footing and foundation wall or monolithic slab and footing according to local code requirements. If the building has a crawl space, place a vapor barrier under the footing and over the ground under the floor (Figure 1-3).

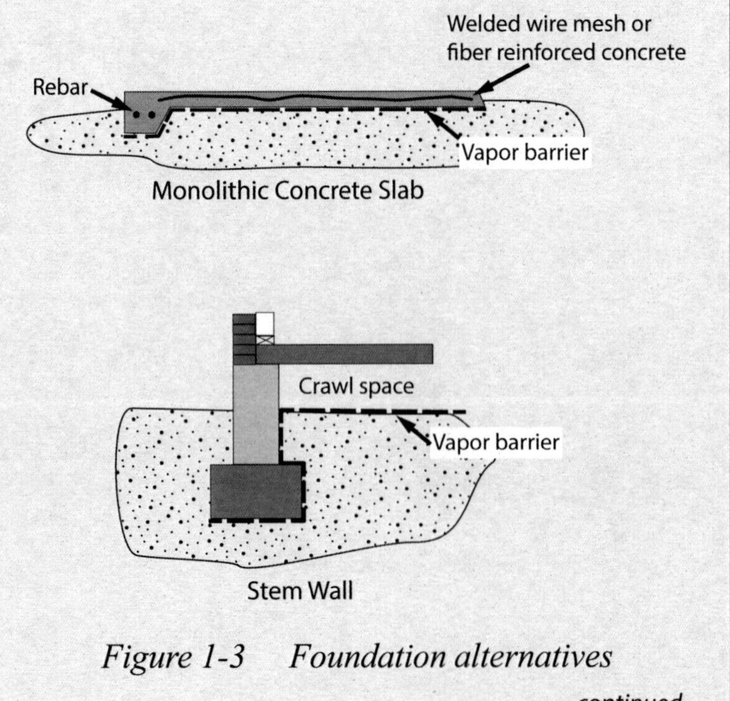

Figure 1-3 Foundation alternatives

continued...

...continued

Eliminate leakage between conditioned and unconditioned space, in particular between living areas, attics, crawl spaces, garages, and in exterior walls.

Figure 1-4 provides details on constructing an exterior frame wall system that will aid in an effective air barrier system, allow for a more complete coverage of the insulation, and reduce framing materials.

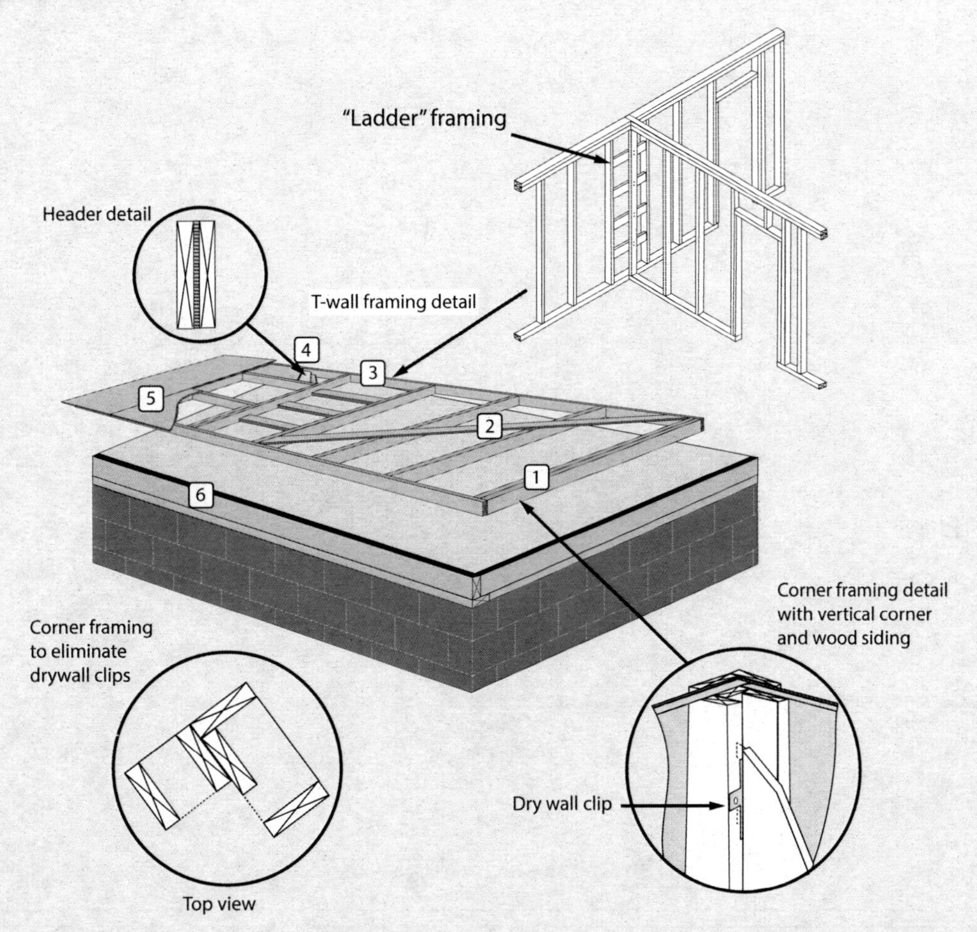

Figure 1-4 Wall framing

1. In corner, use single stud corner, or double stud corner when using wood siding with vertical cornerboard trim. Consider corner framing detail that eliminates drywall clips.
2. Provide adequate bracing near corners and at shear panels.
3. At partition wall (T-wall) intersection, eliminate additional studs for nailing drywall; use "ladder" instead. A ladder consists of 2 × 4 blocking turned on edge and nailed flush with the inside face of the studs 16 inches on center. This will provide nailing for the first stud of the partition while allowing insulation in the exterior wall (see inset).
4. Add 1/2" foam to structural headers.
5. Cover entire wall with 1/2" foam sheathing, including band joists and second top plates.
6. Before lifting wall in place, attach sill sealant material to subfloor.

3. *Continuous insulation system*

Install insulation as continuously as possible between conditioned and unconditioned spaces, including exterior walls; floor systems over unconditioned or exterior spaces; ceilings below unconditioned or exterior spaces (Figure 1-5); and wall areas adjacent to attic space—such as knee walls and attic stairways.

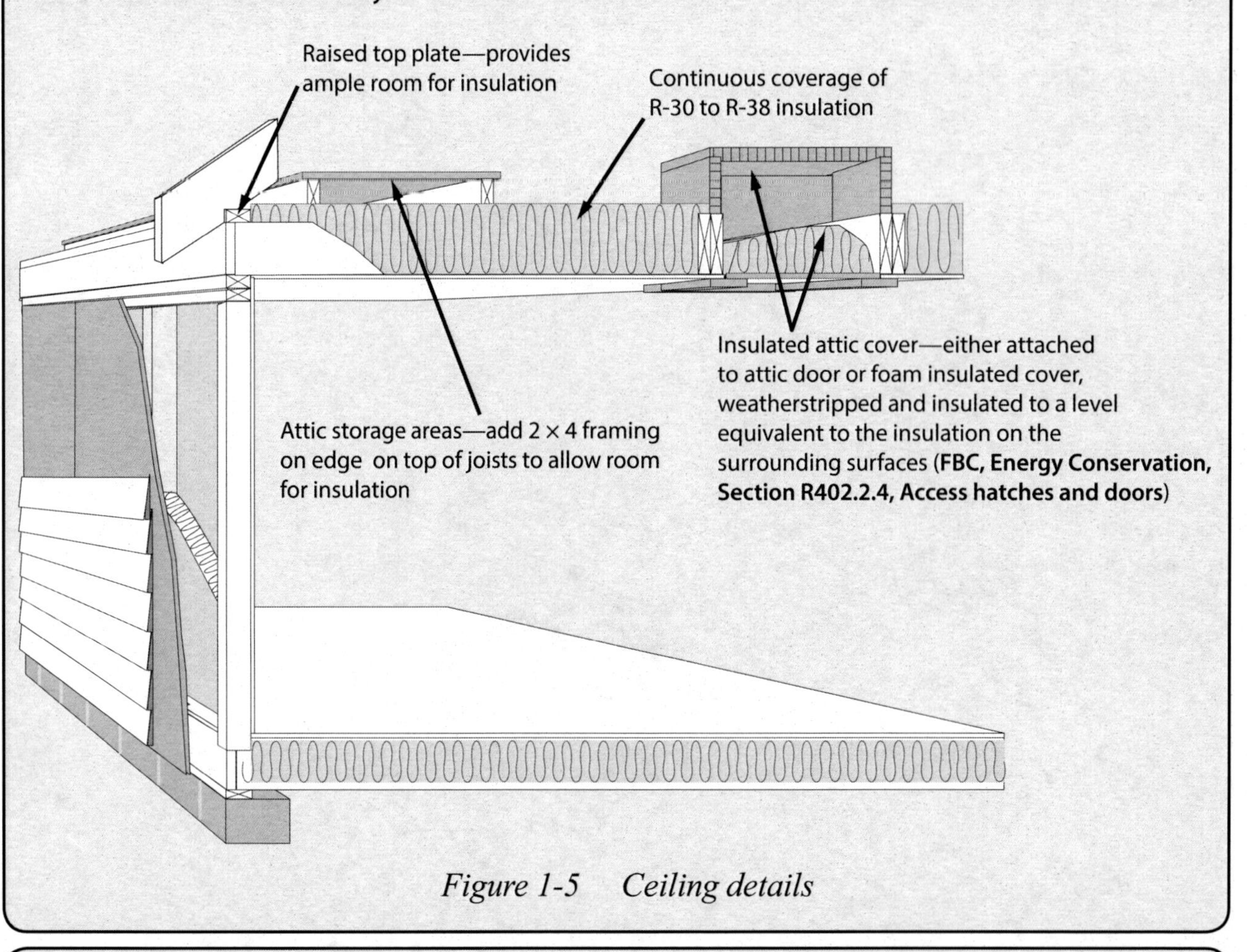

Figure 1-5 Ceiling details

4. *Select and install energy efficient windows*

Design buildings with minimal glass on the southeast and southwest sides.

- Use double-glazed windows with SHGC values of 0.25 or less and U-factors under 0.65 in FBC Climate Zone 1 (south Florida counties). In FBC Climate Zone 2 (the rest of Florida), U-factors should be under 0.40. (See **FBC, Energy Conservation, Table R402.1.2 Insulation and Fenestration Requirements by Component**.)
  - ◊ Note that ENERGY STAR® criteria as of 2015 for the Southern climate zone are SHGC of 0.25 or less, and U-factor of 0.40 or less.
- Consider low-emissivity coatings and other high performance features.
- Shade windows in summertime.

5. *Design cooling and heating system for efficiency (Figure 1-6)*

- Consider multi-split systems
- Select high efficiency equipment designed for local climatic conditions
- Size and install equipment properly
- Avoid negative pressures to prevent backdrafting of combustion appliances
- Install fresh air ventilation systems to bring in outside air when needed
- Accurately account for all latent loads.

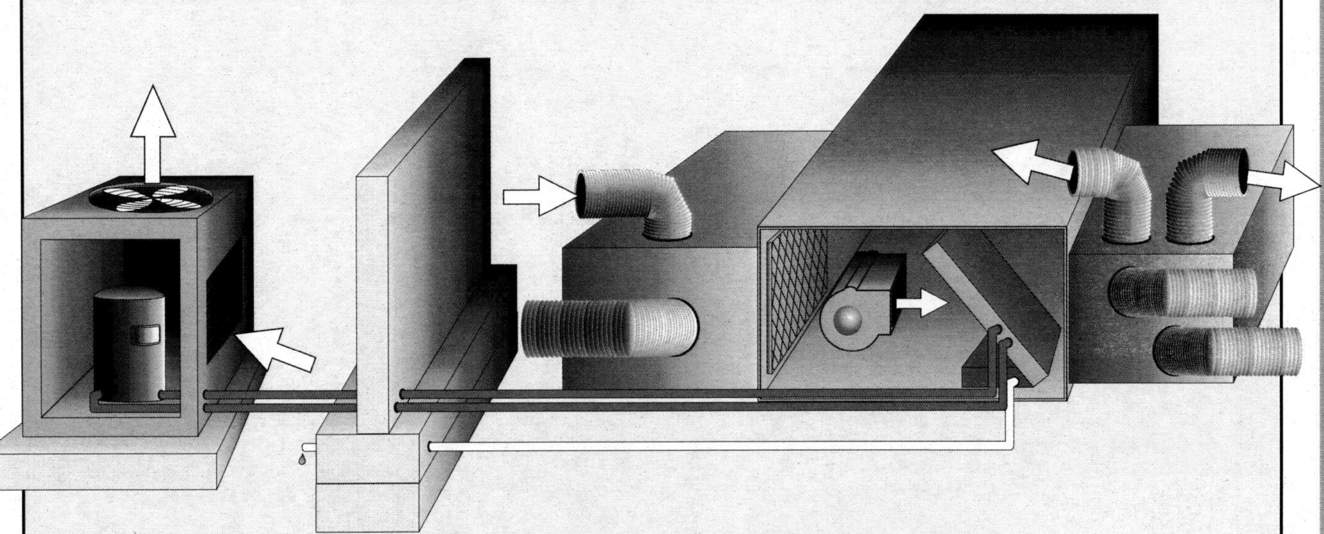

Figure 1-6 HVAC systems

1. Size for heating and cooling load using Manual J techniques (see Chapter 8, "Heating, Ventilation, Air Conditioning").
2. Size latent (dehumidification) load for cooling system.
3. Compare cost and projected energy savings of at least 3 HVAC contractor bids and 3 equipment options:
   - ◊ Minimum efficiency: Unitary cooling equipment and unitary heat pumps shall have Seasonal Energy Efficiency Ratios (SEER) and Heating Seasonal Performance Factors (HSPF) no less than SEER 14 and HSPF 8.2.
   - ◊ High efficiency: SEER 16 cooling, AFUE .95 furnace, HSPF 9.0 heat pump (make sure the Sensible Heating Ratio (SHR) can address the latent loads)
4. Consider automatic zoned system instead of multiple separate systems.

Refer to **FBC, Energy Conservation, Tables C403.2.3 (1), C403.2.3 (2),** and **C403.2.3 (3)** for more details.

continued...

continued...

5. When selecting a contractor, do not just go by price, but consider the following:
   ◊ Reputation for quality
   ◊ Type of duct system
   ◊ Number of returns
   ◊ Sound-muffling components
   ◊ Willingness to ensure and test for airtight ductwork

6. *Seal ductwork*

- Size ductwork to meet the heating and cooling load of each room
- Place ductwork to supply proper airflow; measure airflow to guarantee comfort
- Seal all potential duct leaks—except those in removable components—with mastic or mastic plus fiber mesh, use guidelines in Chapter 9, "Duct Design and Sealing." (Figure 1-7)

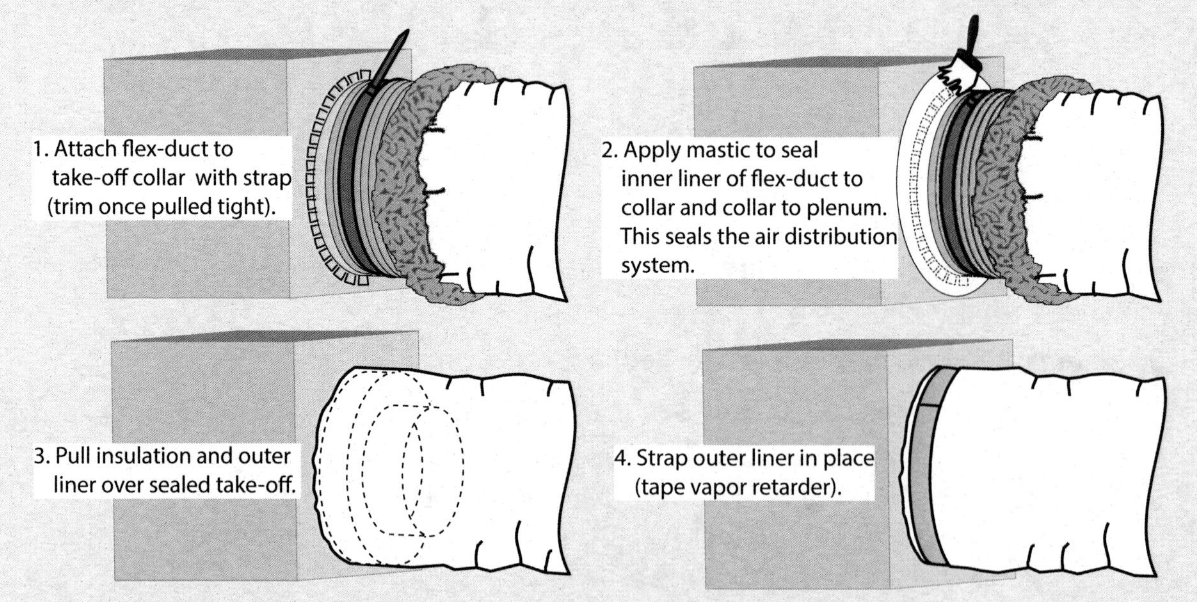

Figure 1-7 Step-by-step duct sealing

◊ Design using Manual D Residential Duct System Concepts.
◊ Install returns or jumper ducts in each room with a closeable door.

continued...

...continued

◊ Test ducts for air tightness (Figure 1-8)

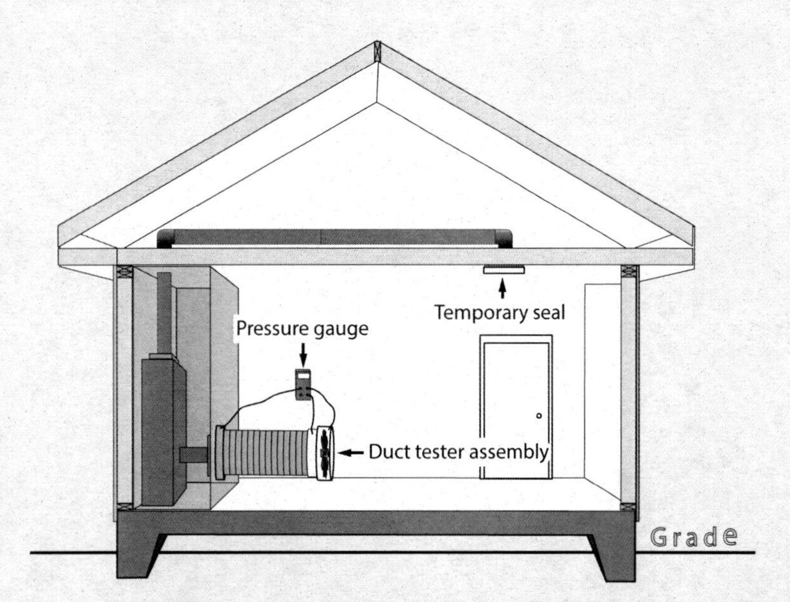

Figure 1-8 Duct testing system

◊ Test home for pressure imbalance problems (Figure 1-9).

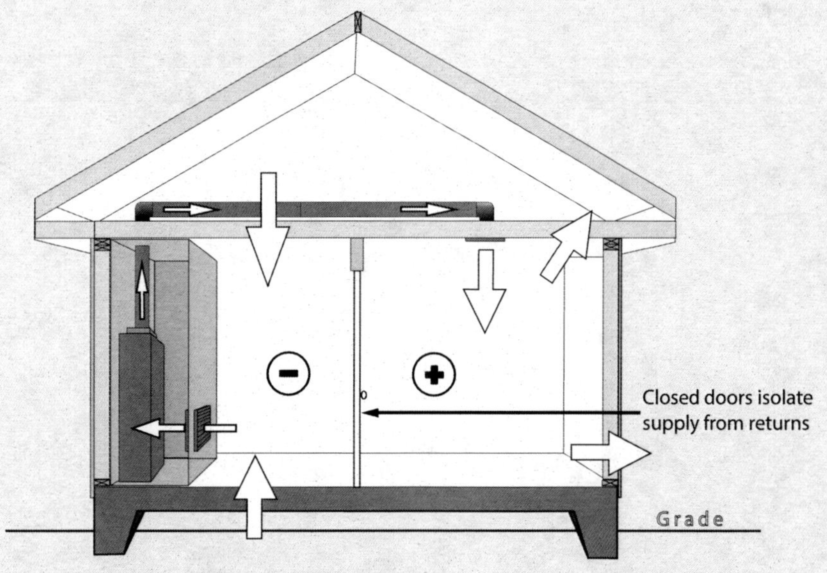

Figure 1-9 Prevent pressure imbalance problems

continued...

continued...

◊ Seal leaks around removable components with UL 181 approved pressure-sensitive tape(Figure 1-10)

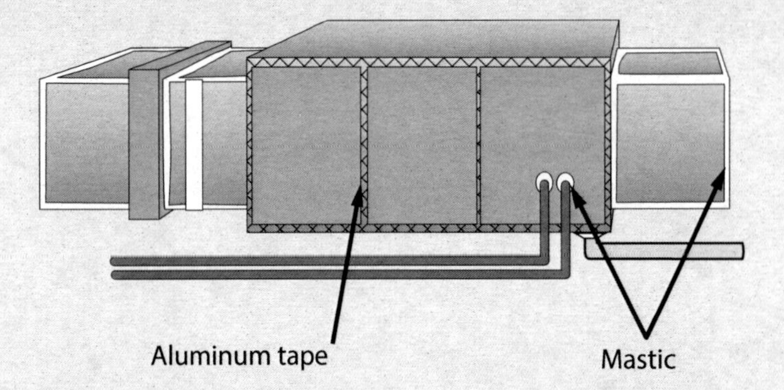

Many air handling cabinets come from the factory with leaks, which should be sealed with duct-sealing mastic.
Removable panels should be sealed with rated aluminum tape.

Figure 1-10 Properly seal air handler

- Provide for return air duct system where appropriate
- Minimize or remove duct system from unconditioned space
- All ducts not inside the building thermal envelope insulated to minimum R-6 (R-8 is a more efficient selection)

7. Seal all building penetrations

- Plumbing penetrations (Figure 1-11)

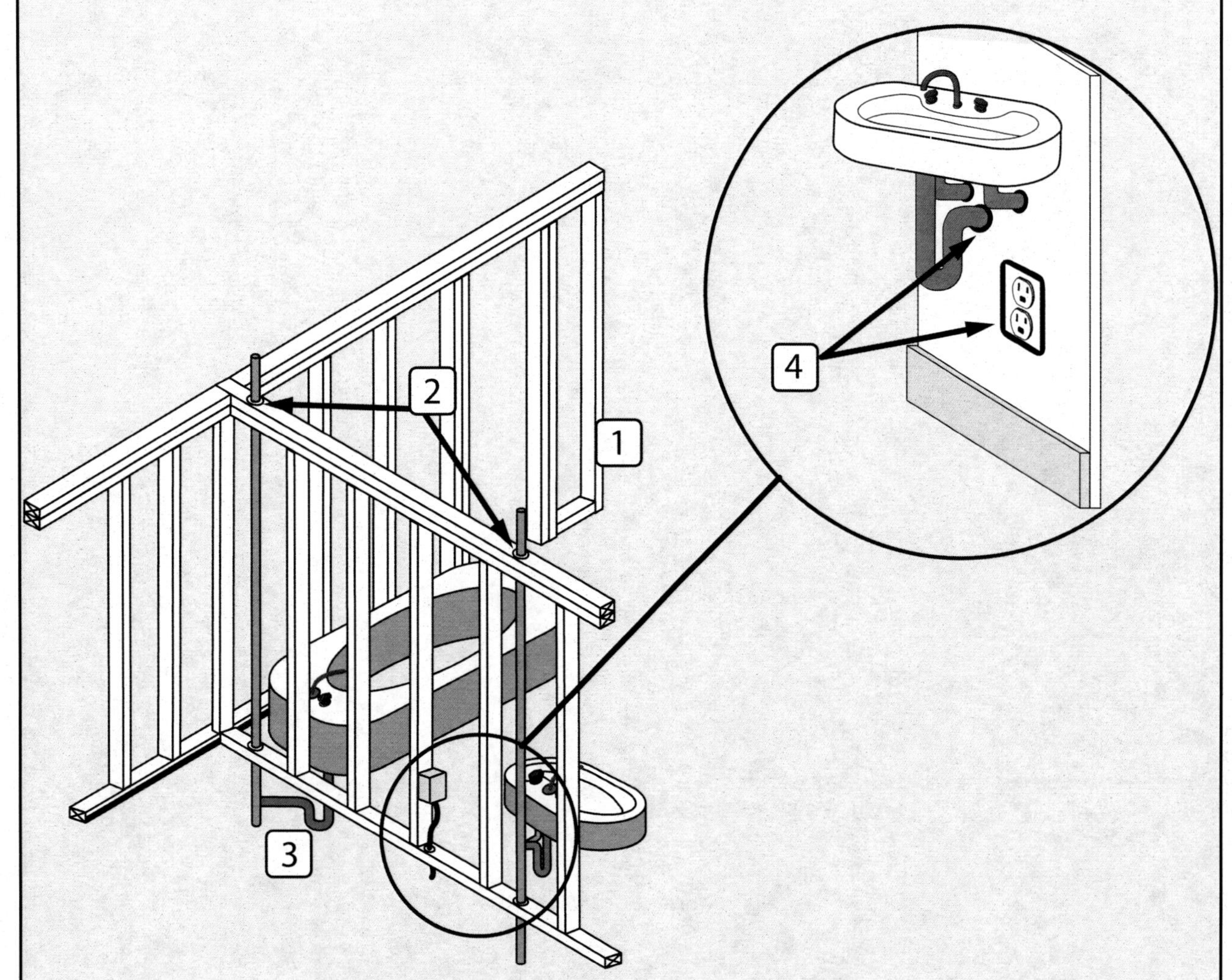

Figure 1-11 Seal plumbing penetrations

1. Locate plumbing on interior walls.
2. Use firestop rated caulk to seal holes into attic.
3. Seal under tubs and showers.
4. Caulk between drywall and piping penetrations.

See **FBC, Energy Conservation, Section R402.4 Air leakage (Mandatory)** and **Section R402.4.1 Building thermal envelope** for details.

continued...

continued...

- Electrical penetrations (Figure 1-12)

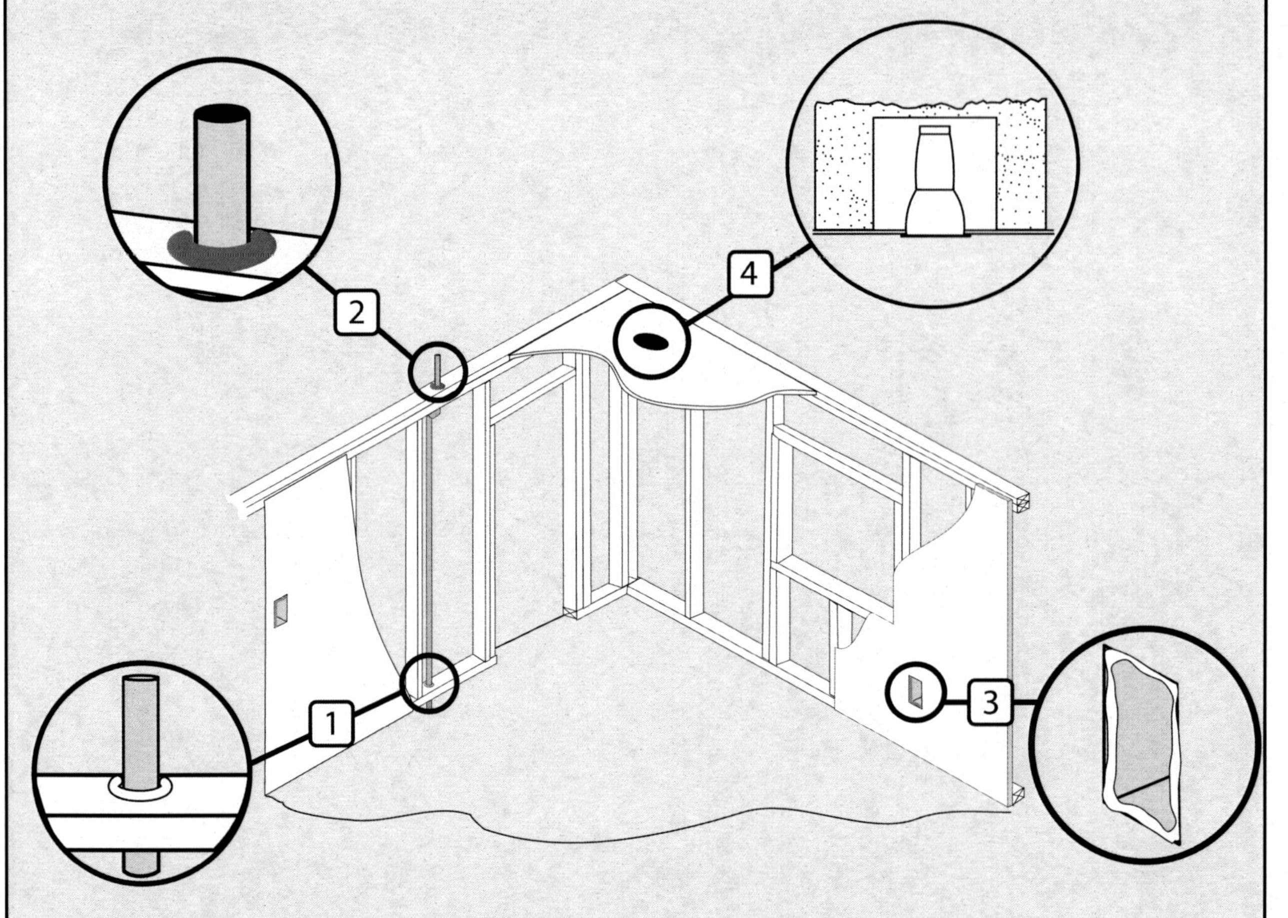

Figure 1-12 Seal electrical penetrations

1. Seal holes through bottom plate.
2. Use firestop rated caulk to seal holes into attic.
3. Caulk between drywall and all electrical boxes, including receptacles, switches and lights.
4. Minimize use of recessed lights; where used, mandatory IC (insulation ceiling) lamps that also have airtight ratings. Always follow manufacturer's specifications for installation. Seal with gasket or calk between housing and the interior wall or ceiling covering (**FBC, Energy Conservation, Section R402.4.5 Recessed lighting**).

8. *Minimize hot water costs*

See Chapter 10, "Domestic Water Heating," for more information.

9. *Choose energy efficient appliances and lighting*

See Chapter 11, "Appliances and Lighting," for more information.

Resources

Note: Web links were current at the time of publication, but can change over time.

Baechler, M. C., & Love, P. M. (2004). *Building America Best Practices Series: Volume 1, Builders and Buyers Handbook for Improving New Home Efficiency, Comfort, and Durability in the Hot and Humid Climate*. U.S. Department of Energy, Energy Efficiency and Renewable Energy, Building Technologies Program. Retrieved from http://apps1.eere.energy.gov/buildings/publications/pdfs/building_america/36960.pdf

Lstiburek, J. (2005). Builder's Guide to Hot-Humid Climates. Building Science Corporation. Retrieved from https://buildingscience.com/bookstore/ebook/ebook-builders-guide-hot-humid-climates

My Florida Home Energy. (n.d.). Retrieved July 28, 2015, from http://www.myfloridahomeenergy.com/

A useful resource with a wide array of information on energy and water efficiency, including The Energy Efficient Home series of fact sheets, available at http://www.myfloridahomeenergy.com/help/library

Southface Energy Institute. (n.d.). Green Home Building Resources. Retrieved July 28, 2015, from http://www.southface.org/learning-center/library/green-home-building-resources

Southface Energy Institute, & Oak Ridge National Laboratory. (2003). *Whole-House Energy Checklist: 50 Steps to Energy Efficiency in the Home* (Technology Fact Sheet). Department of Energy, Office of Energy Efficiency and Renewable Energy, Building Technologies Program. Retrieved from http://www.southface.org/factsheets/WH-Energy%20Checklist%20GO-10099-766.pdf

Tiller, J. S., & Creech, D. B. (1999). *A Builder's Guide to Energy Efficient Homes in Georgia* (Third Edition). Atlanta, GA: Georgia Environmental Facilities Authority. Retrieved from http://www.southface.org/ez/media/georgiabuildersguide.pdf

U.S. Department of Energy (DOE) Energy Saver. Retrieved July 28, 2015, from http://energy.gov/energysaver/energy-saver

U.S. Environmental Protection Agency (EPA) ENERGY STAR. (n.d.). Home Solutions: High Energy Bills. Retrieved July 28, 2015, from http://www.energystar.gov/index.cfm?c=home_solutions.hm_improvement_highenergy

2
WHY BUILD EFFICIENTLY?

Investments in energy (and water) efficient building and remodeling provide short and long term benefits for all.

Homeowners:

- Energy efficient homes have lower monthly utility bills, and usually higher quality construction and fewer maintenance problems.
- Energy efficient homes are more comfortable to live in, have reduced moisture problems, and occupants are likely to have fewer health problems.
- Preferred financing for energy efficient mortgages (EEMs) is available from the Federal Housing Authority (FHA), the Veteran's Administrations (VA), Freddie Mac or Fannie Mae. There are two types—for new or for existing homes. Both allow the cost of the efficiency upgrades to be included in the mortgage. Lenders can reduce the debt-to-income ratio for borrowers based on expected savings on utility costs.
- The owner may be eligible for reduced stormwater fees due to onsite detention and rainfall reuse systems.
- When ready to sell, the home is likely to have a higher resale value.

Builders:

Most homeowners are willing and able to pay a premium price for homes with energy saving features, allowing increased profit to builders.

- Well built homes and customer satisfaction translate to increased customer referrals and a reputation for quality construction for the builder.
- Improved quality subcontractor work equates to higher profits. Building energy efficient labeled homes (such as ENERGY STAR®, Florida Green Building Coalition Green Homes (FGBC), or Leadership in Energy & Environmental Design (LEED), encourages field tests of building components, such as blower door and duct leakage testing and HVAC system inspection. These tests point out potential problems when they can be fixed quickly without litigation and encourage high performance from subcontractors.Identifying and correcting energy performance problems early also results in reduced callbacks from homeowners.

- Some cities and counties have passed ordinances that promote green building, typically voluntary for residential builders and mandatory for government owned facilities. These incentives may include 'fast tracking' of the permit process, priority inspections, a reduction in building permit fees, and free follow-up/secondary submittals for modifications to incorporate additional green infrastructure features.
- Beyond the structure itself, efficient use of water resources may be eligible for local government incentives. Installing a cistern for rainwater harvesting or grey-water capture and reuse (for irrigation or toilet flushing), use of pervious pavements for driveways and patios, and Florida-friendly landscaping are features that may be credited. They are eligible for green building certification credits, and can improve sales potential. In addition, "voluntary" programs, such as Florida Water Star℠ may be required in some municipalities.

Real estate professionals:

- It is easier to sell an energy efficient home. Judging by the most commonly asked questions, 94% of those purchasing new homes are most interested in home maintenance costs and energy features.
- Energy efficient homes are more desirable and therefore, more valuable. Almost 4 out of 5 real estate appraisers surveyed believed that energy efficient homes were worth about 5% more than standard construction. Other studies have found that every $1 decrease in annual energy costs raises the market value of the home by about $20.
- Lower interest rates for energy efficient mortgages increase the pool of those qualifying for home loans by almost 7%.

In addition, buildings that require less energy and water have positive impacts on our natural environment. It should also be noted that federal, state, local, and utility incentives may be available for many energy and water efficient building/remodeling features; see Table 2-1 for more information.

Table 2-1 Categories and features targeted by typical energy incentive programs

Category	Feature	Typical Incentive
Structural	• Attic Insulation • Windows • Reflective Roof • Weatherization	Utility Rebate Programs Manufacturer Rebates
Mechanical	• Duct Sealing • HVAC Service • HVAC Replacement	Utility Rebate Programs Manufacturer Rebates
Appliances	• Water Heater • Refrigerator • Washer • Dryer	Loan Programs Manufacturer Rebates
Lighting	• Compact Fluorescent Lamps (CFLs) • Light Emitting Diodes (LEDs)	Utility and Manufacturer Giveaways
Alternative/Renewable Energy	• Solar Water Heating • Solar Photovoltaic Panels	Federal Tax Credit Utility Rebates Loan Programs
Load Management	• Voluntary (Customer-Driven) • Direct Load Control (Utility-Driven) • On-Site Generation	Tiered Rate Structure Seasonal or Time-of-Use Rates Utility Credit or Rebate

FEDERAL TAX INCENTIVES

Residential

For residential properties, a 30% credit may be available on Federal taxes for renewable energy systems. There is no upper limit on the credit, except for the fuel cells, as noted below.

- Solar electric,
- Solar water heaters,
- Residential wind turbines,
- Geothermal heat pumps—must be ENERGY STAR qualified, and
- Fuel cells (minimum size is 0.5kW, and credit is limited to $500 per 0.5 kW).

Commercial

Fuel Cells and Microturbines

Businesses are eligible for a tax credit on systems that meet the following criteria:

- Fuel cells—the credit is equal to 30% of expenditures, with no maximum credit. However, the credit for fuel cells is capped at $1,500 per 0.5 kilowatt (kW) of capacity. Eligible property includes fuel cells with a minimum capacity of 0.5 kW that have an electricity-only generation efficiency of 30% or higher.
- Microturbines—the credit is equal to 10% of expenditures, with no maximum credit limit stated (explicitly). The credit for microturbines is capped at $200 per kW of capacity. Eligible property includes microturbines up to two megawatts (MW) in capacity that have an electricity-only generation efficiency of 26% or higher.

Appraising Energy Efficiency and Renewable Energy within Buildings

Though Energy Efficient Mortgages (EEMs) have been available for about two decades, it is only in recent years that energy focused green building features have truly begun diffusing into the marketplace. One example is the growing awareness and standardization within the property appraisal and real estate brokerage industries. The following resources provide a glimpse of how appraisers are valuing and how Realtors® are listing green features (web addresses may be found in the Resources section at the end of the chapter):

- Appraisal Institute
  - Green Building Resources: All Things "Green".
  - More Green Resources (including forms)
  - Form 820.04: Residential Green and Energy Efficient Addendum

- National Association of Realtors®: Greening the Multiple Listing Service (MLS)

Of special interest may be the significant resale premiums being seen in homes with solar photovoltaic (PV) modules. Additional resources are included below:

- Appraisal Institute: "Solar Electric Systems Positively Impact Home Values"
- Hoen, Ben, Ryan H. Wiser, Peter Cappers, and Mark A. Thayer. "An Analysis of the Effects of Residential Photovoltaic Energy Systems on Home Sales Prices in California."
- Hoen, Ben, Geoffrey T. Klise, Joshua Graff-Zivin, Mark A. Thayer, Joachim Seel, and Ryan H. Wiser. "Exploring California PV Home Premiums."
- Sandia National Laboratories:
    - News release (January 31, 2012) about PV Value®, an electronic form to standardize appraisals: "Sandia tool determines value of solar photovoltaic power systems."
    - PV Value® : Photovoltaic Energy Valuation Model: updated information from Sandia National Laboratories.
    - The PV Value® tool itself is now being developed and maintained by Energy Sense Finance, LLC, with funding provided by the Department of Energy SunShot Small Business Inovative Research Program.
- *Solar Valuation: An Appraiser's Guide to Solar*: an introductory guide for real estate appraisers on how to accurately value a residential rooftop solar photovoltaic system.

Florida's Support for Energy Efficiency

Statute 255.2575: Energy-efficient and sustainable buildings

The Florida Legislature has recognized the State's leadership role in promoting energy conservation, saving taxpayers money, and raising public awareness of energy rating systems. This statute requires all state, county and local government buildings, educational facilities (K-12, community college and state universities), water management district, and state court buildings be constructed to comply with a sustainable building rating system or national model green building code.

Statute 255.5576: Consideration of energy-efficient materials; high-energy lighting

This requires that the energy efficiency of all materials be considered for any construction or modification of state owned or operated facilities.

Other Resources to Locate Incentives and Rebates

- The Database of State Incentives for Renewables & Efficiency (DSIRE) is an excellent website for locating renewable energy and energy efficiency policies and incentives. It encompasses Federal, State, local government, and utility levels. It is funded by the U.S. Department of Energy and operated by the North Carolina Clean Energy Technology Center at N.C. State University.
- The Efficient Windows Collaborative maintains a list of utilities that offer incentives and rebates for energy efficient window installation and replacement. They vary by utility and include items such as loan programs to install double-pane windows to rebates for window films.
- EPA's ENERGY STAR: Special offers and rebates from ENERGY STAR Partners.
- U.S. Department of Energy, Office of Energy Efficiency & Renewable Energy – Solution Center: Financing Solutions.
- Some manufacturers or retail chains, such as Lowe's and Home Depot, maintain a searchable database of current rebate offers on their websites. They may include items such as ENERGY STAR water heaters and appliances, low flow toilets, and other energy efficient building materials.

Achieving Efficiency

Energy efficient buildings are no accident. Too often, measures that may be easier to market are installed, but key ingredients—such as sealing air leaks (mandatory as found the *Florida Building Code 6th Edition (2017)*) and duct leaks—are insufficient. The result is that buildings fall far short of being efficient, with energy bills higher than necessary, comfort and moisture problems, and owners who are thoroughly dissatisfied—hardly a positive customer relations situation.

Designing and building a structure that uses energy wisely does not mean sacrificing a building's aesthetics or amenities. Buildings of any architectural style can meet the requirements of this book. Any good building design considers the characteristics of a particular site: the local climate, the availability and cost of energy sources, and the occupant's lifestyle.

Building an energy efficient structure does not require special materials or construction skills. However, the quality of basic construction has a major influence on building comfort and energy costs, especially:

- Quality of framing and installation of insulation and windows.
- Attention to detail in sealing areas with potential air leaks.
- Design and installation of the heating and cooling equipment.

- Effectiveness of location and sealing of ducts.

Successful builders of energy efficient homes realize the importance of quality. They also know that achieving low energy costs and greater comfort in today's competitive marketplace requires careful planning throughout the design and construction process.

THE FLORIDA ENERGY CODE

The Florida Energy Efficiency Code for Building Construction was first enacted in 1980 and has been revised and updated several times since. It is now the *Florida Building Code Sixth Edition (2017), Energy Conservation* volume. Florida's code establishes a minimum standard of energy efficiency. This allows the builder or designer flexibility in deciding how to build an efficient building. Prescriptive methods of meeting the code may be used, if the builder prefers, or the code establishes a budget of base points for each building using certain optimized "baseline" features (exterior walls, ceilings, and floors, etc.) and the building's actual areas and configuration. The base points are compared to the designed or "as-built" building using multipliers that reflect energy transfer through the same wall, ceiling and floor areas, and equipment efficiencies of the components actually designed for that building. This is referred to as the Simulated Performance Alternative (Performance) method. The Energy Conservation Code also allows the use of the Energy Rating Index (ERI) Compliance Method. The ERI considers all energy used in the residential building.

This allows the builder or designer considerable versatility. For example, a more efficient heating and cooling system could be chosen instead of double pane windows, or more windows on the north side of the building and fewer on the south or west side of the building.

Code Compliance and Permitting

A building's compliance with the *Florida Building Code Sixth Edition (2017), Energy Conservation* will be based on the climate zone where it is located (**FBC, Energy Conservation, Section C301.1, Table C301.1** and **Section R301.1, Table R301.1, Climate Zones, Moisture Regimes, and Warm-Humid Designations by County**). Once the climate-specific compliance method is chosen, certification is indicated by the forms approved for that method. Figures 2-1 (Residential) and 2-2 (Commercial) on the following pages provide an overview of compliance methods.

Refer to the Code to determine who is authorized to prepare and submit Code compliance forms, as well as jurisdictional and climate zone data. One software package approved for determining compliance is the EnergyGauge (EG) software, available from the Florida Solar Energy Center (contact information is in the next section of this chapter). As part of the *Florida Building Code Sixth Edition (2017),* the Florida Building Commission is charged with the responsibility of approving energy simulation tools (software) for determining compliances with the energy code.

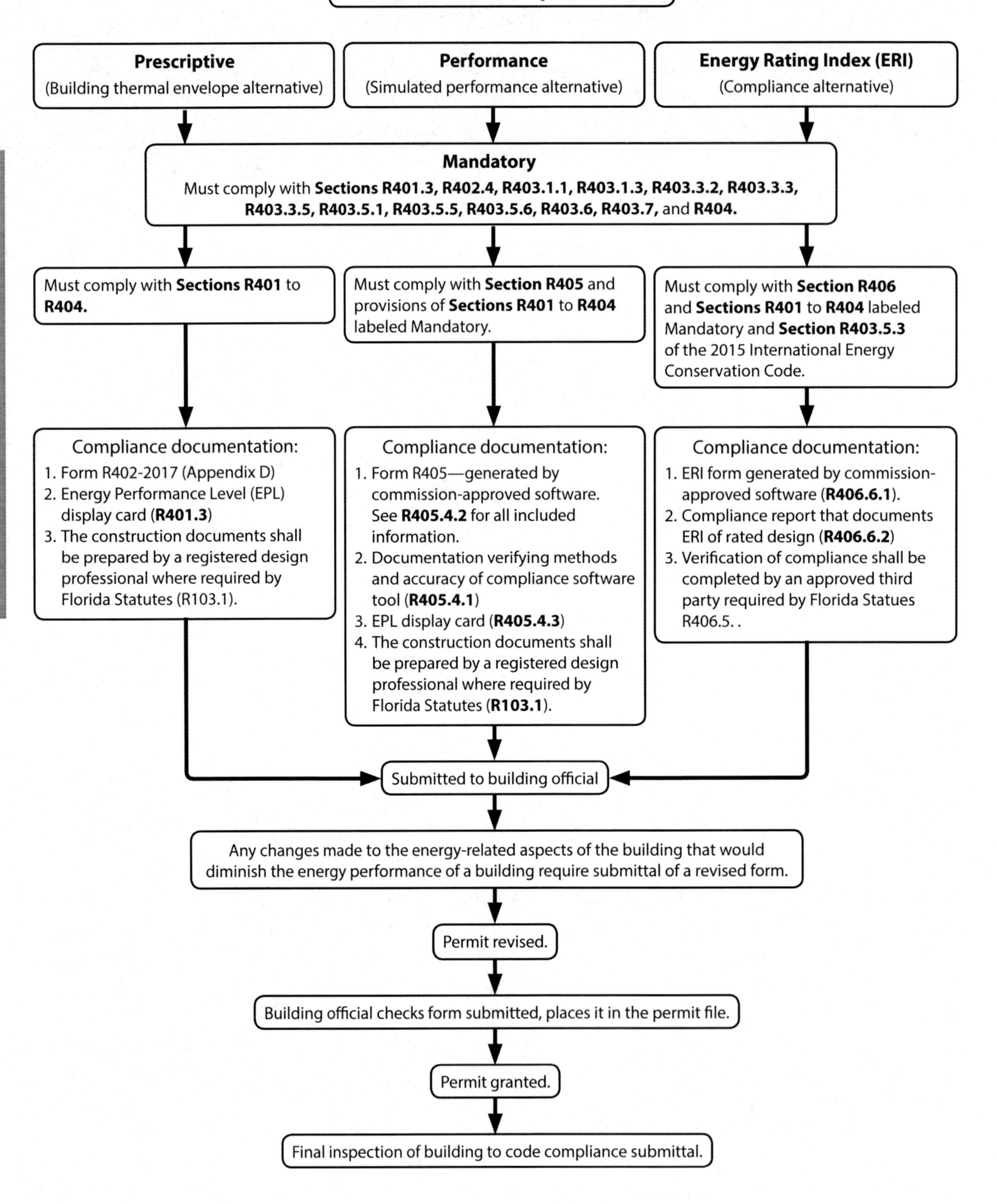

Figure 2-1 Code compliance flow chart - residential

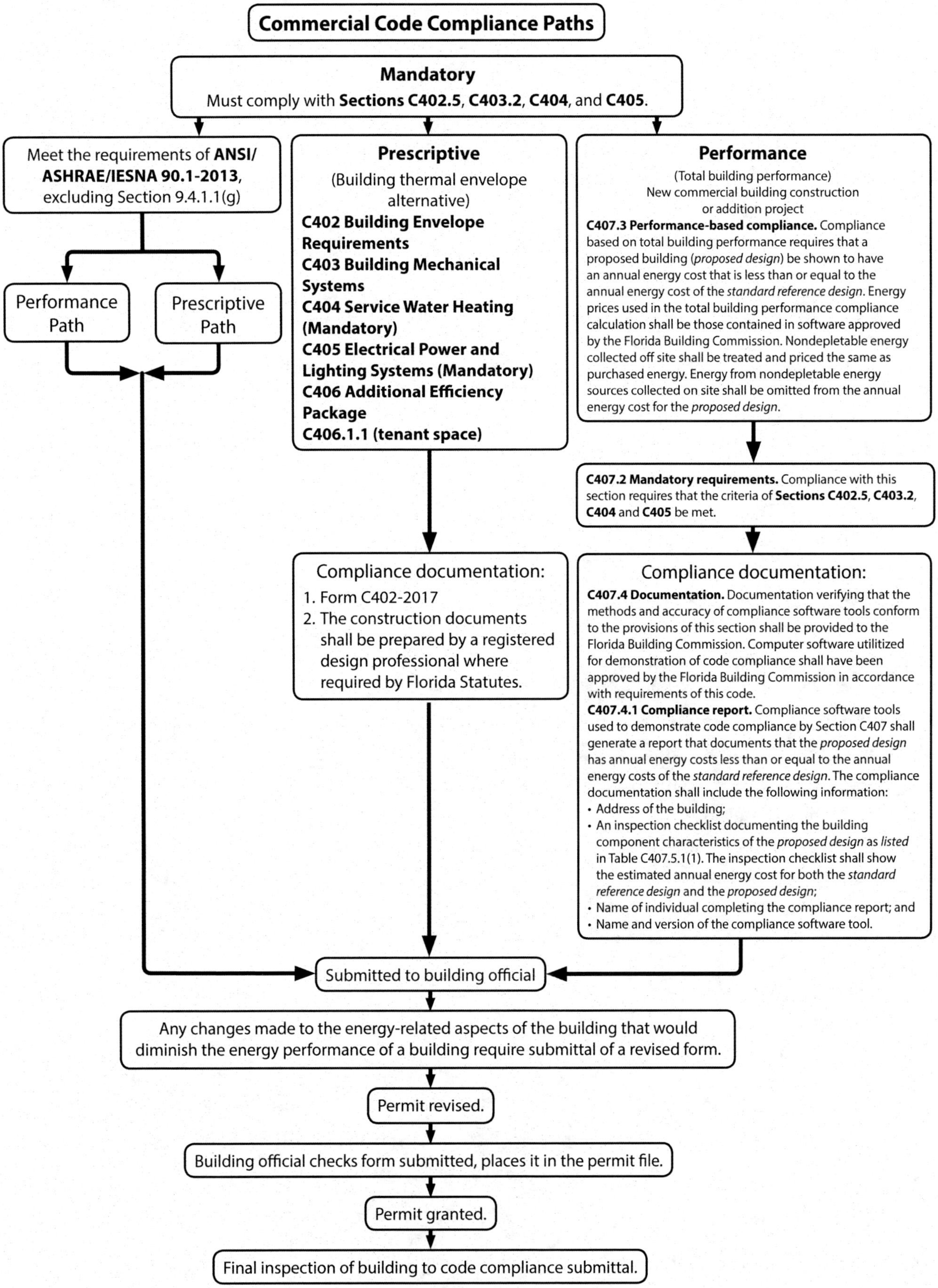

Commercial Code Compliance Paths

Mandatory

Must comply with **Sections C402.5**, **C403.2**, **C404**, and **C405**.

Meet the requirements of **ANSI/ASHRAE/IESNA 90.1-2013**, excluding Section 9.4.1.1(g)

Performance Path

Prescriptive Path

Prescriptive

(Building thermal envelope alternative)

C402 Building Envelope Requirements
C403 Building Mechanical Systems
C404 Service Water Heating (Mandatory)
C405 Electrical Power and Lighting Systems (Mandatory)
C406 Additional Efficiency Package
C406.1.1 (tenant space)

Compliance documentation:

1. Form C402-2017
2. The construction documents shall be prepared by a registered design professional where required by Florida Statutes.

Performance

(Total building performance)
New commercial building construction or addition project

C407.3 Performance-based compliance. Compliance based on total building performance requires that a proposed building (*proposed design*) be shown to have an annual energy cost that is less than or equal to the annual energy cost of the *standard reference design*. Energy prices used in the total building performance compliance calculation shall be those contained in software approved by the Florida Building Commission. Nondepletable energy collected off site shall be treated and priced the same as purchased energy. Energy from nondepletable energy sources collected on site shall be omitted from the annual energy cost for the *proposed design*.

C407.2 Mandatory requirements. Compliance with this section requires that the criteria of **Sections C402.5**, **C403.2**, **C404** and **C405** be met.

Compliance documentation:

C407.4 Documentation. Documentation verifying that the methods and accuracy of compliance software tools conform to the provisions of this section shall be provided to the Florida Building Commission. Computer software utilitized for demonstration of code compliance shall have been approved by the Florida Building Commission in accordance with requirements of this code.

C407.4.1 Compliance report. Compliance software tools used to demonstrate code compliance by Section C407 shall generate a report that documents that the *proposed design* has annual energy costs less than or equal to the annual energy costs of the *standard reference design*. The compliance documentation shall include the following information:

- Address of the building;
- An inspection checklist documenting the building component characteristics of the *proposed design* as *listed* in Table C407.5.1(1). The inspection checklist shall show the estimated annual energy cost for both the *standard reference design* and the *proposed design*;
- Name of individual completing the compliance report; and
- Name and version of the compliance software tool.

Submitted to building official

Any changes made to the energy-related aspects of the building that would diminish the energy performance of a building require submittal of a revised form.

Permit revised.

Building official checks form submitted, places it in the permit file.

Permit granted.

Final inspection of building to code compliance submittal.

Figure 2-2 Code compliance flow chart - commercial

Florida Building Energy-Efficiency Rating System

The Florida Building Energy-Efficiency Rating Act provides for a statewide uniform system for rating the energy efficiency of buildings. This rating system is based on the building's annual energy usage and costs considering local climate conditions, construction practices, and building use.

The energy rating for new and existing *residential* buildings shall be determined using a Florida Building Code approved energy simulation tool (software).

One Florida Building Code approved software package can be obtained from:

Florida EnergyGauge Program
Florida Solar Energy Center
1679 Clearlake Road
Cocoa, Florida 32922-5703
Phone: (321) 323-7255
email: info@energygauge.com
Web site: http://energygauge.com

Financing Energy Efficiency

When it comes to designing, constructing, and living in more energy efficiency buildings, we often pay most of our attention to the best practices in the technologies and systems installed. Yet financing is another critical part of the energy efficiency equation. Properly designed financial products can help a builder market value-added energy related features and help a homeowner overcome the first cost barriers common during decision-making for the new construction or renovation process. Additionally, calculating the return on investment (ROI) you may realize from these building improvements can further assist marketing and risk assessment while providing a sense of confidence in the decision to invest time and money into energy efficiency.

Energy Efficient Mortgages

Since 1995, the U.S. Department of Housing and Urban Development has made Energy Efficient Mortgages (EEMs) available nationwide. These loan products allow the total loan amount to increase in relation to the projected monthly utility bill savings from applicable energy improvements, capped by the lesser of the following three criteria:

- 5% of the value of the property, or
- 115% of the median area price of a single family dwelling, or
- 150% of the conforming Freddie Mac limit.

It is important to note that some details may vary by lender and all borrowers are encouraged to get multiple rate quotes and ask the lenders to compete for their business. Progressive builders should network with EEM approved lenders and inquire about options that may be available to assist builders in offering their clients on-site loan approvals at the time of quoting a construction project. Often, simplicity and streamlining of the loan underwriting and origination process can be more important than loan rates or tenors. For more information about EEMs and to review approved lenders in your area, see the following resources:

- HUD – FHA Energy Efficient Mortgage Program (Overview)
- HUD – FHA Energy Efficient Mortgage: Home Owner Guide
- FSEC – Financing Energy Efficiency: An EEM Handbook
- List of Approved HUD – FHA EEM Lenders

Additionally, if you are not aware of the Home Energy Rating System (HERS), consider visiting the websites of the resources below to learn more about how they work, how they might be utilized in the EEM process, and how you can find certified professionals to serve your needs.

- Home Energy Ratings (Overview)
- List of Certified Building Energy Raters

The SAVE Act

Building on the long history of EEMs, the Sensible Accounting to Value Energy (SAVE) Act (S. 1737) strives to further refine the methods used to internalize energy efficiency into the lending process. More information on this proposed legislation can be found at the website hosted by the Institute for Market Transformation.

EVALUATING ENERGY EFFICIENT PRODUCTS

The energy efficient builder seeks to minimize the lifetime costs of a home rather than the first costs. Making such calculations are often time-consuming and confusing. One of the best ways to determine whether an investment is sound is to compare the annual energy savings with the additional annual mortgage costs to find the Net Annual Savings.

For example, suppose you are wondering whether it is worthwhile for a home to have high efficiency, low-e windows, which use special coatings to reduce heat loss and gain. A builder had planned to install double-glazed units, but is now considering an upgrade to low-e units. He receives the following information from a window dealer:

- Additional window cost = $500
- First year energy savings = $75

He can easily calculate that the payback period on the above investment is just under 7 years. However, he is unsure whether the payback is acceptable. To find the Net Annual Savings, first, he finds the extra mortgage costs for the windows:

1. Mortgage interest rate = 5.0%
   Term of mortgage = 30 years
   Monthly payment per $1,000 (from Appendix 1) = $5.37
2. Annual payment per $1 (multiply the above by 12 and divide by 1,000)
   $5.37 × 12/1,000 = $.064
3. Extra annual payment (multiply the additional cost of the windows by the above factor)
   = $500 × $.064 = $32
4. Net annual energy savings (subtract the annual payment from annual energy savings)
   = $75 – $32 = $43

Since the Net Annual Energy Savings is positive, the investment is sound, especially when considering that energy costs will increase over time, while mortgage costs will remain relatively constant.

It is often useful to calculate the Rate of Return (ROR) for an energy investment. Homeowners can compare the annual percentage return for an energy measure to that earned by their financial investment. The steps for finding the ROR, using the above example, are as follows:

1. Find the payback period (divide the total cost by the annual savings)
   = 500/75 = about 6.7 years
2. Determine the life of the energy measure
   = over 20 years
3. For the payback period and lifetime, find the ROR in Table 2-2
   = more than 20 percent, but less than 23 percent (and it's tax free)

Table 2-2 Rate of Return for Energy Investments (%)

Lifetime of Energy Investments, years

Payback Period, years	5	7	10	12	15	17	20
1.5	71%	75%	77%	77%	77%	77%	77%
2	50%	56%	59%	59%	59%	59%	59%
3	27%	35%	39%	41%	41%	42%	42%
4	15%	24%	29%	31%	32%	32%	33%
5	6%	16%	22%	24%	26%	26%	27%
6	0	11%	18%	20%	22%	22%	23%
7	0	6%	14%	16%	19%	19%	20%
8	0	3%	11%	14%	16%	17%	18%
9	0	0	8%	11%	14%	15%	16%
10	0	0	6%	9%	12%	13%	15%
11	0	0	5%	8%	11%	12%	13%
12	0	0	3%	6%	9%	11%	12%
13	0	0	2%	5%	8%	10%	11%
14	0	0	0	4%	7%	9%	10%
15	0	0	0	3%	6%	8%	9%
16	0	0	0	2%	5%	7%	9%
17	0	0	0	1%	5%	6%	8%
18	0	0	0	0	3%	5%	7%
19	0	0	0	0	3%	5%	7%
20	0	0	0	0	3%	4%	6%

Note: *A zero indicates the rate of return is either negligible or negative. The table assumes energy prices escalate 6.33 percent per year.*

Resources

Note: Web links were current at the time of publication, but can change over time.

Energy Policy Act of 2005, Pub. L. No. 109-58 (2005). Retrieved from http://energy.gov/sites/prod/files/2013/10/f3/epact_2005.pdf

Florida Solar Energy Center. (n.d.). EnergyGauge - Energy and Economic Analysis Software. Main web page: http://energygauge.com/

International Code Council (ICC). (n.d.). *2012 International Energy Conservation Code®*. Retrieved August 3, 2015, from http://shop.iccsafe.org/codes/2012-international-codes.html?p=3

My Florida Home Energy. (n.d.). Retrieved July 28, 2015, from http://www.myfloridahomeenergy.com/

A useful resource with a wide array of information on energy and water efficiency, including The Energy Efficient Home series of fact sheets, available at http://www.myfloridahomeenergy.com/help/library

Residential Energy Services Network (RESNET). (n.d.). Lenders Corner: Energy Mortgage Information for Lenders. Retrieved August 3, 2015, from http://www.resnet.us/professional/lenders

Southface Energy Institute, & Oak Ridge National Laboratory. (1999). *Energy Efficiency Pays: Systems approach cuts home energy waste and saves money* (Technology Fact Sheet). U.S. Department of Energy, Energy Efficiency and Renewable Energy, Office of Building Technology, State and Community Programs. Retrieved from http://www.southface.org/factsheets/EEP-Efficiency_pays%2099-746.pdf

Building Energy Efficient Labeled Homes

Florida Green Building Coalition (FGBC). (n.d.). Retrieved August 3, 2015, from http://www.floridagreenbuilding.org/home

U.S. Environmental Protection Agency (EPA) ENERGY STAR. (n.d.). ENERGY STAR Certified New Homes. Retrieved August 3, 2015, from http://www.energystar.gov/index.cfm?c=new_homes.hm_index

U.S. Green Building Council. (n.d.). Leadership in Energy & Environmental Design (LEED). Retrieved August 3, 2015, from http://www.usgbc.org/leed

Appraising Energy Efficiency and Renewable Energy within Buildings

Appraisal Institute. (n.d.). Retrieved July 26, 2015, from http://www.appraisalinstitute.org/

Green Building Resources - Education Resources (n.d.). Retrieved July 26, 2015, from http://www.appraisalinstitute.org/education/education-resources/green-building-resources/

More Green Resources (n.d.). Retrieved July 26, 2015, from http://www.appraisalinstitute.org/education/education-resources/green-building-resources/more-green-resources/

Residential Green and Energy Efficient Addendum (Form 820.04) (January 2013). Appraisal Institute. Retrieved from http://www.appraisalinstitute.org/assets/1/7/ai-residential-green-energy-effecient-addendum.pdf

Solar Electric Systems Positively Impact Home Values. (October 13, 2013). News Release. Retrieved from http://www.appraisalinstitute.org/solar-electric-systems-positively-impact-home-values-appraisal-institute-/

Hoen, B., Klise, G. T., Graff-Zivin, J., Thayer, M., Seel, J., & Wiser, R. (2013). *Exploring California PV Home Premiums* (No. LBNL-6484E) (p. 41). Lawrence Berkeley National Laboratory. Retrieved from http://emp.lbl.gov/sites/all/files/lbnl-6484e.pdf

Hoen, B., Wiser, R. H., Cappers, P., & Thayer, M. A. (2011). *An Analysis of the Effects of Residential Photovoltaic Energy Systems on Home Sales Prices in California* (No. LBNL-4476E). Berkeley, CA: Lawrence Berkeley National Laboratory. Retrieved from http://eetd.lbl.gov/node/49019

National Association of Realtors®. (n.d.). Greening the MLS. Retrieved July 26, 2015, from http://www.greenresourcecouncil.org/green-resource-council-info/greening-mls

Sandia National Laboratories:

News Release (2012, January 31). Sandia tool determines value of solar photovoltaic power systems. Retrieved from https://share.sandia.gov/news/resources/news_releases/pv-value-tool/#.Vb-Qovnij2d

PV Value®. (n.d.). Retrieved July 26, 2015, from http://energy.sandia.gov/energy/renewable-energy/solar-energy/photovoltaics/solar-market-tranformation/pv-value/

PV Value® Photovoltaic Energy Valuation Model. (n.d.). Retrieved July 26, 2015, from https://www.pvvalue.com/

Solar Valuation: An Appraiser's Guide to Solar. (2012). SunPower Corporation. Retrieved from http://us.sunpower.com/sites/sunpower/files/media-library/white-papers/wp-residential-real-estate-appraisers-guide-accurately-valuate-residential-rooftop-solar-electric-pv.pdf

Other Resources to Locate Incentives & Rebates

Efficient Windows Collaborative. (2015, June 26). Incentives and Rebates for Energy-Efficient Windows Offered Through Utility and State Programs. Retrieved from http://www.efficientwindows.org/UtilityIncentivesWindows.pdf

North Carolina Clean Energy Technology Center at N.C. State University. (n.d.). Database of

State Incentives for Renewables & Efficiency®. Retrieved August 3, 2015, from http://www.dsireusa.org/

U.S. Department of Energy, Office of Energy Efficiency & Renewable Energy. (n.d.). State and Local Solution Center. Retrieved August 3, 2015, from http://energy.gov/eere/slsc/state-and-local-solution-center

U.S. Environmental Protection Agency (EPA) ENERGY STAR. (n.d.). Special Offers and Rebates from ENERGY STAR Partners. Retrieved August 3, 2015, from https://www.energystar.gov/rebate-finder

Energy Efficient Mortgages

Florida Solar Energy Center. (n.d.). *Financing Energy Efficiency: An EEM Handbook*. Retrieved from http://www.fsec.ucf.edu/en/consumer/buildings/homes/ratings/eem/index.htm

U.S Department of Housing and Urban Development (HUD):

Energy Efficient Mortgage Homeowner Guide - HUD. Retrieved August 3, 2015, from http://portal.hud.gov/hudportal/HUD?src=/program_offices/housing/sfh/eem/eemhog96

HUD FHA Insured Energy Efficient Mortgages. Retrieved August 3, 2015, from http://portal.hud.gov/hudportal/HUD?src=/program_offices/housing/sfh/eem/energy-r

HUD FHA Lender List. Retrieved August 3, 2015, from http://www.hud.gov/ll/code/llslcrit.cfm

Home Energy Rating System

Florida Solar Energy Center:

Certified Building Energy Raters. Retrieved August 3, 2015, from https://securedb.fsec.ucf.edu/engauge/engauge_search_rater

Home Energy Ratings. Retrieved August 3, 2015, from http://www.fsec.ucf.edu/en/consumer/buildings/homes/ratings/

The SAVE Act

The SAVE Act. Retrieved August 3, 2015, from http://www.imt.org/finance-and-real-estate/save-act

3
Siting and Passive Design Features

The initial decision of how to site and orient the house can have far-reaching impacts on the overall energy efficiency and thermal comfort of the home. Passive design includes non-mechanical means of using solar energy, wind patterns, and topography to heat homes in the winter and cool them in the summer months. Simple building elements, such as overhangs and awnings that take advantage of the natural setting, can reduce energy use. In addition, careful siting can help reduce the impacts of construction on soils, water, and the plant and animal community. A building that blends into its surroundings can have a higher value than buildings that do not. You can help your clients make wise decisions by pointing out the total savings over a ten- or twenty-year period that will result from good siting and passive design features.

In Florida, there are two principal factors in siting and passive design: access to solar radiation and ventilation.

Solar Radiation

Let's look at solar radiation first. The impact of solar radiation is a critical issue because heating and cooling account for a large part of the energy use in both residential and commercial buildings (Figure 3-1). Exposure to the sun is required only during the winter months in hot, humid regions where heating requirements are small. Shading and air movement are the most important factors during the rest of the year.

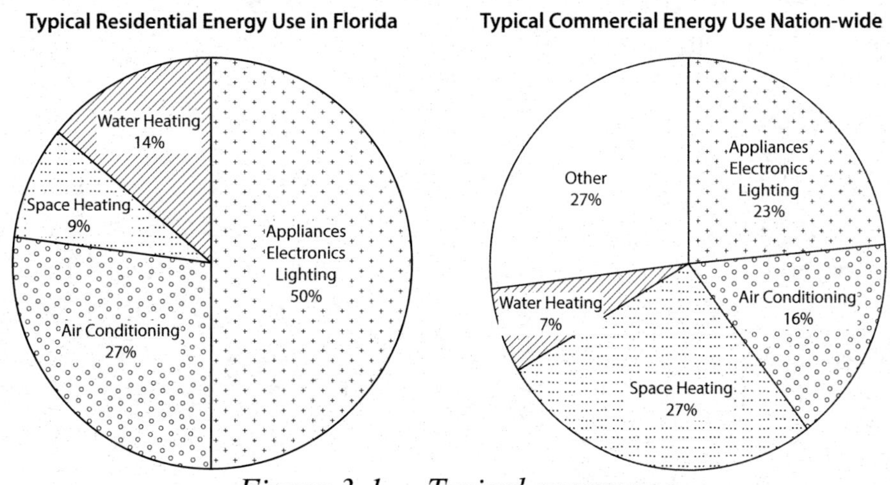

Figure 3-1 Typical energy use
Source: U.S. Energy Information Administration

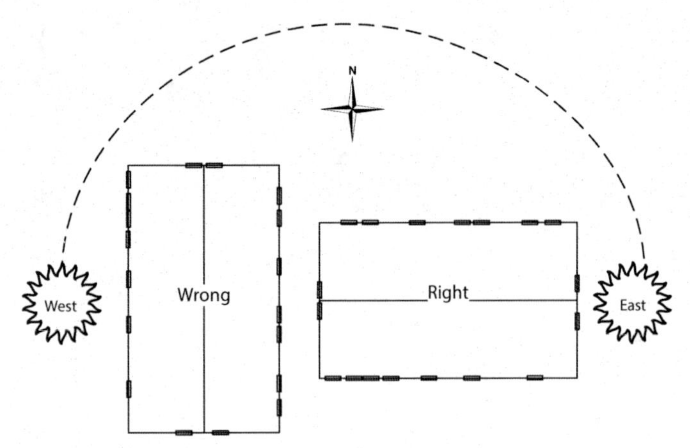

Figure 3-2 *Buildings should be oriented so that the long axis is east-west, to minimize the wall surface area exposed to the sun's rays.*

With this in mind, the long axis of a building should be oriented within 15º of the east-west axis of the sun's path across the sky (Figure 3-2). This reduces overheating because the short east and west sides are exposed to the sun for most of the day during the summer. The long side of the building is exposed to the low-angle rays of the winter sun. Calculate the east-west axis from true north. True north is shown on topographic maps, but not by a compass. Conventional compasses point toward magnetic north, which can vary as much as 20º from true north. Apps are available on the iPhone and other mobile devices that provide indicators for both magnetic and true north.

Heat loss or gain through the "skin," or exterior wall, of the house is greater for elongated than compact forms of the same volume. Compact designs gain less heat in the summer (and lose less heat in the winter) through their skin, since a compact shape has a smaller surface area than an elongated one with the same volume, and therefore less area where the outside temperature wants to balance the inside temperature. Houses designed so that rooms are clustered to reduce skin area will result in less heat loss and gain in the different seasons. They also use less building materials and make better use of the same amount of space.

For example, see Figure 3-3, where the two designs have equal floor areas—1500 sq ft—yet significantly different wall areas—1600 sq ft vs. 1900 sq ft, assuming 10-ft exterior walls. Compact design can apply to other features as well. Long, thin designs with many roof pitches (lines) are less energy efficient than square, single roofline homes. Two stories are generally better than one to achieve the same number of square feet. A building should be at least one and one-half times as long as it is wide, with the longer side of the building oriented to make maximum use of solar radiation, since a combination of active and passive cooling systems are usually used in Florida.

In homes, kitchens and dining rooms should be on the north or east sides or in the center of the building, helping to prevent the heat generated from cooking being added to the heat from solar radiation on the west side. Primary living spaces, such as living and family rooms, should be on the south side of homes to provide for year-round moderate temperature control where low sun angles can provide passive solar heating. Rooms that can tolerate temperature swings, such as closets, utility rooms and storage rooms, can act as insulating buffers to the living spaces.

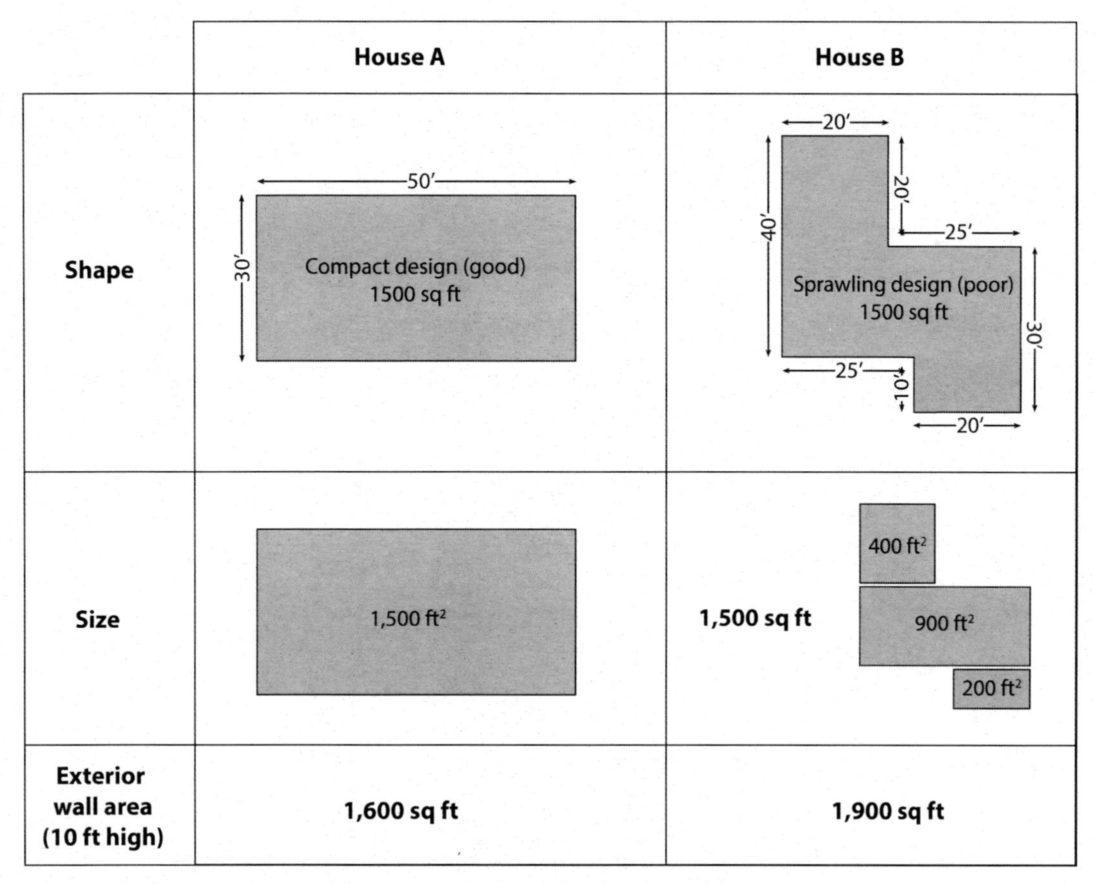

Figure 3-3 *Compact house designs, such as the one above, are more energy-efficient to build, heat, and cool.*
(House designs courtesy of Hyun-Jeong Lee, University of Florida)

Predominant window areas of a building should face north or south to reduce heat gain from solar radiation during the summer. One 6 × 8 foot clear glass area on a west wall can require more air conditioning to offset heat gain than all the rest of the wall. (Ideally, glass areas should not exceed about 10% of the wall area in single-story houses and about 6% in two-story houses.) The actual percentage of glass area shall not exceed the maximum acceptable percentage specified for the compliance package chosen. Window efficiencies can have an impact on allowable glass to floor areas. Solar screens or tinted windows (i.e. glazing) can reduce heat gain from glass areas significantly. Clear glass has a shading coefficient of 1. A solar screen or film with a shading coefficient of 0.2 would reduce direct solar radiation by 80%. Low-e (for low-emissive) glass allows for passage of visible light but reflects infrared energy back toward the warm side of the glass, providing greater thermal comfort. Double-pane windows are better insulators against cold and heat than single-pane glass.

Storm windows are another alternative. Their depth depends on the site's latitude. You need to know the sun's pattern over the site and the height of the window-to-roof line to determine the correct overhang depth.

Providing outdoor shade for windows during hot summer months can lessen indoor heat load and increase thermal comfort. The two principal approaches incorporate architectural elements and landscaping. Architectural elements such as awnings, roof overhangs, and shading devices of horizontal lattices or louvers placed over extensive window areas can be designed specifically for the latitude of house and size of windows. Covered porches over doors and windows on the east and west sides of homes do not compromise passive solar gain during winter months and can help lessen heat load during the summer.

In many areas, the best trees are those that have a dense canopy in the summer and an open, leafless canopy in the winter. For example, deciduous trees on the east, west, and south sides of a building provide shading in the summer, reducing cooling bills by as much as 40%. Yet, they still allow the sun's rays to penetrate in the winter. In South Florida, where winters are very mild and short, evergreens are better because they provide shading year-round. *Enviroscaping to Conserve Energy: Trees for South Florida* provides information on species selection and landscape design. This publication can be found at your local County Extension Service Office or through the University of Florida's Institute of Food and Agricultural Sciences' (IFAS) Web site (see the Resources section at the end of the chapter). Also available on the IFAS Web site is *Energy Efficient Homes: Landscaping*, as well as publications for determining shade patterns in North, Central, and South Florida.

Of growing concern in large cities of Florida is the **urban heat island (UHI)** effect; that is, when an area of a city is significantly warmer than its surrounding rural areas due to building materials and human activities such as traffic. Asphalt, concrete and other surface materials of buildings, surface parking lots and roadways absorb heat during the day and radiate that heat back into the atmosphere in the evenings, making the temperature difference larger at night than during the day. UHI can affect energy use by having to mechanically cool homes and buildings for longer periods of time during the day and night. However, choice of building materials can help mitigate UHI. **Albedo** is the term that indicates the fraction of solar radiation reflected back into the atmosphere in proportion with the heat energy absorbed for a given surface. A low albedo means a surface reflects a small amount of the incoming radiation and absorbs the rest. In comparison, high albedo materials reflect a large percentage of the solar radiation back into the atmosphere. Simple material choices, such as a roof painted a light reflective color, can help reduce ULI and energy use as well as provide more comfortable living during hot summer months. Figure 3-4 shows albedo ratings of various building and landscape elements.

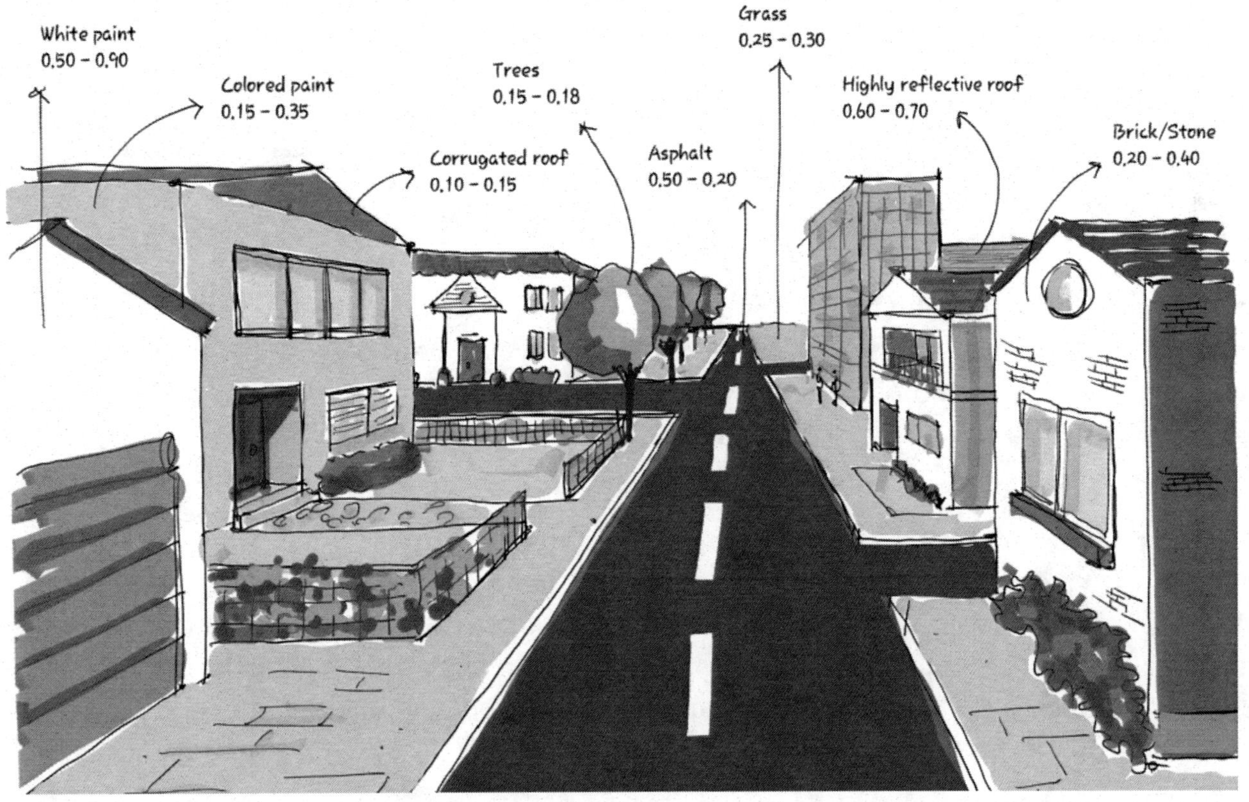

Figure 3-4 Albedo measures of typical residential building materials and vegetation
(Figure by Mahshad Kazem-Sadeh)

Ventilation

Now let's look at the second important factor, ventilation. Natural ventilation brings in outdoor air, which passes directly over people to increase cooling from evaporation on the skin. However, a completely passive system is rarely possible even where this is the main cooling strategy because it requires consistent, relatively high wind velocities with low humidity (see Figure 3-5), conditions not found in most of Florida. Nonetheless, buildings that permit maximum use of natural ventilation during the fall and spring reduce air conditioning needs significantly.

Natural ventilation is most appropriate when indoor temperatures and humidity are above the outdoor level. Natural ventilation is not recommended if the outside dew point temperature exceeds 60ºF. It will only increase the humidity indoors and contribute to the growth of mold and mildew. Well-designed buildings use the prevailing wind direction to cool in the summer, but protect buildings from winter winds. The meteorological station at the nearest airport can give you information about local wind patterns so that you can judge how well a building in the area is designed to take advantage of natural ventilation.

Ideal orientation for solar access and ventilation sometimes conflict. Solar radiation is more important because it affects the heating and cooling of the building more. Nonetheless, there are key siting solutions that can augment energy efficiency through natural ventilation. Siting of homes in a subdivision can be organized to preserve each home's access to breezes. Compared to multi-story buildings, single story homes can be spaced closer together to avoid wind shadows. If the houses are staggered, the wind flow around one house can help provide airflow for the adjacent house.

Protecting against prevailing north winds during the winter is important in cooler parts of the state. A long south side is desirable not only to prevent excessive heat gain from solar radiation, but also to promote cross-ventilation for incoming breezes from the north. Rooms oriented towards the prevailing breeze increase the effectiveness of cross-ventilation for cooling.

Topographic features can also help make buildings more energy efficient. Hills can protect buildings from cold winter winds in cooler regions. In much warmer regions, houses situated at the top of slopes maximize wind exposure and, if oriented east, decrease solar exposure in the hot afternoons. Natural topographic features can also be used to reduce danger of flooding and to increase the kinds of vegetation in the landscape.

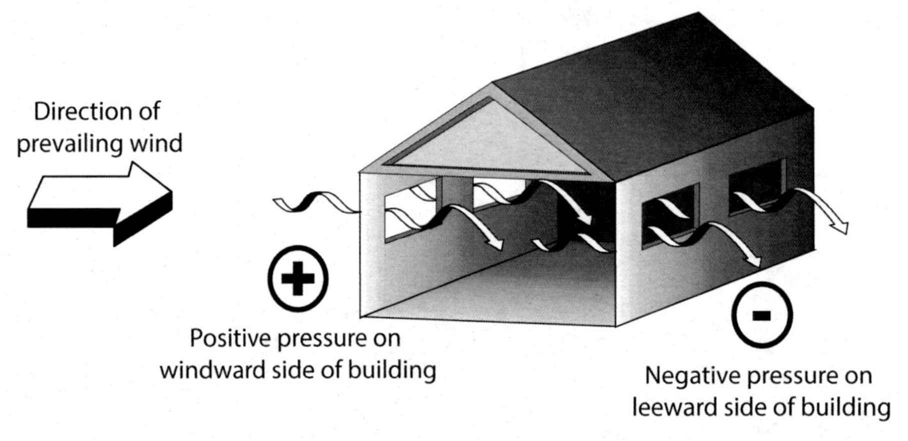

Figure 3-5 Cross ventilation depends on using prevailing winds

Cross ventilation allows air to flow from a strong positive pressure area to a negative pressure area in the opposite wall (Figure 3-5). The rate at which air flows through a house, carrying heat with it, is a function of the area of fenestration (inlets and outlets such as windows, doors, vents), the wind speed, the direction of the wind relative to the openings, and the temperature difference outside air and indoor air. Correct window placement depends on the prevailing wind direction, and cross ventilation is less successful where there is no strong prevailing wind direction. Where there are sufficient prevailing winds, window areas should be split about equally between the windward and leeward sides of a building for good cross ventilation. Window area should be about 15% of the floor area to both maximize cross ventilation and to prevent excessive heat gain when using air conditioning. It is key that the window placement allows air to move pass the occupants to be cooled. Windows low on the wall allow viewing while sitting or standing. If the occupants will be sitting most of the time, the windows should be about 3 feet above the floor, ending no higher than 6.5 feet from the floor.

A low window is especially important when hoppers or jalousies are used because they tend to deflect air upward. Casement or pivoting windows deflect the air stream from side to side. They can act as fin walls when they swing outwards. Hoppers or jalousies deflect wind vertically. They can be used to deflect rain while still admitting air. However, they may also direct the air over the occupants' heads. Horizontal or strip windows are often the best choice to ventilate large areas.

Fin walls can greatly increase the air that enters a building from windows on the same wall. They are used when the wall does not directly face into the prevailing winds. Fin walls must be placed correctly, as seen in Figure 3-6, to be effective. They keep air from entering the room if they are placed incorrectly.

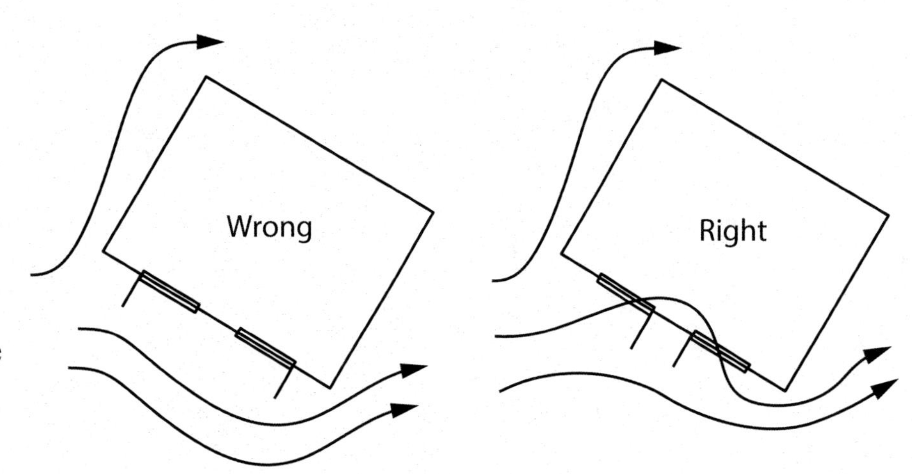

Figure 3-6 Incorrect and correct placement of fin walls

More open plans are better than plans with many interior partitions because partitions resist airflow and decrease total ventilation. Transoms—windows or vents above doors—allow for some cross-ventilation and yet maintain privacy between adjacent rooms.

In areas with hot days and temperate climates at night, air movement can be slow, in which case stack ventilation may be a useful design strategy. Roof vents and cupolas can lower attic temperatures and help ventilate buildings if the wind speed in an area is adequate. The vacuum produced by wind blowing across the cupola will exhaust the rising hot air from the house. In homes cooled by stack ventilation, warm air rises and exits through openings at the top of the rooms, with air replaced by the cooler air entering through low openings and vents in the rooms. The rate at which the air moves through the house depends upon the vertical distance between the openings or vents, their size, and the difference between outside temperature and average inside temperature over the height of the room.

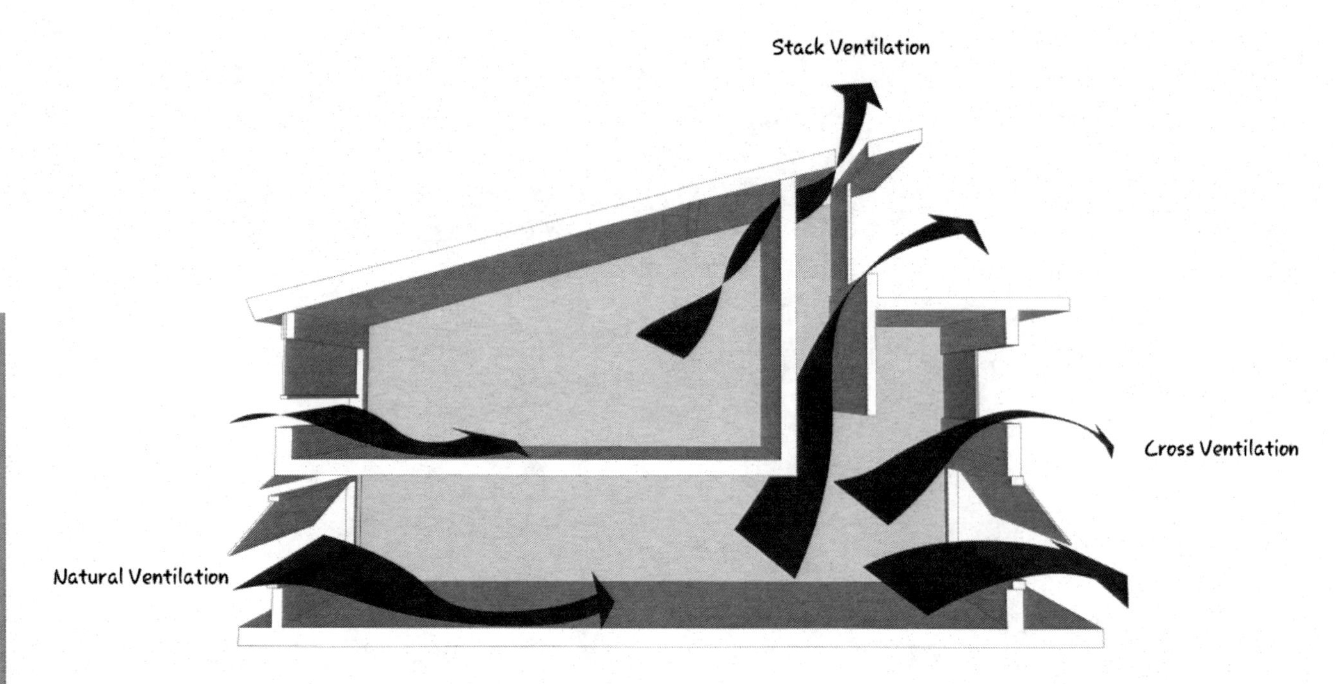

Figure 3-7 Example of both stack and cross-ventilation (Figure by Mahshad Kazem-Sadeh)

Stack- and cross-ventilation require different room organization in the homes. Linked room arrangements are appropriate for cross-ventilation because each space presents a large part of its skin area to the prevailing winds. Bunched room arrangements allow all the rooms to share the same ventilating stack. (See Figure 3-7.)

Wind is a liability in the winter. Both wind speed and flow affect heat transmission through the building envelope. Vegetation and other barriers can reduce wind speed and flow at sites where reducing heat loss is important. Or, they can be used to increase heat loss at sites where excessive heat buildup is the more important problem. Trees, for example, can be a windbreak or they can funnel cool breezes into a building. Plants near buildings can help prevent cool winds from easily escaping around the sides of the building (Figure 3-8).

THE USE OF VEGETATION IN EFFECTIVE HOME DESIGN

In northern Florida

- Plant deciduous vegetation; in summer, the plants will shade the home, but the loss of leaves in winter allows the sun's warming rays to reach the building walls.
- Plant trees and shrubs (or build walls) that protect the building from cooling winds in the winter.

In central and southern Florida

- Plant evergreen vegetation where protection from the sun's rays is a year-round issue.
- Trees and shrubs or walls should direct breezes toward the building where cooling is critical (Figure 3-8).
- Wind resistance should be an important consideration in the choice of tree species, especially those planted near buildings.

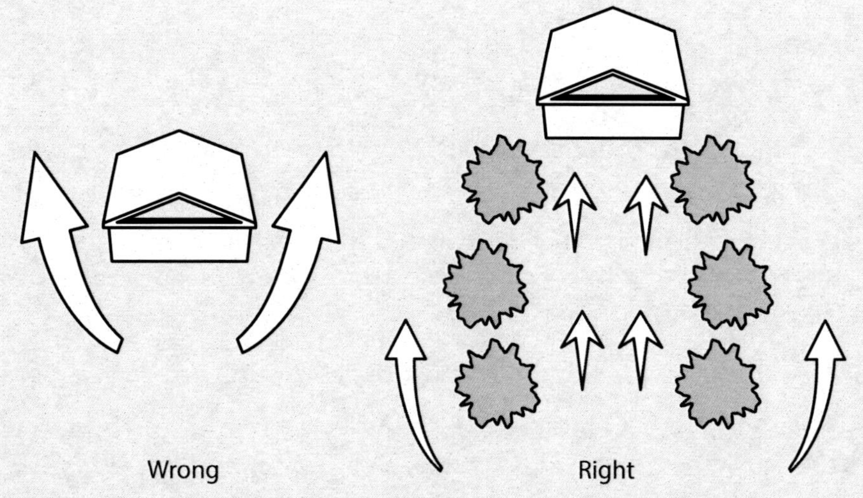

Figure 3-8 When desired, vegetation can help prevent cool breezes from escaping around the sides of the building.

KEY PASSIVE DESIGN FACTORS IN FLORIDA HOME DESIGN: A CHECKLIST

Orientation

- The long side of a building should face south and be oriented within 15° of the east-west axis.
- Buildings can be oriented to take advantage of prevailing winds during the fall and spring.
- Take advantage of natural topographic features to protect buildings from the sun.

Configuration

- Building design should be compact rather than sprawling.
- Roof vents and cupolas can help ventilate buildings.

Floor plan

- Fewer interior partitions are preferable to many small rooms.
- Transoms above doors aid in ventilation.
- Primary living spaces should be located on the south side.
- Kitchens and dining rooms should be situated on the north or east side, or at the center of the house.
- Closets, utility rooms, and storage rooms can act as buffers to living spaces.

Windows

- Predominant window areas should be on the north or south walls.
- Window areas should be split about equally between the windward and leeward sides of a building. This placement allows cross ventilation from prevailing winds.
- Consider placing fin walls on walls with windows to help direct air into the building.
- Unshaded/untinted glass areas should be 10% or less of the total wall area in single story homes, and 6% or less in two-story homes.
- Consider the advantages of:
  a. solar screens
  b. energy-efficient windows
  c. low-e glass
  d. interior shading
  e. roof overhangs, awnings and outdoor shading devices over windows

Building Materials

- HIgh albedo building materials and colors on exterior surfaces, particularly roofs, walls and walkways
- Trees and other vegetation planted in locations around buildings, parking lots and walkways.

Hurricanes Andrew, Hugo, Charlie, Ivan, Frances, and Jeanne showed that some trees resist wind damage better than others. One of the most durable is the Florida state tree, the Sabal Palm. Preliminary findings indicate that live oak, Southern magnolia, crape myrtle, dogwood, American holly and gumbo limbo are among the species that appear to have the greatest wind resistance. Least wind resistant species include sand, loblolly, and spruce pines, laurel and water oaks, avocado, citrus species, bottlebrush, and queen palm. Place these species well away from buildings to avoid possible damage.

Visit your local County Extension office to determine plant materials suitable for your site (see Resources, to find the location of your local office).

Resources

Note: Web links were current at the time of publication, but can change over time.

Energy Efficiency and Renewable Energy Clearinghouse. (2001). *Passive Solar Design for the Home*. DOE/GO-102001-1105. U.S. Department of Energy (DOE). Retrieved from http://www.nrel.gov/docs/fy01osti/27954.pdf

Broschat, T. K., Meerow, A. W., & Black, R. J. (2013). Enviroscaping to Conserve Energy: Trees for South Florida (No. EES42). Gainesville, FL: University of Florida. Retrieved from http://edis.ifas.ufl.edu/eh142

DelValle, T. B., Bradshaw, J. P., Larson, B. C., & Ruppert, K. C. (2015). Energy Efficient Homes: Landscaping (No. FCS3281). Gainesville, FL: University of Florida. Retrieved from http://edis.ifas.ufl.edu/fy1050

Fairey, P., & Chandra, S. (1984). *Principles of Low Energy Building Design in Warm, Humid Climates*. Cocoa, FL: Florida Solar Energy Center.

My Florida Home Energy. (n.d.). Retrieved July 28, 2015, from http://www.myfloridahomeenergy.com/

A useful resource with a wide array of information on energy and water efficiency, including The Energy Efficient Home series of fact sheets, available at http://www.myfloridahomeenergy.com/help/library

Ruppert, K. C., Miller, C. R., Swanson, C., & Lee, H.-J. (2015). Passive Solar Orientation (Fact Sheet). Gainesville, FL: University of Florida. Retrieved from http://www.myfloridahomeenergy.com/help/library/choices/orientation/

National Oceanic and Atmospheric Administration (NOAA). (n.d.). National Centers for Environmental Information (NCEI) (formerly known as National Climatic Data Center). Retrieved August 4, 2015, from https://www.ncdc.noaa.gov/

Sheinkopf, K. G. G., Sonne, J. K., & Vieira, R. K. (1992). *Energy-Efficient Florida Home Building Manual*. Cocoa, FL: Florida Solar Energy Center. Retrieved from http://www.fsec.ucf.edu/En/publications/html/FSEC-GP-33-88/index.htm

UF/IFAS Extension: Solutions for Your Life. (n.d.). Local Offices. Retrieved August 7, 2015, from http://sfyl.ifas.ufl.edu/map/index.shtml

Web sites, maps, and driving directions to local Extension Offices, Research and Education Centers, and Research and Demonstration Sites for the state of Florida.

U.S. Department of Energy (DOE) Energy Saver:

(n.d.-a). Reducing Your Electricity Use. Retrieved August 11, 2015, from http://energy.gov/energysaver/articles/reducing-your-electricity-use

(n.d.-b). Small Solar Electric Systems. Retrieved August 11, 2015, from http://energy.gov/energysaver/articles/small-solar-electric-systems

The Building as a System

We sometimes think of buildings as independent structures, placed on an attractive lot, and lived in without regard to the world around. Yet, most buildings have problems—some simply minor nuisances, others life-threatening:

- Mold on walls, ceilings, and furnishings
- Mysterious odors
- Excessive cooling and heating bills
- High humidity
- Rooms that are never comfortable
- Decayed structural wood and other materials
- Termite or other pest infestations
- Fireplaces that do not draft properly
- High levels of formaldehyde, radon, or carbon monoxide

These problems occur because of the failure of the building to properly react to the outdoor or indoor environment. The building should be designed to function well amid fluctuating temperatures, moisture levels, and air pressures.

Health and Comfort Factors

The following factors define the quality of the living environment. If kept at desirable levels, the building will provide comfort and healthy air quality.

- Moisture levels—often measured as the relative humidity (RH). High humidity causes discomfort and can promote growth of mold and organisms such as dust mites.
- Temperature—both dry bulb (measured by a regular thermometer) and wet bulb, which indicates the amount of moisture in the air. The dry bulb and wet bulb temperatures can be used to find the relative humidity of the air.
- Air quality—the level of pollutants in the air, such as formaldehyde, radon, carbon monoxide, and other detrimental chemicals, as well as organisms such as mold and dust mites. The key determinant of air quality problems is the strength of the source of pollution.
- Air movement—the velocity at which air flows in specific areas of the building. Higher velocities make occupants more comfortable in summer, but less comfortable in winter.
- Structural integrity—the ability of the materials that make up the building to create a long-term barrier between the exterior and interior.

Concepts

Heat Flows in Buildings

Health and comfort factors are determined considerably by how readily heat moves through a building and its exterior envelope. The next page explains the three primary modes of heat transfer.

HOW HEAT MOVES

Conduction

- The transfer of heat through solid objects, such as the ceiling, walls, and floor of a home.
- Insulation (and multiple layers of glass in windows) reduce conduction losses.

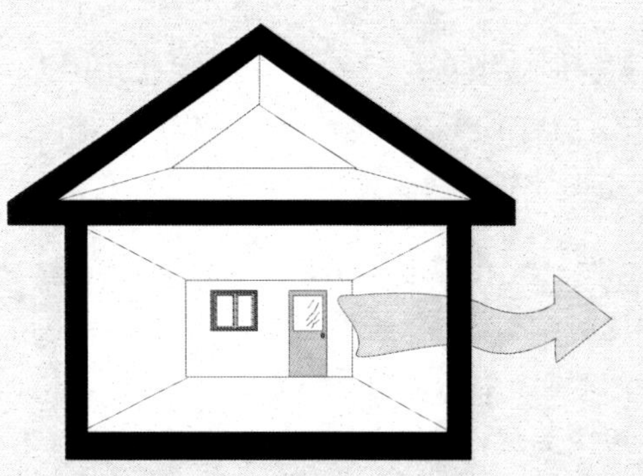

Figure 4-1 Conduction

Convection

- The flow of heat by currents of air.
- As air becomes heated it rises; as it cools, it becomes heavier and sinks.
- The convective flow of air into a home is known as *infiltration*; the outward flow is called *exfiltration*. In this book, these air flows are known generally as *air leakage*.

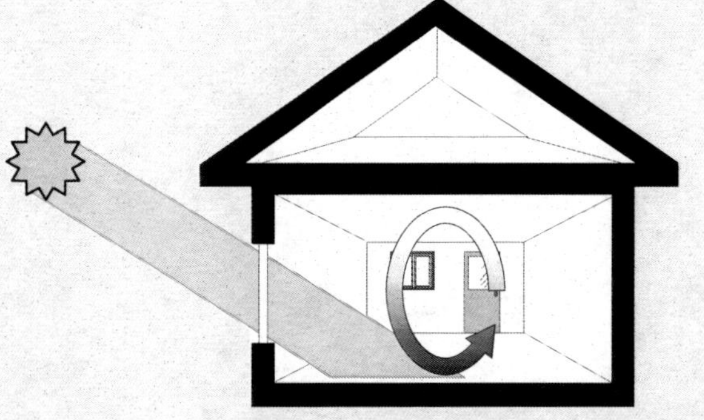

Figure 4-2 Convection

Figure 4-3 Air leakage

Radiation

- The net movement of radiant energy from warmer to cooler objects across empty spaces (like warmth from a fireplace).
- Examples include radiant heat traveling from:
  a. inner panes of glass to outer panes in double-glazed windows in winter
  b. roof deck to attic insulation during hot, sunny days
- Radiant energy exchange can be minimized by installing reflective barriers; examples include radiant heat barriers in attics and low-emissivity coatings for windows.

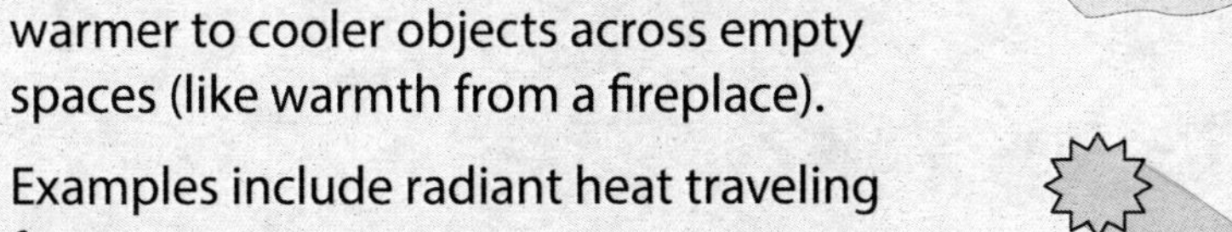

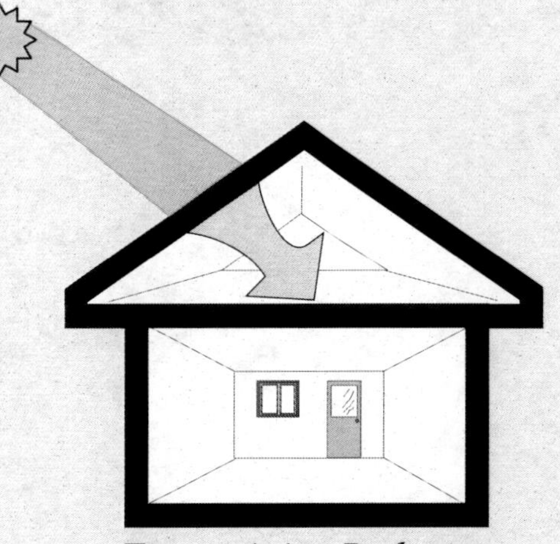

Figure 4-4 Radiation

Air Leaks and Indoor Air Quality

Both building professionals and building occupants have concerns about indoor air quality. It is important to understand that few studies on the subject have shown a strong relationship between indoor air quality and the air tightness of a building.

The major factor affecting indoor air quality is the level of the pollutant causing the problem. Most experts feel the solution to poor indoor air quality is removing the source of the pollution. Designing a leakier building may help lessen the intensity of the problem, but it will not eliminate it, nor necessarily create a healthy living situation.

Air leaks often bring in air quality problems from attics, the outside, and crawl spaces (Figure 4-5), such as:

- Mold
- Radon from crawl spaces and under-slab areas
- Humidity
- Pollen and other allergens
- Dust and other particles

The best solution to air quality problems is to construct a building as tightly as possible and install an effective ventilation system that can bring in fresh outside air (not crawl space or attic air). Ventilation system design options and indoor air quality are described in greater detail in Chapter 8, "Heating, Ventilation, Air Conditioning (HVAC)".

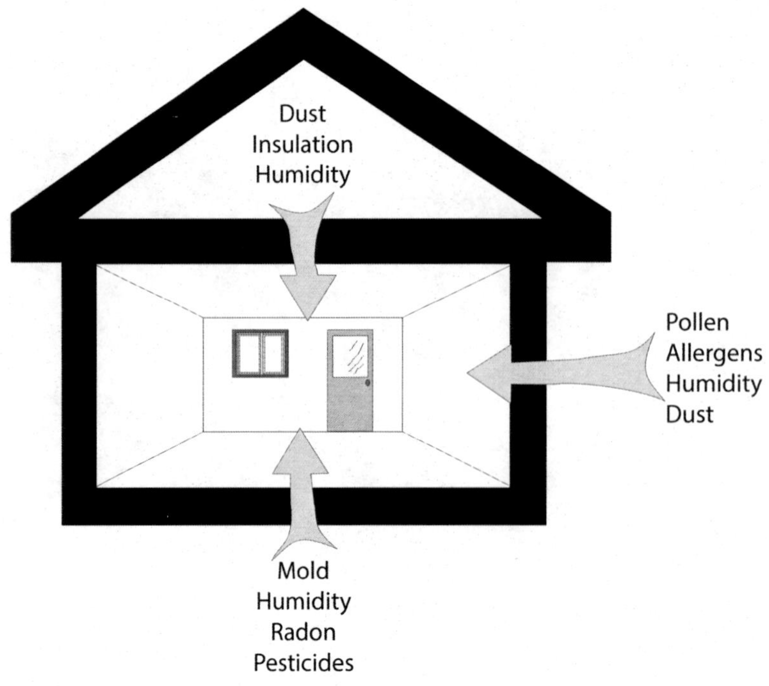

Figure 4-5 Air quality problems from "fresh" air

Relative Humidity and Water Condensation

Air is made up of gases such as oxygen, nitrogen, and water vapor. The amount of water vapor that air can hold is determined by its temperature. Warm air holds more vapor than cold air. The amount of water vapor in the air is measured by its relative humidity (RH).

$$\textbf{RH} = \frac{\text{the amount of water vapor in the air at a given temperature}}{\text{the maximum amount of water vapor that air can hold at that temperature}}$$

At 100% RH, water vapor condenses into a liquid. The temperature at which water vapor condenses is its dew point (Figure 4-6).

The dew point of air depends on its temperature and relative humidity. A convenient tool for examining how air temperature and moisture interact is the Psychrometric Chart (see next page). Preventing condensation involves reducing the RH of the air or increasing the temperatures of surfaces exposed to the air.

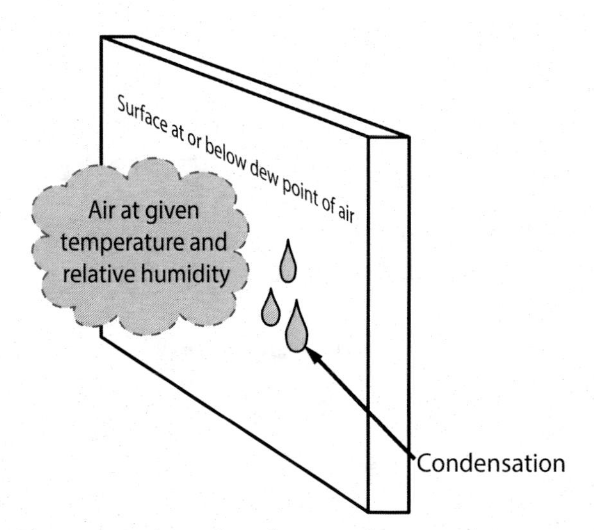

Figure 4-6 Conditions for condensation

Effect of Relative Humidity

People respond dramatically to changes in relative humidity:

- At lower RH, we feel cooler as moisture evaporates more readily from our skin.
- At higher levels, we may feel uncomfortable, especially at temperatures above 78° F.
- Dry air can often aggravate respiratory problems.
- Molds grows best in air over 70% RH.
- Dust mites prosper at over 50% RH.
- Wood decays when the RH is near or at 100%.
- People are most comfortable at 30% to 50% RH.

MOISTURE AND RELATIVE HUMIDITY

A psychrometric chart aids in understanding the dynamics of moisture control. A simplified chart shown in Figure 4-7 relates temperature and moisture. Note that at a single temperature, as the amount of moisture increases (moves up the vertical axis), the relative humidity of the air also increases. At the top curve of the chart, the relative humidity reaches 100%—air can hold no additional water vapor at that temperature (called the *dew point*) so condensation can occur.

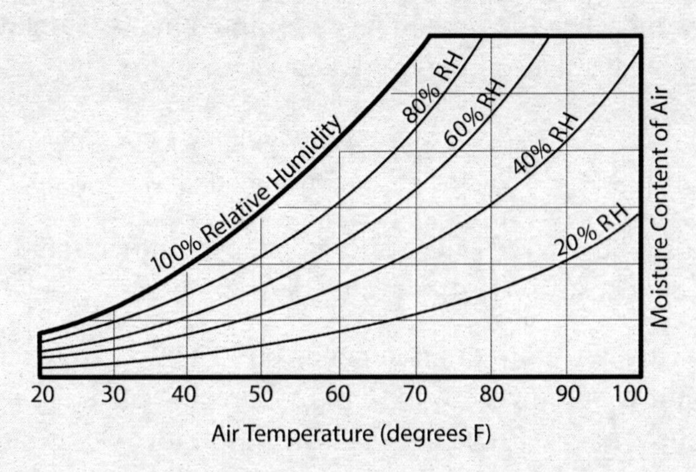

Figure 4-7 Psychrometric chart

Winter Condensation in Walls

In a well built wall, the temperature of the inside surface of the sheathing will depend on the insulating value of the sheathing, and the indoor and outdoor temperatures. When the temperature is 35°F outside, and 70°F at 40% relative humidity inside:

- The interior surface of plywood sheathing will be around 39°F
- The interior surface of insulated sheathing would be 47°F

The psychrometric chart can help predict whether condensation will occur:

1. In Figure 4-8, find the point representing the indoor air conditions (70°F) and 40% RH.
2. Draw a horizontal line to the 100% RH line.
3. Next, draw a vertical line down from where the horizontal line intersects the 100% RH line.

In the example, condensation would occur if the temperature of the inside surface of the sheathing were at 44°F. Thus, under the temperature conditions in this example, water droplets may form on the plywood sheathing, but not on the insulated sheathing.

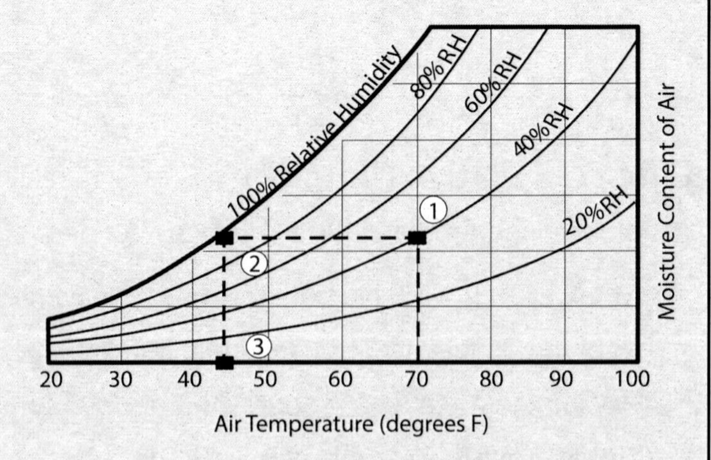

Figure 4-8 Winter dewpoint temerature inside walls

continued…

continued…

Summer Condensation in Walls

Figure 4-9 depicts a similar case in summer. If the interior air is 75°F, and outside air at 95°F and 40% relative humidity enters the wall cavity, will condensation occur? Using the psychrometric chart we find that the dew point of the outside air leaking into the wall cavity would be about 67°F. Since the drywall temperature is greater than the dew point, condensation should not form.

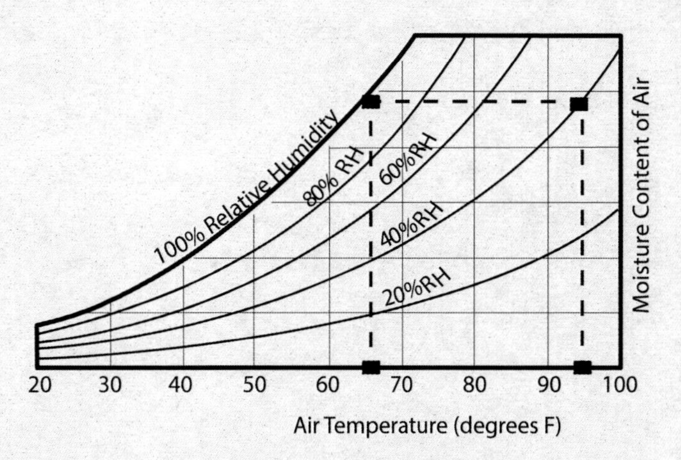

Figure 4-9 Psychrometric chart

Summer Condensation in Walls—Hot, Rainy Days

Afternoon rain is a frequent occurrence throughout the sub-tropical Floridian peninsula. Let's look at another example: what happens when rain has soaked a building and sunlight strikes a wall after the shower is over? In particular, let's examine the effect this has on a stucco-covered frame wall. The stucco will store water until the sun comes out. Sunlight striking the wall heats the water to its vapor state. This results in a temperature at the outer wall surface of 100° F or more and 100% RH (point 1). Water vapor at this location will condense all the way through the wall cavity. This is shown graphically in Figure 4-10, moving along the saturation line from point 1 to point 2 (the same state as the room air).

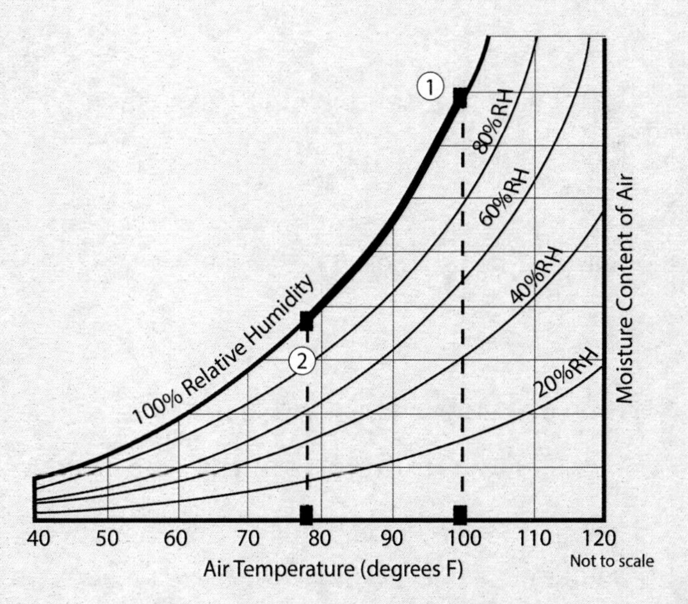

Figure 4-10 Psychrometric chart

4: The Building as a System

SYSTEMS IN A BUILDING

Whether the health and comfort factors of temperature, humidity, and air quality remain at comfortable and healthy levels depends on how well the building works as a system. Every building has systems that are intended to provide indoor health and comfort:

- Structural system
- Moisture control system
- Air barrier system
- Thermal insulation system
- HVAC system

Structural System

The purpose of this book is not to show how to design and build the structural components of a building, but rather to describe how to maintain the integrity of these components. Key problems that can affect the structural integrity of a building include:

- Erosion
- Roof leaks
- Water absorption into building systems
- Excessive relative humidity levels
- Fire
- Summer heat build-up
- Wind

Structural recommendations

To prevent these structural problems, the designer and builder should:

- Ensure the roof is watertight to prevent rainwater intrusion. Provide diverters or gutters and downspouts to direct water away from entries and walkways. If gutters are used, make sure that they are properly designed to handle Florida's frequent and heavy rainfall. (The owner should clean them regularly.) Gutters should be attached so that water cannot enter the fascia or spike penetrations, which should be carefully sealed.
- Divert ground water away from the building through proper grading and install effective gutters and downspouts.
- Ensure soffits have been properly installed per manufacturer's instructions.

- Seal penetrations that allow moisture to enter the building envelope via air leakage. Use firestopping sealants to close penetrations that are potential sources of "draft" during a fire.
- Prevent air from washing over attic insulation.
- Install a series of capillary breaks that keep moisture from migrating through foundation systems into wall and attic framing.

Moisture Control System

Buildings should be designed and built to provide comfortable and healthy levels of relative humidity. They should also prevent both liquid water and water vapor from migrating through building components.

The moisture control system includes quality construction to shed water from the building and its foundation, vapor and air barrier systems that hinder the flow of air infiltration and water vapor, and cooling and heating systems designed to provide comfort throughout the year.

There are four primary modes of moisture migration into buildings. Each of these must be controlled to preserve comfort, health, and building durability.

Bulk moisture transport

- The flow of moisture through holes, cracks, or gaps (Figure 4-11)
- Primary source is rain
- Causes include:
  - Poor flashing
  - Inadequate drainage
  - Poor quality weatherstripping or caulking around joints in building exterior (such as windows, doors, and bottom plates)

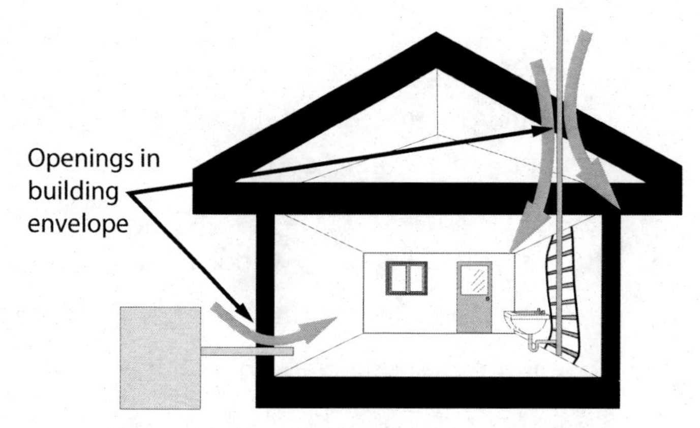

Figure 4-11 Bulk moisture transport

- Solved through quality construction with durable materials
- Most important of the four modes of moisture migration. (Bulk moisture leaks represent pathways for liquid water to immediately move into building cavities. Other methodologies—vapor, capillary action, air transport—are more time-dependent and take longer to create a significant problem.)

4: THE BUILDING AS A SYSTEM

Capillary action

- Wicking of water through porous materials or between small cracks (Figure 4-12)
- Primary sources are from rain or ground water
- Causes include:
  - water seeping between overlapping pieces of exterior siding
  - water drawn upward through pores or cracks in concrete slabs
  - water migrating from crawl spaces into attics through foundation walls and wall framing
- Solved by completely sealing pores or gaps, increasing the size of the gaps (usually to a minimum of 1/8 inch), or installing a waterproof, vapor barrier material to form a capillary break

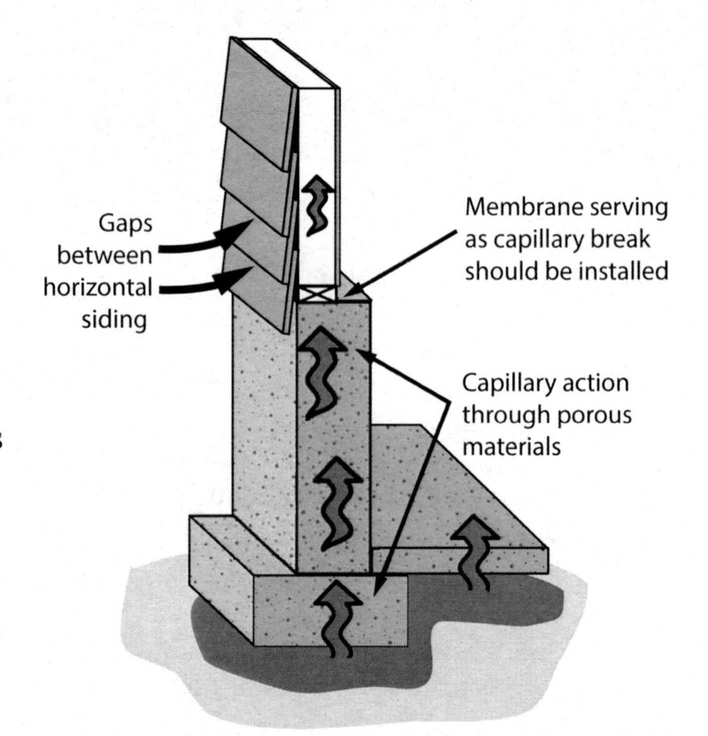

Figure 4-12 Capillary action

Air transport

- Unsealed penetrations and joints between conditioned and unconditioned areas allow air containing water vapor to flow into enclosed areas. Air transport can bring 50 to 100 times more moisture into wall cavities than vapor diffusion.
- Primary source is water vapor in air
- Causes include air leaking through holes, cracks, and other leaks between:
  - interior air and enclosed wall cavities
  - interior air and attics
  - exterior air and interior air, adding humidity to interior air in summer
  - crawl spaces and interior air
- Solved by creating an Air Barrier System

Vapor diffusion

- Water vapor in air moves through permeable materials (those having Perm ratings over 1, see Table 4-1).
- Primary source is water vapor in the air
- Causes:
  - exterior moisture moving into the building in summer
  - interior moisture permeating wall and ceiling finish materials
  - moist crawl space air migrating into the building

MOISTURE PROBLEM EXAMPLE

The owner of a residence complains that her ceilings are dotted with mildew. On closer examination, an energy auditor finds that the spots are primarily around recessed lamps located close to the exterior walls of the building.

What type of moisture problem may be causing the mildew growth, which requires relative humidities over 70% to grow optimally? In reality, any of the forms of moisture transport could cause the problem:

Bulk moisture transport—the home may have roof leaks above the recessed lamps.

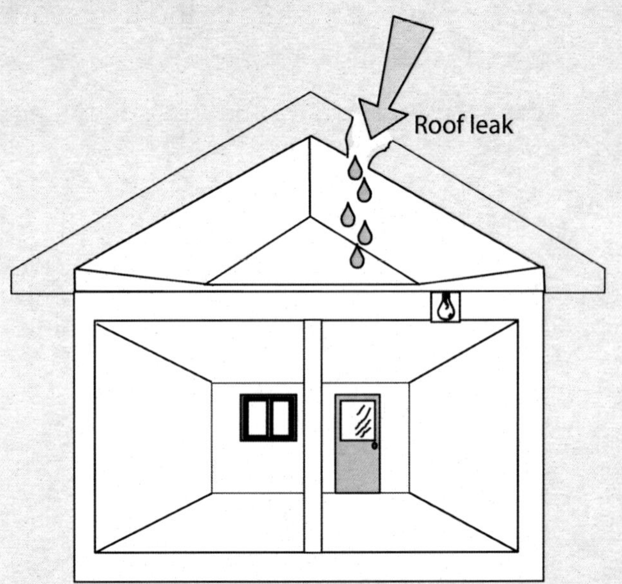

Figure 4-13 Bulk moisture transport

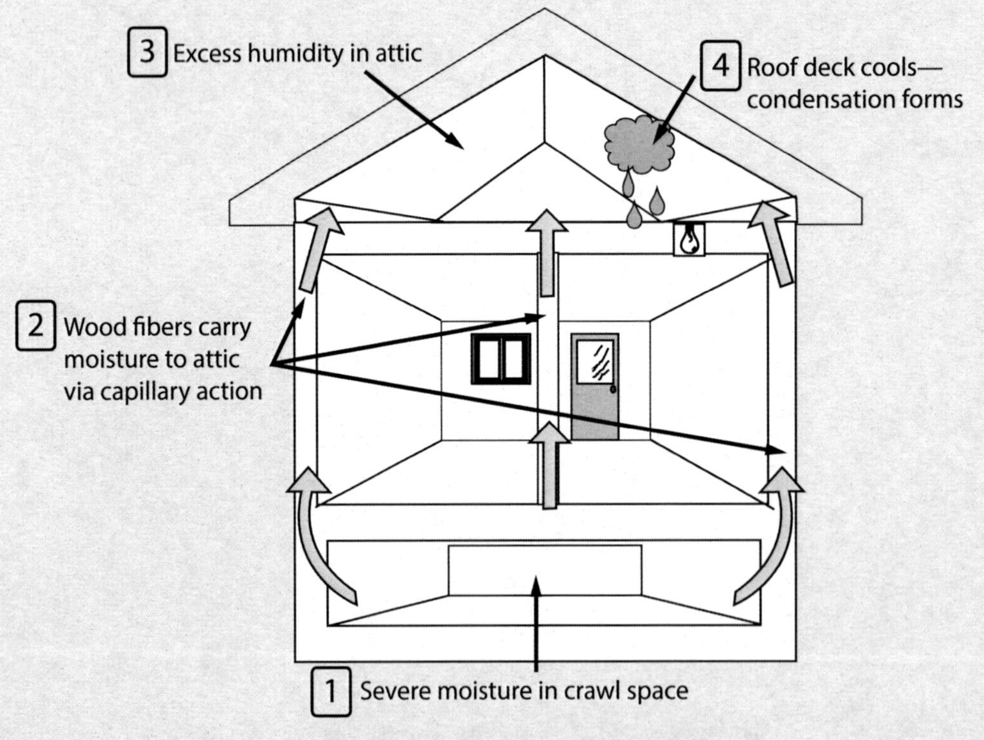

Figure 4-14 Capillary action

Capillary action—the home may have a severe moisture problem in its crawl space or under the slab. Moisture travels up the slab via capillary action into the framing lumber, and all the way into the attic. If the attic air becomes sufficiently moist, it may condense on the surface of the cool roof deck and drip onto the insulation and drywall below.

continued…

continued...

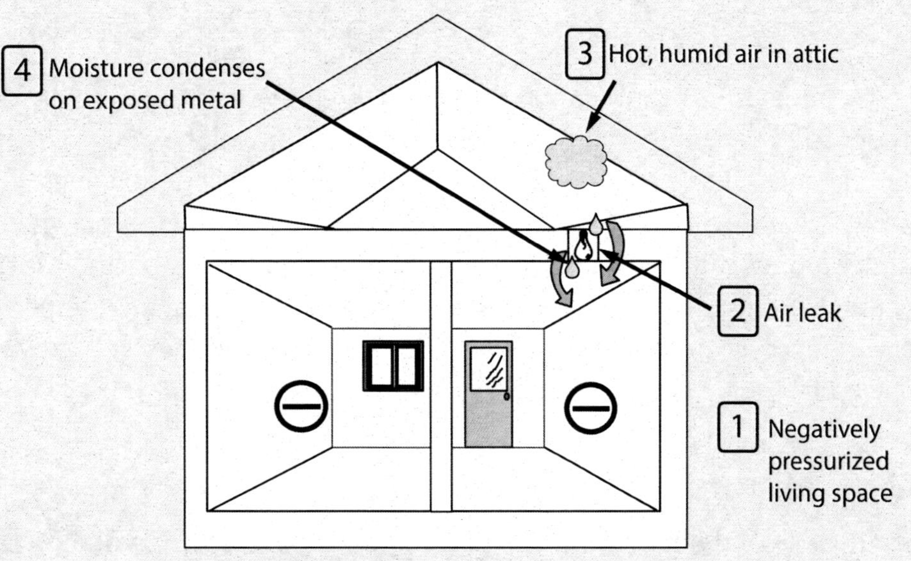

Figure 4-15 Air transport

Air transport—most recessed lighting cans are quite leaky; if the living space is negatively pressurized, hot humid air can be drawn from the attic to the living space. As the air contacts the metal can in the air-conditioned space, water vapor condenses on the exposed surface. *This is the most likely explanation.*

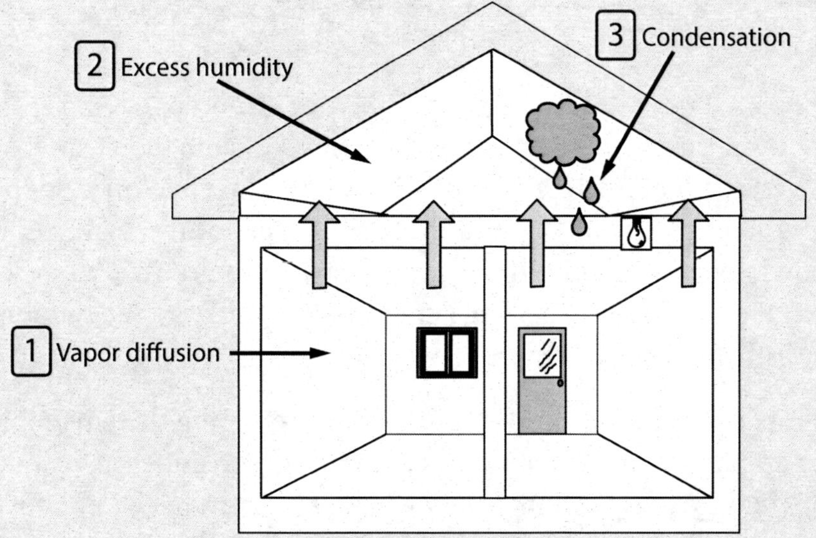

Figure 4-16 Vapor diffusion

Vapor diffusion—the least likely explanation.

Focus on: Vapor Diffusion Retarders (Barriers) and Air Barriers

Vapor diffusion retarders, air retarders, and air/vapor retarders all relate to the interaction of temperature and moisture in and around the building envelope. A vapor barrier or vapor diffusion retarder (VDR) is a material that reduces the rate at which water vapor can move through a material. The older term "vapor barrier" may still be used, however, this is incorrect since it implies that the material stops all of the moisture transfer. Since everything allows some water vapor to diffuse through it to some degree, the term "vapor diffusion retarder" is more accurate. No matter what you call them, they have become an important building issue for most regions. The following information describes what they are, how they work, and when to use them.

The Thermal-Moisture Dynamic

Water vapor moves in and out of a building basically in three ways: with air movement, by diffusion through materials, and by heat transfer. Of the these three, air movement is the dominant force because, like most fluids, air naturally moves from a high pressure area to a lower one by the easiest path possible. This is generally through any available hole in the building envelope. Moisture transfer by air currents is very fast (in the range of several hundred cubic feet of air per minute) and accounts for more than 98% of all water vapor movement in building cavities. Thus it's very important that unintended paths that it may follow be carefully and permanently sealed. The other two driving forces are much slower processes and most common building materials slow moisture diffusion to a large degree, although never stop it completely.

In decades past, buildings did not need to restrict the flow of airborne moisture, since when the building cavities got wet they also generally dried quickly due to the "leaky" construction methods that allowed air to move freely through the building envelope. So the water vapor movement really didn't matter much until the introduction of thermal insulation. When insulation is added, the temperature of the water vapor can drop very quickly since it is being isolated from the heat of the building (in the winter) or from the outdoors in the summer if the building is being air-conditioned.

Whether from the indoors or outdoors, airborne water vapor entering the envelope of the building through holes around plumbing pipes, ductwork, wiring, and electrical outlets are some of the less obvious, yet important, points where air can move in and out of the thermal envelope. During the winter in Northern climates, any warm air entering the walls from the house cools and condenses its water vapor inside building cavities. In the South, humid air does much the same except it comes from the outdoors and condenses inside the wall cavities during the cooling season.

The laws of physics govern how moist air reacts within various temperature conditions. This behavior is technically referred to as "psychrometrics." A psychrometric chart is used by professionals to determine at what temperature and moisture concentration water vapor begins to condense. This is called the "dew point." By understanding how to find the dew point, you will better understand how to avoid moisture problems in a building. (Refer back to "Moisture and Relative Humidity" examples earlier in this chapter.)

Relative humidity (RH) refers to the amount of moisture contained in a quantity of air compared to the maximum amount of moisture the air could hold at the same temperature. As air warms, its ability to hold water vapor increases. As air cools this capacity decreases. For example according to the psychometric chart: air at 68°F with 0.216 ounces of water (H_2O) per pound of air has a 100% RH. The same air at 59°F reaches 100% RH with only 0.156 ounces of water per pound of air. The colder air holds about 28% less moisture than the warmer air does. The moisture that the air can no longer hold condenses on the first cold surface it encounters (the dew point.) If this surface is within an exterior wall cavity, wet insulation and framing will be the result.

Types of Vapor Diffusion Retarders

Vapor diffusion retarders (VDRs) are typically available as membranes or coatings. Membranes are generally thin, flexible materials, but also include thicker sheet materials sometimes termed "structural" vapor diffusion retarders. Materials such as rigid insulation, reinforced plastics, aluminum, and stainless steel are relatively resistant to water vapor diffusion. (Refer to Table 4-1.) These types of vapor diffusion retarders are usually mechanically fastened and then sealed at the joints.

Thinner membrane types of VDRs come in rolls or as integral parts of building materials. A common example of this is aluminum- or paper-faced fiber glass roll insulation. Foil-backed wallboard is another type commonly used. Polyethylene, a plastic sheet material, is the most commonly used VDR in very cold climates.

Most paint-like coatings also retard vapor diffusion. While it was once believed that only coatings with low perm ratings (see next section) constituted the only effective VDR, it is now believed that any paint or coating is effective at restricting most water vapor diffusion in milder climates such as in Florida.

Perm Ratings

The ability of a material to retard the diffusion of water vapor is measured by units known as "perms" or permeability. A perm at 73.4°F (23°C) is a measure of the number of grains of water vapor passing through a square foot of material per hour at a differential vapor pressure equal to one inch of mercury (1" W.C.) Any material with a Perm rating of less than 1.0 is considered a vapor retarder.

Table 4-1 Perm Ratings of Different Materials*

Material	Perm Rating
Aluminum foil (.35 mil)	0.05
Polyethylene plastic (6 mil)	0.06
Plastic-coated insulated foam sheathing	0.4 to 1.2
Asphalt-coated paper backing on insulation	0.40
Vapor barrier paint or primer	0.45
Plywood with exterior glue	0.70
Drywall (unpainted)	50.0
Drywall (painted—non-vapor retarder latex paint)	3.0 or more

** Rating of 1 or less qualifies as a vapor barrier (retarder).*

The Florida Building Code defines *vapor retarder class* as

> a measure of the ability of a material or assembly to limit the amount of moisture that passes through that material or assembly. Vapor retarder class shall be defined using the desiccant method of ASTM E 96 as follows:
>
> **Class I**: 0.1 perm or less.
>
> **Class II**: $0.1 < \text{perm} \leq 1.0$ perm.
>
> **Class III**: $1.0 < \text{perm} \leq 10$ perm.

A good rule to remember is: To prevent trapping any moisture in a cavity, the cold-side material's Perm rating should be at least five times greater than the value of the warm-side.

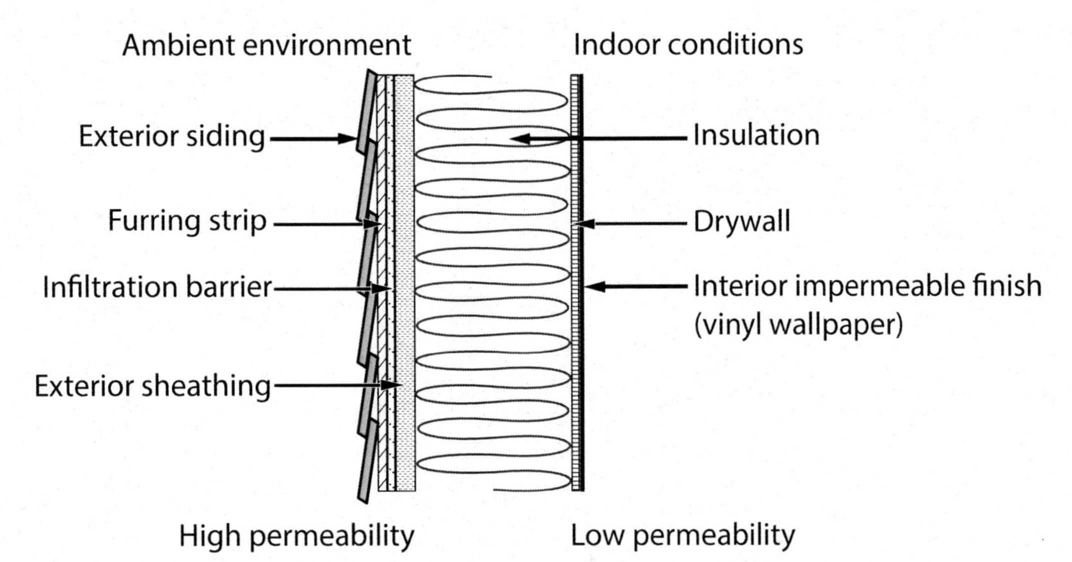

Figure 4-17 Wall cavity detail - typical

In Florida the outside ambient conditions are hot and humid. Figure 4-17 show a typical frame constructed exterior wall:

- *Exterior siding*: This example shows a cementitious fiber-reinforced overlapped siding. This product should be caulked and sealed at the edges.
- *Furring strip*: The idea of a secondary drainage plane is becoming more popular. The furring strip allows for a gap where any water that gets past the siding can drain to, without entering the wall cavity.
- *Infiltration barrier*: Commonly known as housewraps, these materials block air infiltration while still allowing water vapor to pass through. Unless otherwise noted, these products are ***not*** vapor barriers.
- *Exterior sheathing*: Typically oriented strand board (OSB) or plywood.
- *Insulation*: The building cavity is filled with an insulation material suited to the application.
- *Drywall*: The interior surface of the wall is usually covered with a gypsum board or drywall product.
- *Paint or vinyl wallpaper*: Many times a well intentioned builder or homeowner will cover the inside surface of a wall with vinyl wall paper. In our hot-humid climate, the vinyl wallpaper can act as a vapor barrier. With the outdoor climate driving water vapor into the wall cavity, the vinyl wallpaper acts as a barrier which prevents the wall from breathing. The trapped moisture sets up a favorable environment for mold.

A well designed wall cavity will be constructed so that drying occurs to the inside. The wall in Figure 4-18 is constructed so the inside wall is covered with a permeable paint layer instead of an impermeable vinyl wallpaper. Since the Florida climate offers few opportunities when moisture moves from the structure to the outside, the best design choice is to design a wall that has low permeability within the outside materials and high permeability within the inner layers. In this manner, the drying effect of the air conditioning keeps moisture from collecting in the wall cavity.

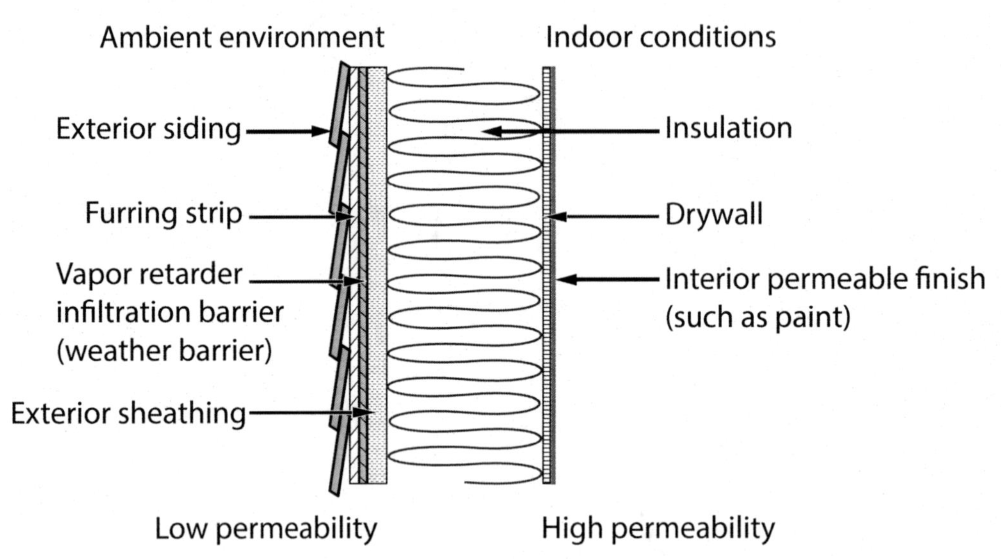

Figure 4-18 Wall cavity detail - well-designed

The concrete masonry unit (CMU) can and does pass some water vapor in Florida's hot-humid climate. This is why the interior insulation must be permeable or semi-permeable (see Figure 4-19). Drywall covered with paint is also permeable, which allows any water vapor to pass through and be removed by the structure's HVAC system.

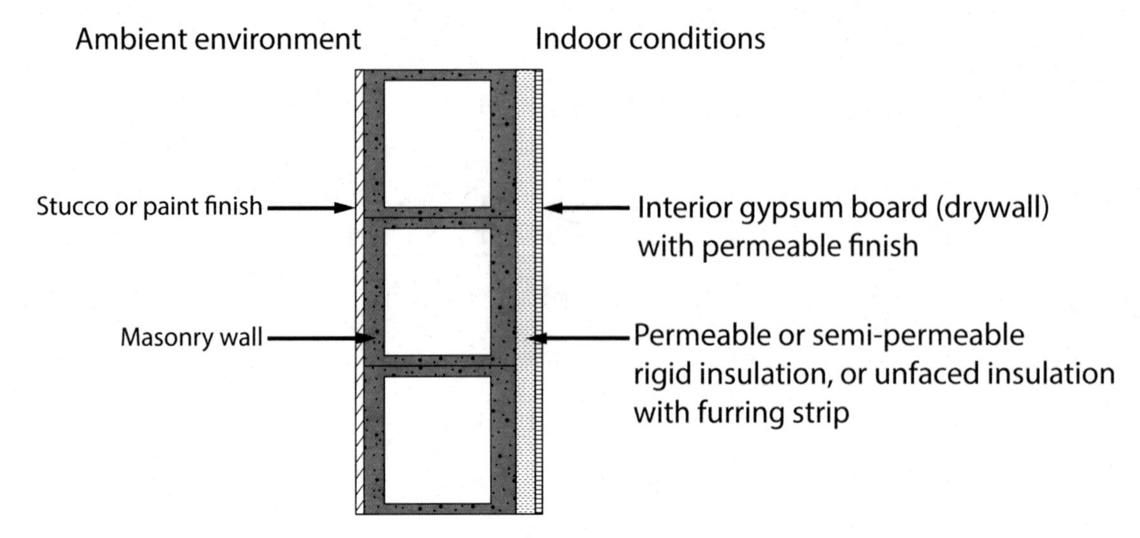

Figure 4-19 Wall cavity detail - concrete block

Installing Vapor Diffusion Retarders

It is important for VDRs to minimize condensation or moisture problems in the following areas of a building: walls, ceilings, and floors; under concrete slabs; and in crawl spaces. A continuous VDR with reliable air sealing is very important if you have a house constructed on a concrete slab. Use a VDR with a perm value of less than 0.50 if you also have a high water table. In moderate heating dominated climates—those with less than 4,000 Heating Degree Days—which includes all of Florida, materials like painted gypsum wallboard and plaster wall coatings impede moisture diffusion to acceptable levels and no further VDR is needed in the walls or ceiling.

(Note: Keep in mind that the American Society of Heating, Refrigerating and Air-Conditioning Engineers [ASHRAE] *Fundamentals Handbook* does not recommend VDRs at all in the walls and ceilings in Florida.)

Air Barriers

Air barriers are intended to block random (unintended) air movement through building cavities. Air barriers can be made of almost anything. A continuous air barrier is an important feature in energy-efficient design not only for the energy it can save but also because the water vapor carried by the air is the primary way moisture related damage gets started in structural cavities. As the water vapor cools it condenses and can lead to structural damage, rotting wood, or mold growth. Air barriers reduce this problem by stopping much of the air movement but still allowing what water vapor that does get in to diffuse back out again.

Some common materials used for this purpose are: "housewrap," plywood, drywall (gypsum) board and foamboard. Many of these materials are also used for insulation, structural purposes, and finish surfaces. What to choose and how to use it depends mainly on where you are building and the climate.

The most common air barrier material in use today is "housewrap." Some wraps have better weathering or water repelling properties than others. All come in a variety of sizes for different purposes and are made of fibrous spun polyolefin plastic, matted into sheets and rolled up for shipping. Sometimes, they also have other materials woven or bonded to them to make them more resistant to tearing.

Housewraps are usually wrapped around the exterior of a house during construction. Sealing all of the joints with "housewrap tape" is a good practice that improves the wrap's performance about 20%. All housewrap manufacturers have a special tape for this purpose.

In wet climates housewrap sometimes reacts poorly with certain kinds of wood siding. Lignin (a natural occurring substance in many species of wood) is water-soluble and acts as a detergent. Like all detergents, it decreases surface tension and so destroys the housewrap's ability to repel water. Field research has shown that wood lignin makes it easier for liquid water to pass into the wall. Certain types of wood siding like redwood, cedar, and manufactured hardboard siding seem to accelerate the problem. The updated article, "Housewraps, Felt Paper and Weather Penetration Barriers" by Paul Fisette, is a good overview of research on this topic (see Resources section at the end of the chapter).

Air/Vapor Retarders

An air/vapor retarder attempts to combine water vapor and the air movement control with one material. This method is most appropriate for wet Southern climates like Florida, where keeping humid outdoor air from entering the building cavities is critical during the cooling season.

In addition to air leakage resistance, permeance, and moisture resistance, two other material characteristics are worth considering: ultraviolet (UV) sunlight resistance and strength. All major housewrap brands have a manufacturer's rated UV exposure time ranging from 120 days to more than one year. Some products are manufactured with antioxidants and UV stabilizers, while others are naturally more resistant by their composition. In the field, however, covering the housewrap as quickly as practical is recommended, as some UV degradation will occur even over a short period, and other unrelated damage to the membrane can be avoided.

Strength of the housewrap can be critical, as wind conditions or adverse job site handling can tear or puncture the material during and after installation. Even small holes can negatively affect overall performance. The inherent strengths of housewrap can be judged on three levels: tensile strength, tear strength, and burst strength. Respectively, these are the material's ability to withstand damage from pulling and stretching; withstand tearing at nail and staple locations; and to withstand separation of material fibers, fabrics, or films. Unfortunately, testing procedures and standards vary between manufacturers, so product comparison is difficult.

The material is generally placed around the perimeter of the building just under the exterior finish, or it may actually be the exterior finish. In many cases it's constructed of one, or a combination of, the following: builder's foil, foamboard insulation, and other exterior sheathings. The key to making this method work effectively is to permanently and carefully seal all of the seams and penetrations, including around windows, doors, electrical outlets, plumbing stacks, and vent fans.

Missed gaps of any size not only increase energy use, but also increase the risk of moisture damage to the house especially during the cooling season. An air/vapor retarder should also be carefully inspected after installation before other work covers it. If small holes are found, you can repair them with caulk or polyethylene or foil tape. Areas with larger holes or tears should be removed and replaced. Patches should always be large enough to cover the damage and overlap any adjacent wood framing.

Note: More information on this topic is found in Chapter 5, "Air Leakage—Materials and Techniques."

Air Barrier System

Sealing against air leakage is mandatory, as stated in the **FBC, Energy Conservation Code, Section R402.4 Air leakage (Mandatory)**. Air leakage can be detrimental to the long term durability of buildings. It can also cause a substantial number of other problems, including:

- High humidity levels in summer and dry air in winter
- Allergy problems
- Radon entry via leaks in the floor system
- Mold growth
- Drafts
- Window fogging or frosting
- Excessive heating and cooling bills
- Increased damage in case of fire

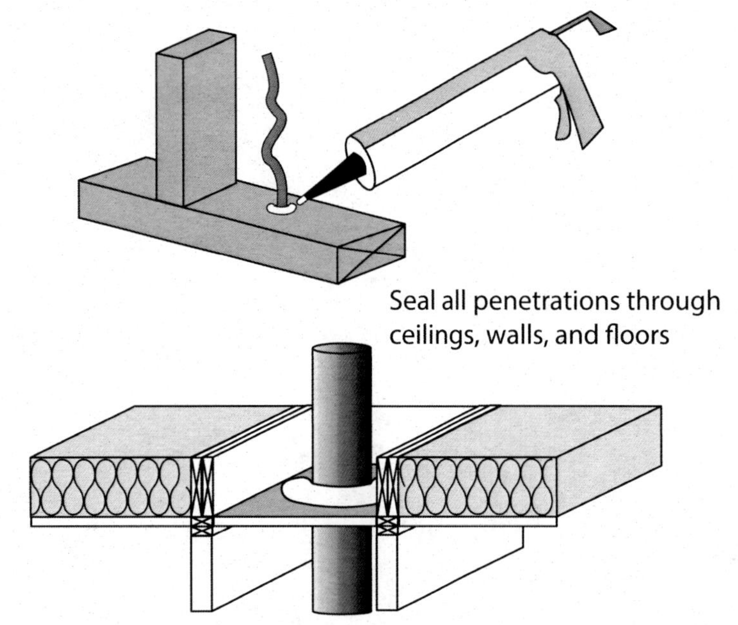

An air barrier system may sound formidable (Figure 4-20), but it is actually a simple concept—seal all leaks between conditioned and unconditioned spaces with durable materials. Achieving success can be difficult without diligent efforts, particularly in buildings with multiple stories and changing roof lines.

Air barriers may also help a building meet local fire codes. One aspect of controlling fires is preventing oxygen from entering a burning area. Most fire codes have requirements to seal air leakage sites.

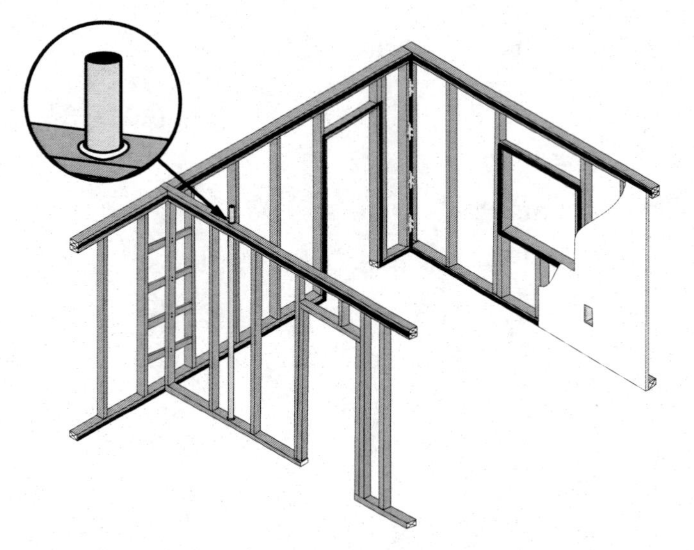

Figure 4-20 Air barrier system requirements

Air sealing reduces air infiltration and prevents water vapor in the air from entering the wall. About 50 cups of water can enter through a single ½-inch hole in a year.

Chapter 5, "Air Leakage—Materials and Techniques," describes air barrier systems—which can be effective with proper installation. They are one of the key features of an energy efficient building. The basic approach is:

- Seal all air leakage sites between conditioned and unconditioned spaces:
  - caulk or otherwise seal penetrations for plumbing, electrical wiring, and other utilities
  - seal junctions between building components, such as bottom plates and band joists between conditioned floors
  - consider air sealing insulating materials, such as plastic foam
- Seal bypasses—hidden chases, plenums, or other air spaces through which attic or crawl space air leaks into the building.
- Install a continuous air barrier material such as the airtight drywall approach or continuous air barrier system.

Thermal Insulation System

Thermal insulation and energy efficient windows are intended to reduce heat loss and gain due to conduction. As with other aspects of energy efficient construction, the key to a successfully insulated building is quality installation.

Substandard insulation not only inflates energy bills, but may create comfort and moisture problems. Key considerations for effective insulation include:

- Install R-values equal to or exceeding those indicated in the Florida Energy Conservation Code.
- Do not compress insulation.
- Provide full insulation coverage of the specified R-value; gaps dramatically lower the overall R-value and can create areas subject to condensation.
- Air seal and insulate knee walls and other attic wall areas.
- Support insulation so that it remains in place, especially in areas where breezes can enter or rodents may reside or disturb it.

More information on insulation can be found in Chapter 6 of this publication, "Insulation—Materials and Techniques."

HVAC System

The heating, ventilation, and air conditioning (HVAC) system is designed to provide comfort and improved air quality throughout the year, particularly in summer and winter. Energy efficient buildings, particularly passive solar designs, can reduce the number of hours during the year when the HVAC system is needed.

These systems are often not well designed and may not be installed to perform as intended. As a consequence, buildings often suffer higher heating and cooling bills and have more areas with discomfort than necessary. Poor HVAC design often leads to moisture and air quality problems, too.

One major issue concerning HVAC systems is their ability to create pressure imbalances in the building. The side-bar on the following page shows that duct leaks can create serious problems. In addition, even closing a few doors can create situations that may endanger human health. (See "Carbon Monoxide Disaster," later in this chapter, for more detail.)

Pressure imbalances can increase air leakage, which may draw additional moisture into the building. Proper duct design and installation helps prevent pressure imbalances from occurring.

HVAC systems must be designed and installed properly, and maintained regularly by qualified professionals to provide continued efficient and healthy operation.

DUCT LEAKS AND INFILTRATION

Forced-air heating and cooling systems should be *balanced*—the amount of air delivered through the supply ducts should be equal to that drawn through the return ducts. If the two volumes of air are unequal, pressure imbalances may occur in the home, resulting in increased air leakage and possible health and safety problems. The *Florida Building Code Sixth Edition (2017), Energy Conservation* prohibits using wall, ceiling, and floor cavities for supply and return air in residential air distribution systems (**FBC, Energy Conservation, Section R403.3.5 Building cavities (Mandatory)**).

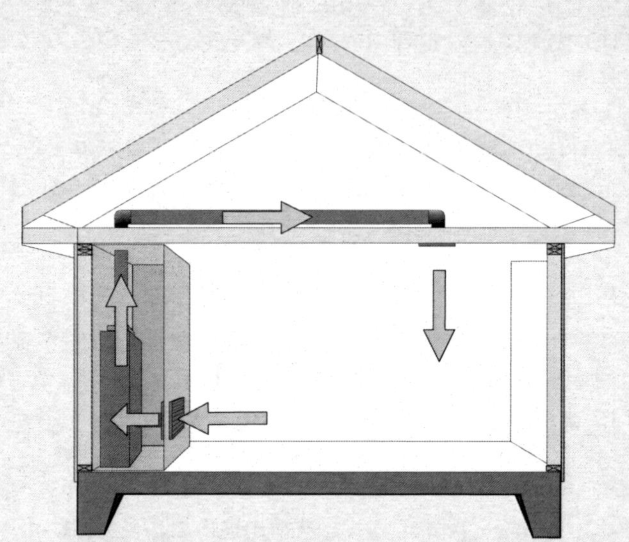

Figure 4-21 Balanced air distribution

If *supply ducts* in unconditioned areas have more leaks than return ducts:

- Heated and cooled air will escape to the outside, increasing energy costs.
- Less air volume will be "supplied" to the building, so the pressure inside the building may become negative, thus increasing air infiltration.
- The negative pressure can actually *backdraft* flues—pull exhaust gases back into the home from fireplaces and other combustion appliances. The health effects can be deadly if flues contain substantial carbon monoxide.

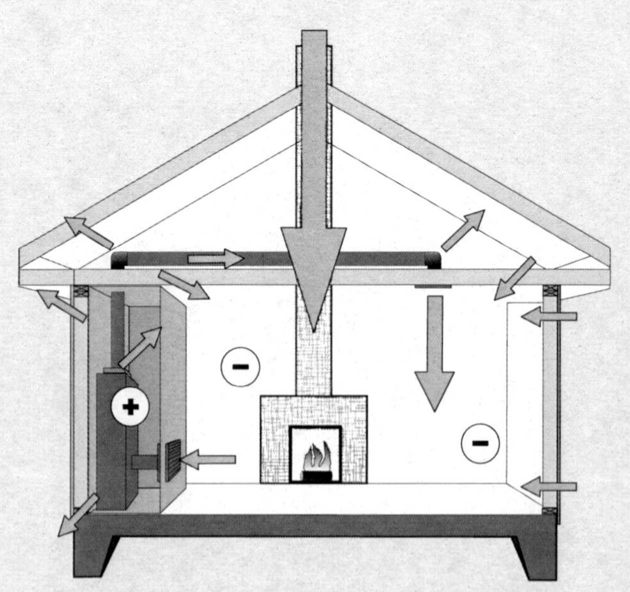

Figure 4-22 Air leaks in supply ducts

continued…

continued...

If *return ducts* in unconditioned spaces leak:

- The home can become pressurized, thus increasing air leakage out of the envelope.
- Hot, humid air is pulled into system ducts in summer; cold air is drawn into the ducts in winter.
- Human health may be endangered if ducts are located in areas with radon, mold, or toxic chemicals from soil termite treatments, paints, cleansers, and pesticides.
- If combustion appliances are located near return leaks, the negative pressure created by the leaks can be great enough to backdraft flues and chimneys.

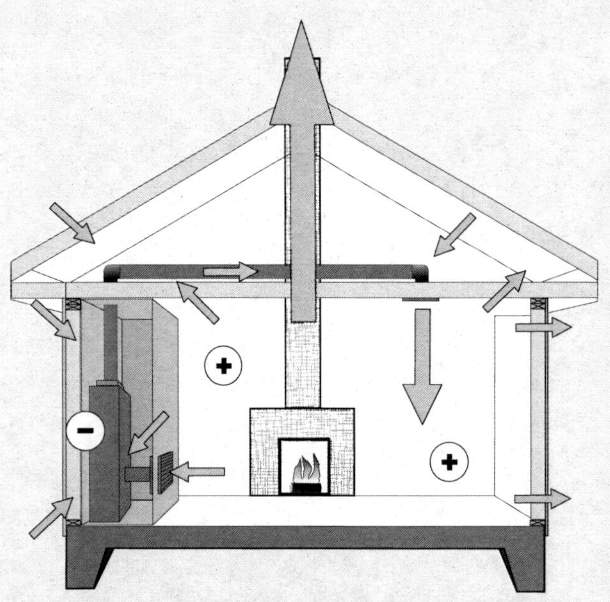

Figure 4-23 Air leaks in return system

Pressure differences can also result in homes with tight ductwork if the home only has one or two returns. When interior doors are closed it may be difficult for the air in these rooms to circulate back to the return ducts. The pressure in the closed-off rooms increases, and the pressure in rooms open to the returns decreases.

Installing multiple returns, "jumper" ducts that connect closed-off rooms to the main return, and undercutting doors to rooms without returns will alleviate these problems. See Chapter 8, "Heating, Ventilation, Air Conditioning," and Chapter 9, "Duct Design and Sealing," of this publication for further details.

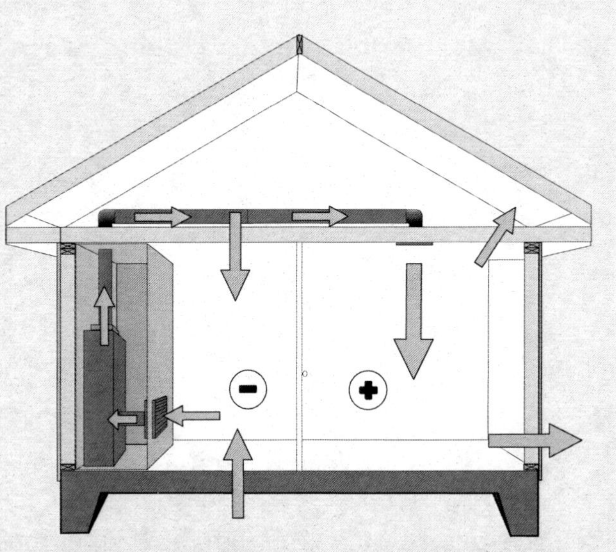

Figure 4-24 Return blocked by door

WALL MOISTURE EXAMPLE

The following pages describe two examples of building science problems due to common failures of the home's systems. These problems can be minimized through careful attention to the construction techniques described in this book.

A homeowner notices that paint is peeling on the exterior siding near the base of a bathroom wall. The drywall interior has mildew and the baseboard paint is peeling as well. What happened?

1. The interior of the wall has numerous air leaks—an air barrier system failure.
2. The bathroom has no return air duct, its door is usually closed and is not undercut at the bottom. Therefore, when the heating and cooling system operates the room becomes pressurized. This is an HVAC system failure.
3. The bath fan is installed improperly and does not exhaust moist air—another HVAC system failure.
4. When air leaks into the wall, it carries substantial water vapor, thus the failure of the air barrier and HVAC systems has led to a moisture control system failure.
5. The interior wall has a polyethylene vapor barrier/retarder, which was installed by a builder using Northern building construction techniques. The exterior wall has CDX plywood sheathing, which also can serve as a vapor barrier/retarder.
6. When the air leaks carry water vapor into the wall cavity, the two vapor barriers/retarders hinder drying—a moisture control system failure.
7. In winter, the inner surface of the plywood sheathing will be several degrees cooler than foam sheathing would have been. Thus, the plywood-sheathed wall has more potential for condensation—a thermal insulation system failure.
8. As the water vapor condenses on the sheathing, it runs down the wall and pools on the bottom plate of the wall. Now the following problems occur:
   - The water threatens to cause structural problems by rotting the wall framing.
   - It wets the drywall, causing mold to grow.
   - It travels through the unsealed back surfaces of the wood siding and baseboard, causing the paint to peel when it soaks through the wood.
   - The multiple failures of the building systems create a potential structural disaster.

To solve this moisture problem the builder must address all of the failures. If only one aspect is treated, the problem may even worsen.

continued...

continued...

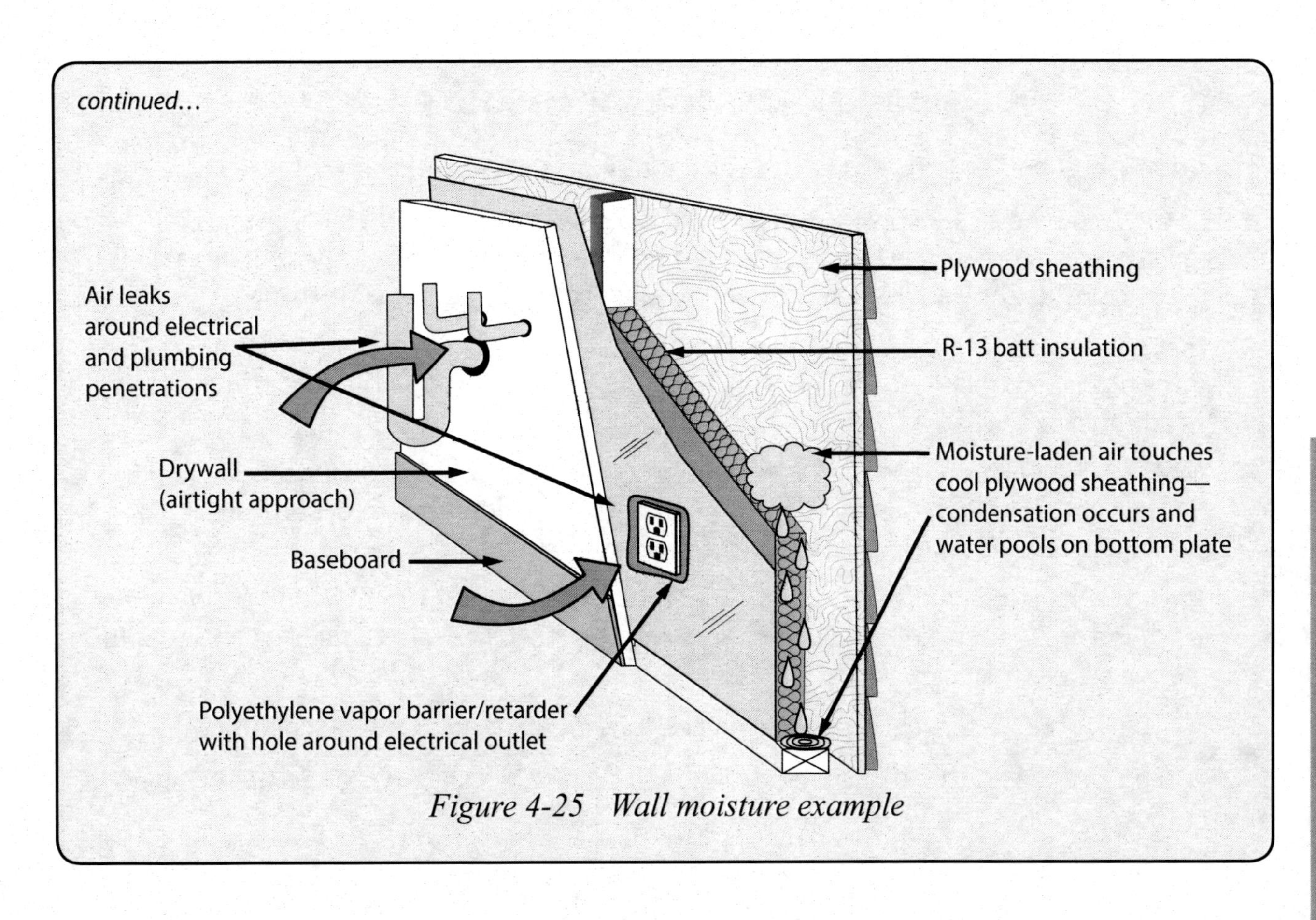

Figure 4-25 Wall moisture example

CARBON MONOXIDE DISASTER

1. A home has been built to airtight specifications—an air barrier system success.
2. However, the home's ductwork was not well sealed—a HVAC system failure. It has considerably more supply leakage than return leakage which creates a strong negative pressure inside the home when the heating and cooling system operates.
3. The homeowners are celebrating winter holidays. With overnight guests in the home, many of the interior doors are kept closed. The home has only a single return in the main living room.
4. When the system operates, the rooms with closed doors become pressurized, while the central living area with the return becomes significantly depressurized. Because the building is very airtight, it is easier for these pressure imbalances to occur.
5. The home has a beautiful fireplace without an outside source of combustion air. When the fire in the unit begins to dwindle, the following sequence of events could spell disaster for the household:
   - The fire begins to smolder and produces considerable carbon monoxide.
   - Because the fire's heat dissipates, the draft pressure, which draws gases up the flue, decreases.
   - The reduced output of the fire causes the thermostat to turn on the heating system. Due to the duct problems, the blower creates a relatively high negative pressure in the living room.
   - Because of the reduced draft pressure in the fireplace, the negative pressure in the living room causes the chimney to backdraft—the flue gases are drawn back into the home. They contain carbon monoxide and can now cause severe, if not fatal, health consequences for the occupants.

This example is extreme, but similar conditions occur in a number of Florida homes each year. Comparable conditions can also occur with gas-fired hot water heaters—and all fossil fuel-burning equipment. The solution to the problem is not to build leakier homes — they can experience similar pressure imbalances. Instead, eliminate the causes of pressure imbalances, as described in detail in Chapter 8, "Heating, Ventilation, Air Conditioning," and install an external source of combustion air for the fireplace.

continued...

continued...

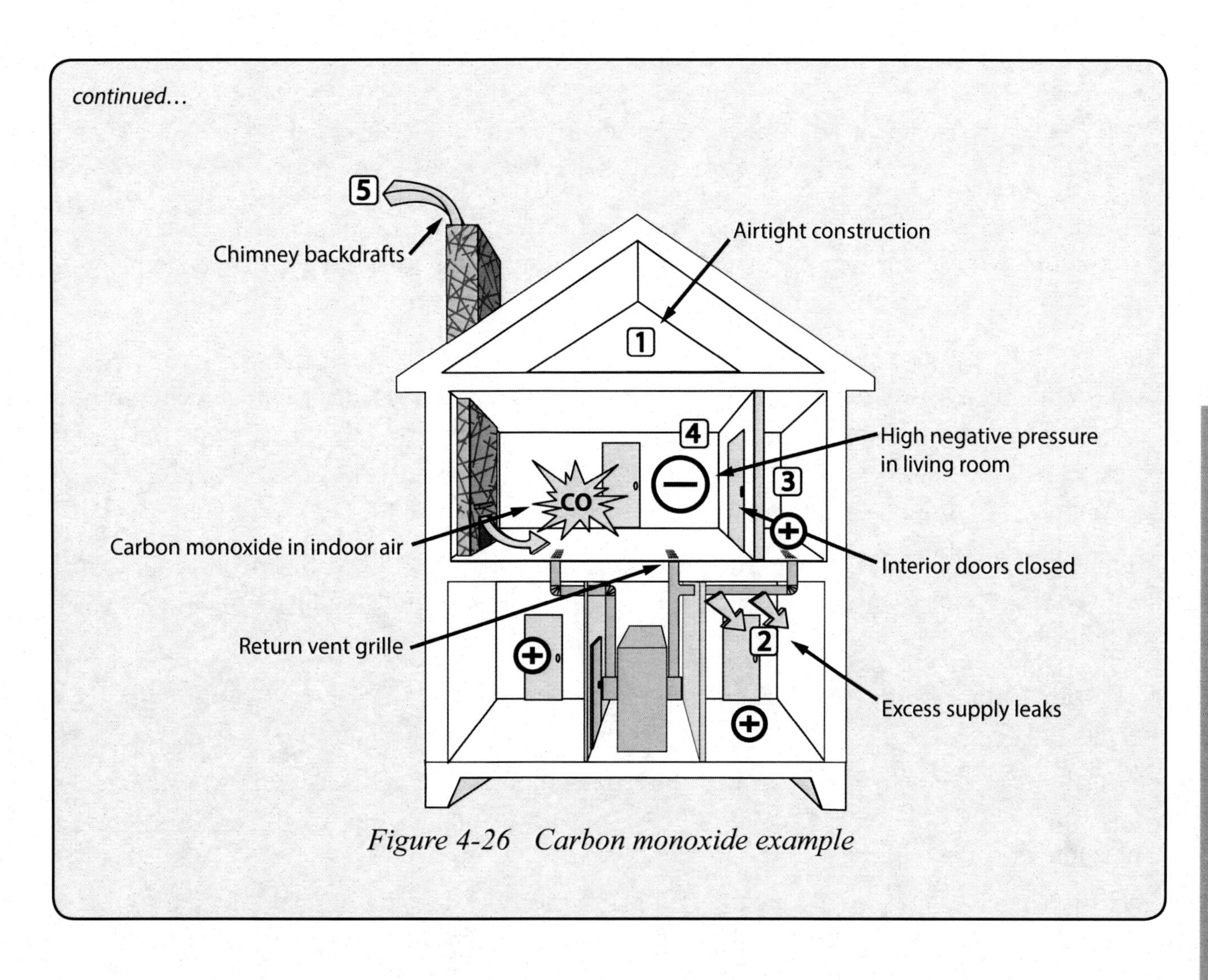

Figure 4-26 Carbon monoxide example

Resources

Note: Web links were current at the time of publication, but can change over time.

American Society of Heating, Refrigerating and Air Conditioning Engineers (ASHRAE). (2017). *2017 ASHRAE Handbook—Fundamentals (2017 Edition)*. Atlanta, GA: American Society of Heating, Refrigerating, and Air-Conditioning Engineers, Inc. Retrieved from https://www.techstreet.com/ashrae/lists/ashrae_handbook.tmpl?ashrae_auth_token=

Fisette, P. (2001; webpage updated 2007). *Housewraps, Felt Paper and Weather Penetration Barriers*. University of Massachusetts Amherst, Building and Construction Technology Program. Retrieved from http://bct.eco.umass.edu/publications/by-title/housewraps-felt-paper-and-weather-penetration-barriers/?q=bmatwt/publications/articles/housewraps_feltpaper_weather_penetration_barriers.html

Lstiburek, J. (2005). Builder's Guide to Hot-Humid Climates. Building Science Corporation. Retrieved from https://buildingscience.com/bookstore/ebook/ebook-builders-guide-hot-humid-climates

My Florida Home Energy. (n.d.). Retrieved December 20, 2017, from http://www.myfloridahomeenergy.com/

A useful resource with a wide array of information on energy and water efficiency, including The Energy Efficient Home series of fact sheets, available at http://www.myfloridahomeenergy.com/help/library

NAHB Research Center, Southface Energy Institute, & Oak Ridge National Laboratory. (2000). *Weather-Resistive Barriers: How to select and install housewrap and other types of weather-resistive barriers* (Technology Fact Sheet). U.S. Department of Energy, Office of Building Technology, State and Community Programs, Energy Efficiency and Renewable Energy. Retrieved from http://www.southface.org/factsheets/WRB-Weather-resist-barriers%2000-769.pdf

Southface Energy Institute, & Oak Ridge National Laboratory. (2000). *Combustion Equipment Safety: Provide Safe Installation for Combustion Appliances* (Technology Fact Sheet). U.S. Department of Energy, Office of Building Technology, State and Community Programs, Energy Efficiency and Renewable Energy. Retrieved from http://www.southface.org/factsheets/CES-Combustion-safety%2000-784.pdf

Wilson, A. (2003, July). "Moisture Control in Buildings: Putting Building Science in Green Building." *Environmental Building News*, 12(7). Retrieved from https://www.buildinggreen.com/feature/moisture-control-buildings-putting-building-science-green-building

5 Air Leakage—Materials and Techniques

Air leakage is a major problem for both new and existing buildings and can:

- Contribute to over 30% of the cooling and heating costs
- Create comfort and moisture problems
- Pull pollutants such as radon and mold into buildings
- Serve as easy access for insects and rodents

To reduce air leakage effectively requires a *continuous air barrier system*—a combination of materials linked together to create a tight **building envelope** (Figure 5-1). An effective building envelope should form both a continuous air barrier and an insulation barrier (insulation is discussed in detail in Chapter 6). An air barrier minimizes air currents inside the cavities of the building envelope which helps maintain insulation R-values.

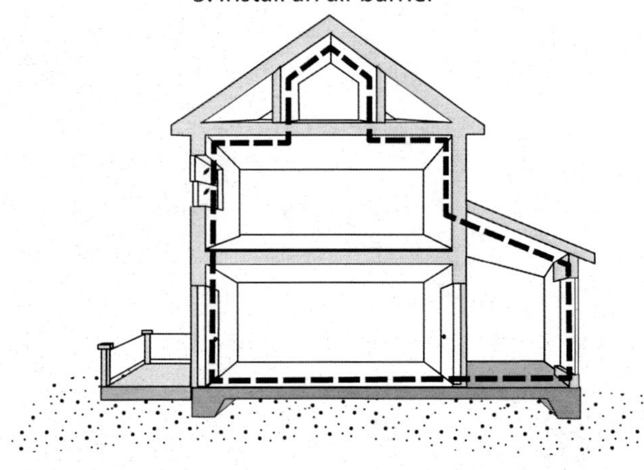

Figure 5-1 Creating a continuous air barrier system

The air barrier should seal all leaks through the building envelope—the boundary between the conditioned portion of the building and the unconditioned area. Most standard insulation products are not effective at sealing air leakage. The R-value for these products may drop if air leaks through the material.

The builder should work with his or her own crew and subcontractors to seal all penetrations through the envelope. Then, continuous material should be installed around the envelope. It is critical in the air sealing process to use durable materials and install them properly. Controlling air leakage is mandatory in the *Florida Building Code 2017, Energy Conservation,* under **Section R402.4 Air leakage (Mandatory)**.

Infiltration Control

Section R402.4 Air leakage (Mandatory) and **Table R402.4.1.1 Air Barrier and Insulation Installation** of the *Florida Building Code, Energy Conservation* describe the prescriptive requirements for air infiltration in:

- All joints, seams and penetrations
- Openings between window and door assemblies and their respective jambs and framing
- Utility penetrations
- Dropped ceilings or chases adjacent to the thermal envelope
- Knee walls
- Site-built windows, doors and skylightings
- Walls and ceilings separating a garage from conditioned space
- Behind tubs and showers on exterior walls
- Common walls between dwelling units
- Attic access openings
- Rim joist junction
- HVAC register boots
- Shower/tub on exterior walls
- Other sources of infiltration (fireplace)

An infiltration barrier shall provide a continuous air barrier from the foundation to the top plate of the ceiling and shall be sealed at the foundation, the top plate, at openings in the wall plane (windows, doors, etc.) and at the seams between sections of infiltration barrier material. When installed on the interior side of the walls, such as with insulated face panels with an infiltration barrier, the infiltration barrier shall be sealed at the foundation or subfloor. This prevents wind from circulating air within the insulation. If properly sealed at the seams and ends, plywood and builder's felt will serve as an infiltration barrier, but not as a moisture retardant. A vapor retarder will essentially stop moisture transmission or diffusion. Common vapor retarders include 6-mil polyethylene sheet and aluminum foil-backed paper or boards. Contrary to northern construction practices, a vapor retarder, including vinyl wall coverings, installed next to the conditioned space is not recommended. Otherwise, water may condense on the vapor retarder surface within the wall cavity when the inside temperature is below the outside dew point in the summer. This could wet and degrade insulation, deteriorate wall components, and contribute to mold and mildew. Vapor retarders are not recommended on the conditioned side of walls in Florida buildings. In fact, the ASHRAE 2013 *Fundamentals Handbook* does not recommend vapor retarders at all in Florida (on walls or ceilings). It should be noted that vinyl wallpaper acts like a vapor barrier. Placement on the conditioned side of the wall may cause moisture problems.

Air Barriers

Housewraps serve as exterior air barriers and help reduce air leakage through outside walls. Most products block only air leakage, not vapor diffusion, so they are not vapor retarders. (See the section 'Focus On: Vapor Diffusion Retarders (Barriers) and Air Barriers' in Chapter 4, "The Building as a System," for more detailed information on this topic.)

Typical products are rolled sheet materials that can be affixed and sealed to the wall between the sheathing and exterior finish material (Figure 5-2). For best performance, a housewrap must be sealed with caulk or tape at the top and bottom of the wall and around any openings, such as for windows, doors, and utility penetrations, and be installed per manufacturer's specifications.

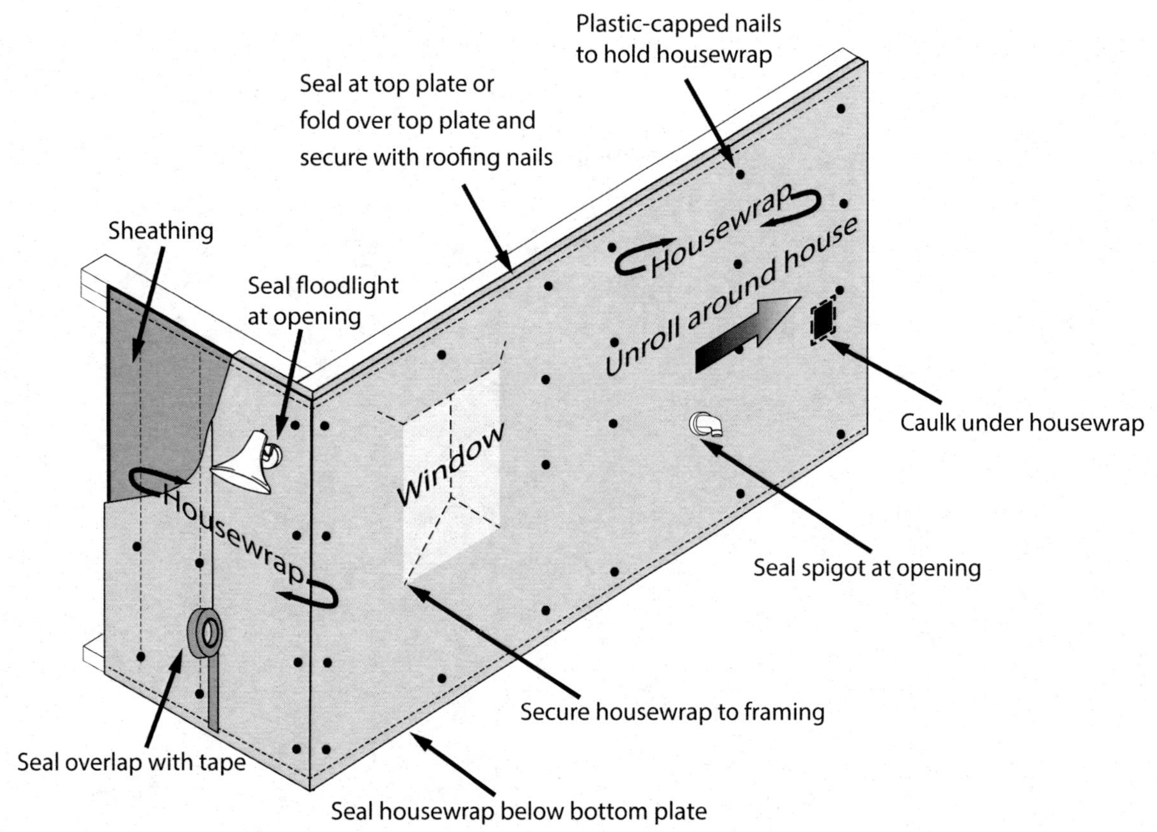

Figure 5-2 Housewrap installation details

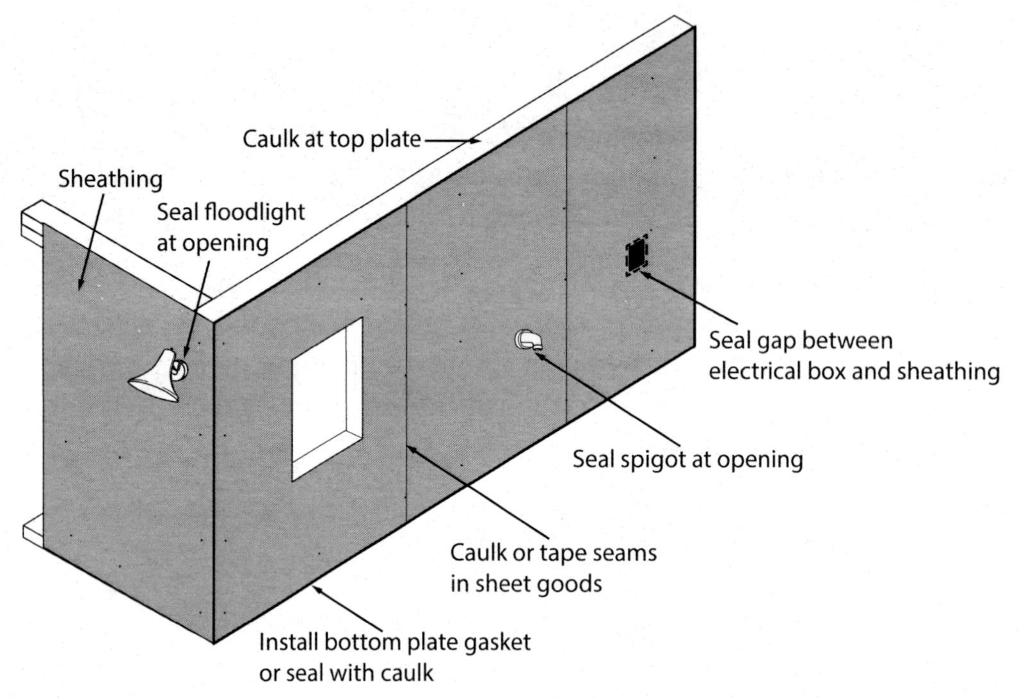

Figure 5-3 Exterior air sealing of sheathing

A housewrap can help reduce air leakage through exterior walls, but by itself is not a continuous air barrier for the entire envelope, and hence is not a substitute for the airtight drywall approach. Housewraps are recommended primarily as further insurance against air leakage and, because they can block liquid water (bulk water) penetration, can help protect the building from moisture damage. In some instances, the exterior sheathing may be used as an outside air barrier (Figure 5-3). Careful sealing of all seams and penetrations, including windows, is required (Figure 5-4).

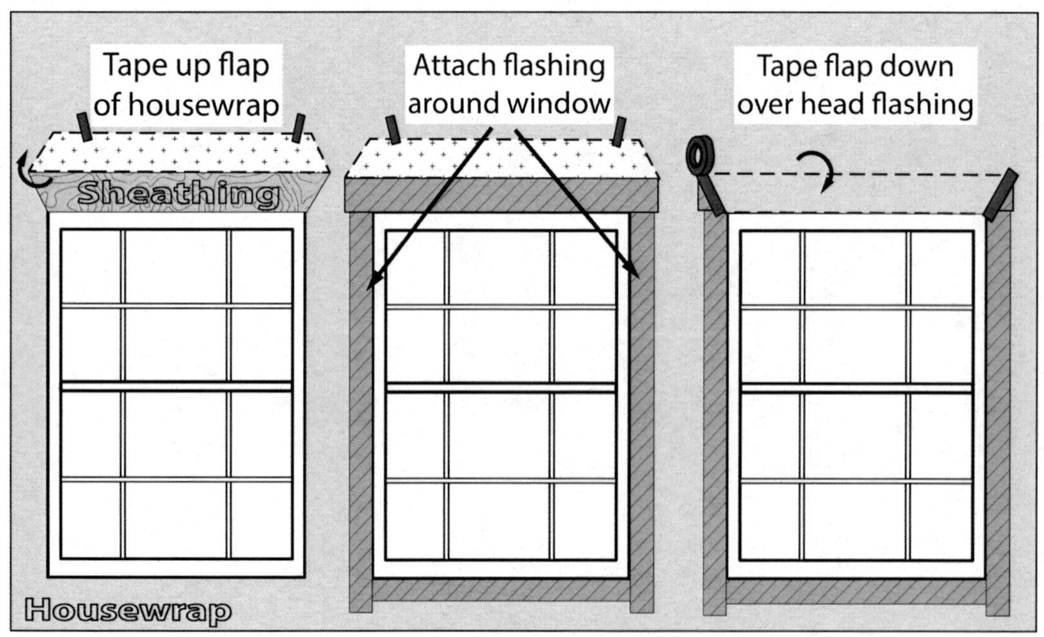

Figure 5-4 Window taping detail

Materials

Most air barrier systems rely on a variety of caulks, gaskets, weather-stripping, and sheet materials, such as plywood, drywall, and housewraps.

Use a combination of these different air sealing materials:

- **Caulk**—Use to seal gaps less than ¼". Select grade (interior, exterior, high temperature) based on application.
- **Spray foam**—Expands to fill large cracks and small holes. It can be messy; consider new, water-based foams. **Not** recommended near flammable applications (flue vents, etc.). Also not permitted around PVC pipe. May be prohibited around windows by window manufacturer.
- **Gaskets**—Can be applied under the bottom plate before an exterior wall is raised, or used to seal drywall to framing.
- **Housewrap**—Installed between the sheathing and exterior finish material. Must be sealed with tape or caulk to form an airtight seal. Resists liquid water and is *not* a vapor retarder.
- **Sheet goods (plywood, drywall, rigid foam insulation)**—These are solid materials which form the building envelope. Air will only leak at the seams or through unsealed penetrations.
- **Sheet metal**—Used with high temperature caulk for sealing high temperature components, such as flues, to framing.
- **Polyethylene plastic**—Inexpensive material for airsealing that also stops vapor diffusion. Must have all edges and penetrations sealed to be effective air barrier. This material is recommended for use under slabs. Incorrect application in wall cavities can lead to mold and mildew problems.
- **Weatherstripping**—Used to seal moveable components, such as doors and windows.

SEAL PENETRATIONS AND BYPASSES

The first step for successfully creating an air barrier system is to seal all of the holes in the building envelope. Too often, builders concentrate on air leakage through windows, doors, and walls, and ignore areas of much greater importance. Many of the key sources of leakage—called *bypasses* (Figure 5-5)—are hidden from view behind soffits for cabinets, bath fixtures, dropped ceilings, chases for flues and ductwork, recessed lighting fixtures, or insulation. Attic access openings and whole house fans are also common bypasses. Sealing these bypasses is critical to reducing air leakage in a building and maintaining the performance of insulation materials. Table 5-1 provides examples of commonly used sealants for various types of leaks.

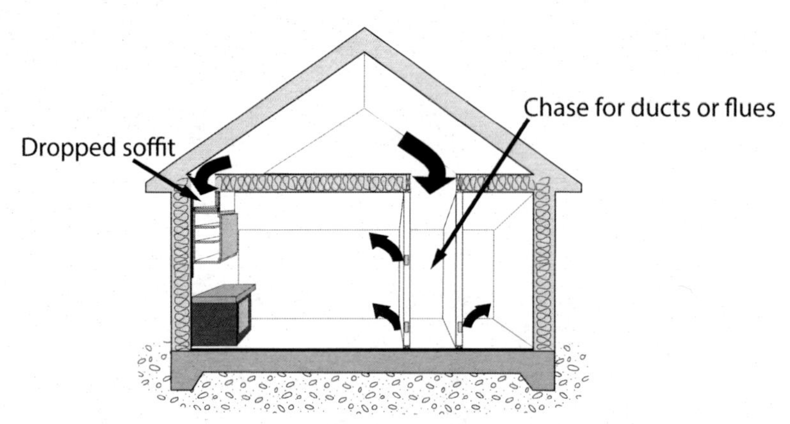

Figure 5-5 Air leakage through bypass

Table 5-1 Leaks and Sealants

Type of Leak	Commonly Used Sealants
Thin gaps between framing and wiring, pipes or ducts through floors or walls	40-year caulking; one-part polyurethane is recommended
Leaks into attics, cathedral ceilings, wall cavities above first floor	Firestop caulking, foam sealant (latex or polyurethane)
Gaps, or cracks or holes over 1/8 inch in width not requiring firestop sealant	Gasket, foam sealant, or stuff with fiber glass or backer rod, and caulk on top
Open areas around flues, chases, plenums, plumbing traps, etc.	Attach and caulk a piece of plywood or foam sheathing material that covers the entire opening • Seal penetrations • **If a flue requires a non-combustible clearance, use a noncombustible metal collar, sealed in place, to span the gap**
Final air barrier material system	Install airtight drywall approach or other air barrier

The guidelines that follow in Figure 5-6 show important areas that should be sealed to create an effective air barrier. The builder must clearly inform subcontractors and workers of these details to ensure that the task is accomplished successfully. Where appropriate, FBC references are provided.

1. *Slab Floors*—If a house is to be constructed on a concrete slab, a vapor retarder of plastic sheeting should be placed under the slab. Without a vapor retarder, moisture will migrate from the ground through the porous slab and into the house. If a house is to be built off-grade with an unvented crawl space, a sheet of 6-mil polyethylene plastic, overlapped 6 inches and taped at joints, should always be placed directly on the ground under the house to prevent moisture from moving upward from the soil.

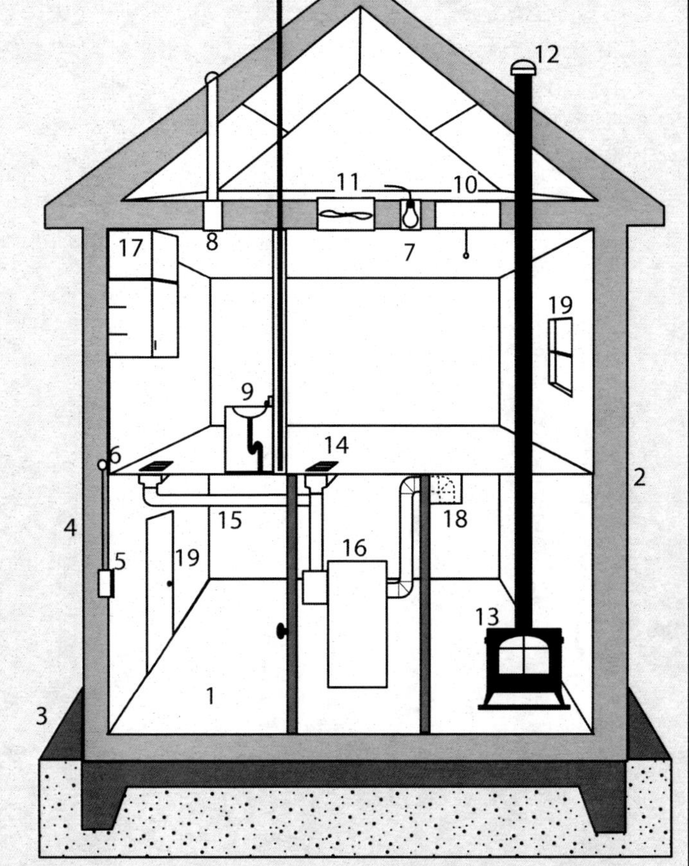

Figure 5-6 Typical home air leakage sites

1. *Floor Joist*—seal sill plates in basements and unvented crawl spaces. Caulk or gasket rim or band joists between floors in multi-story construction.
2. *Bottom Plate*—use either caulk or gasket between the plate and subflooring.
3. *Electrical Wiring*—use wire-compatible caulk or spray foam to seal penetrations.
4. *Electrical Boxes*—use approved caulk to seal wiring on the outside of electrical boxes. Seal between the interior finish material and boxes.
5. *Electrical Box Gaskets*—caulk foam gaskets to all electrical boxes in exterior and interior walls before installing coverplates.
6. *Recessed Light Fixtures*—consider using surface-mounted light fixtures rather than recessed lights. When used, specify airtight models rated for insulation contact sealed to drywall. Ensure fixtures meet appropriate fire codes and manufacturer specifications.
7. *Exhaust Fans*—seal between the fan housing and the interior finish material. Choose products with tight-fitting backdraft dampers.
8. *Plumbing*—locate plumbing in interior walls, and minimize penetrations. Seal all penetrations with sealant or caulk.
9. *Attic Access in Conditioned Spaces*—weatherstrip attic access openings. For pull-down stairs, use latches to hold the door panel tightly against the weatherstripping. Cover the attic access opening with an insulated box (insulated to an equivalent level to the insulation on surrounding surfaces).

continued...

continued...

10. *Whole House Fan*—use a panel made of rigid insulation or plastic to seal the interior louvers.
11. *Flue Stacks*—install a code-approved flue collar and seal with fire-rated caulk (Figure 5-7).
12. *Combustion Appliances*—closely follow all codes for firestopping measures, which reduce air leakage as well as increase the safety of the appliance. Make certain all combustion appliances, such as stoves, inserts, and fireplaces, have an outside source of combustion air and tight-fitting dampers or doors.

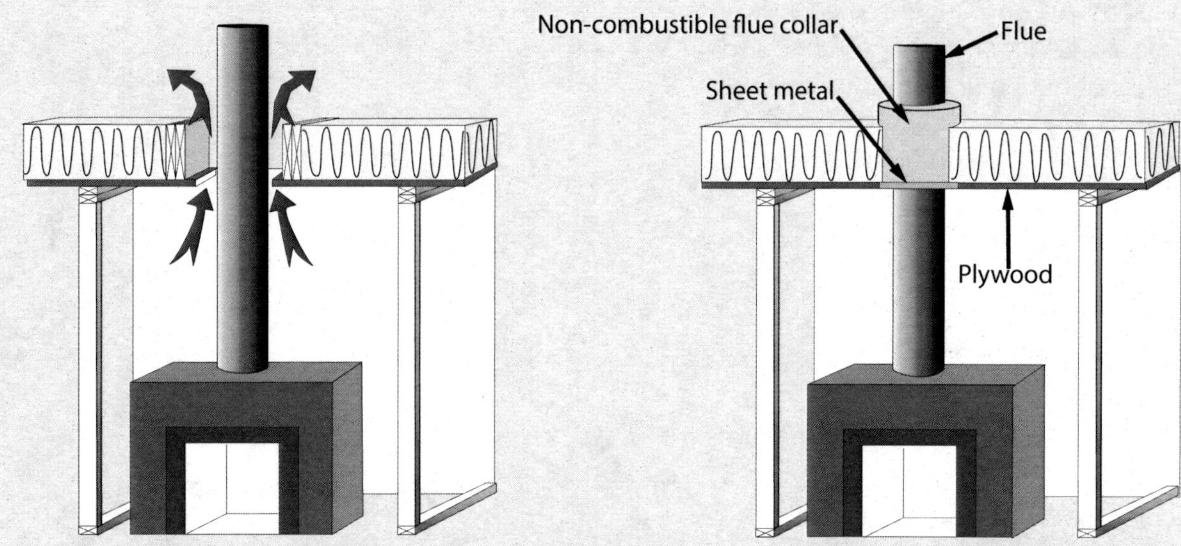

Figure 5-7 Sealing bypasses for flues

13. *Return and Supply Registers*—seal all boots connected to registers or grilles to the interior finish material.
14. *Ductwork*—seal all joints in supply and return duct systems with mastic. Mechanically attach duct systems to prevent dislocation and massive leakage. See **Table C403.2.7.2 Duct System Construction and Sealing** in the Florida Energy Conservation Code.
15. *Air Handling Unit* (for heating and cooling system)—seal all cracks and unnecessary openings with mastic according to **Table C503.2.7.2 Duct System Construction and Sealing** in the Florida Energy Conservation Code**.** Seal service panels with UL 181 listed and labeled tape.

continued...

continued...

16. *Dropped Ceiling Soffit*—use sheet material and sealant to stop air leakage from attic into the soffit or wall framing, then insulate (See Figure 5-8 for more details).
17. *Chases* (for ductwork, flues, etc.)—prevent air leakage through these bypasses with sheet materials and sealants (See Figures 5-7 and 5-9 for more details).
18. *Windows and Sliding Glass Doors*—Must meet allowable air infiltration rates found in the Florida Energy Conservation Code, **Section R402.4.3 Fenestration air leakag**e. (Exceptions: site-built windows, and skylights.)

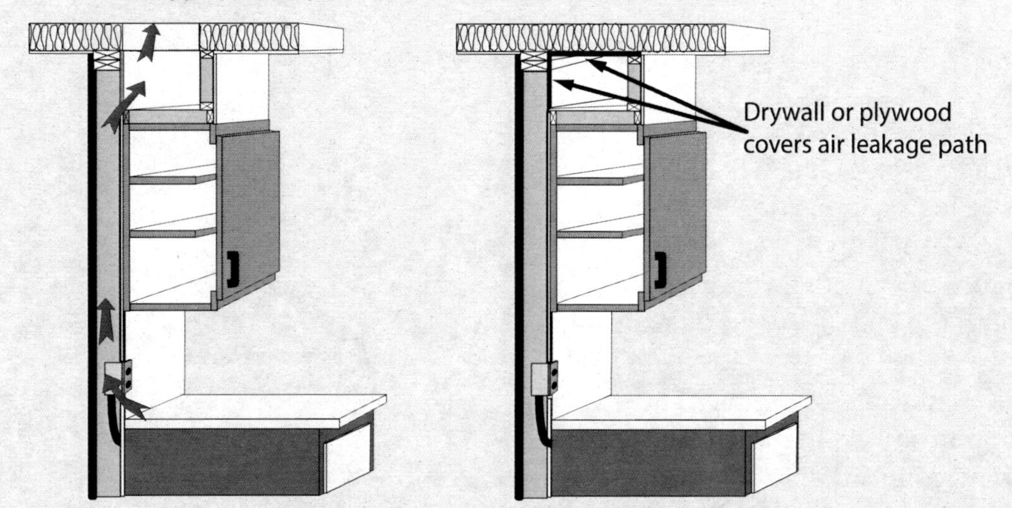

Figure 5-8 Sealing ceiling soffit

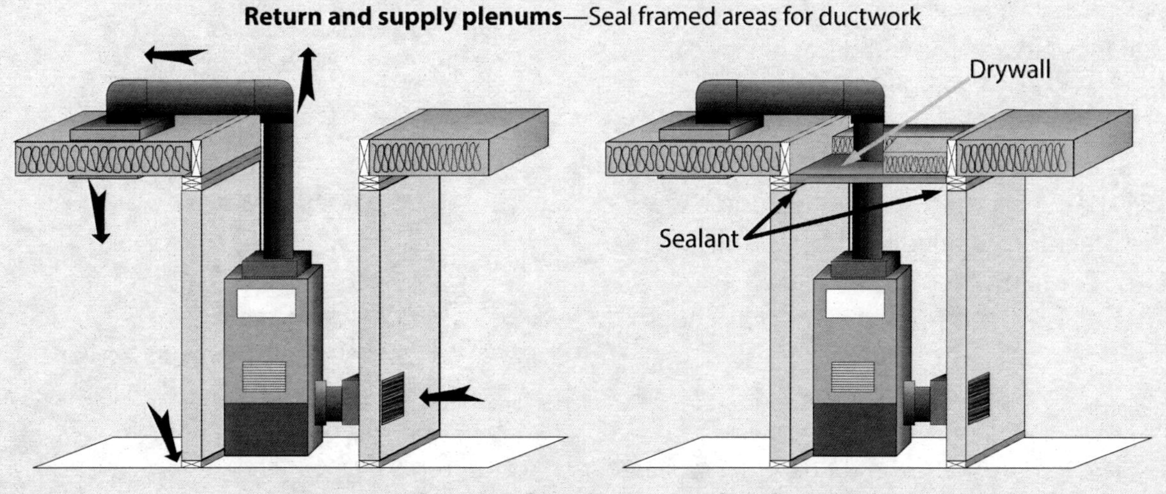

Figure 5-9 Sealing bypasses for ductwork

AIR LEAKAGE DRIVING FORCES

Requirements for air leakage to occur:

- *Holes*—the larger the hole, the greater the air leakage. Large holes have higher priority for air sealing efforts.
- *Driving force*—a pressure difference that forces air to flow through a hole. Holes that experience stronger and more continuous driving forces have higher priority.

The common driving forces are:

- *Wind*—caused by weather conditions.
- *Mechanical blower*—induced pressure imbalances caused by operation of fans and blowers.
- *Stack effect*—upward air pressure due to the buoyancy of air.

Wind is usually considered to be the primary driving force for air leakage. When the wind blows against a building, it creates a high pressure zone on the windward areas. Outdoor air from the windward side infiltrates into the building while air exits on the leeward side. Wind acts to create areas of differential pressure which cause both infiltration and exfiltration. The degree to which wind contributes to air leakage depends on its velocity and duration.

On average, wind in the Southeast creates a pressure difference of 10 to 20 Pascals on the windward side.

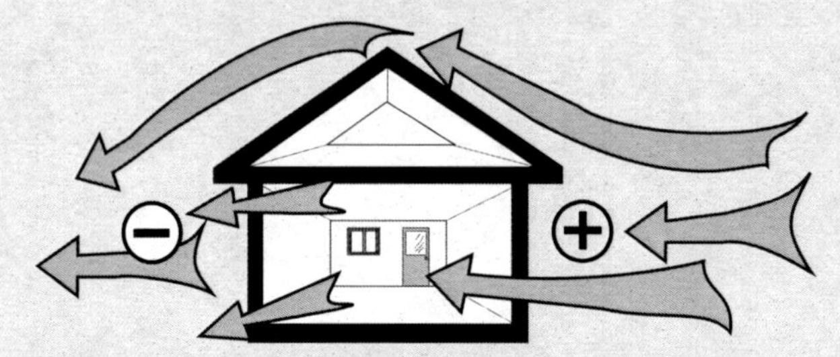

Figure 5-10 Wind-driven infiltration

Poorly designed and installed forced-air systems can create strong pressure imbalances inside the building, which can triple air leakage whenever the cooling and heating system operates. In addition, unsealed ductwork located in attics and crawl spaces can draw pollutants and excess moisture into the building. Correcting duct leakage problems is critical when constructing an energy efficient building.

Leaks in supply and return ductwork can cause pressure differences of up to 30 Pascals. Exhaust equipment such as kitchen and bath fans and clothes dryers can also create pressure differences.

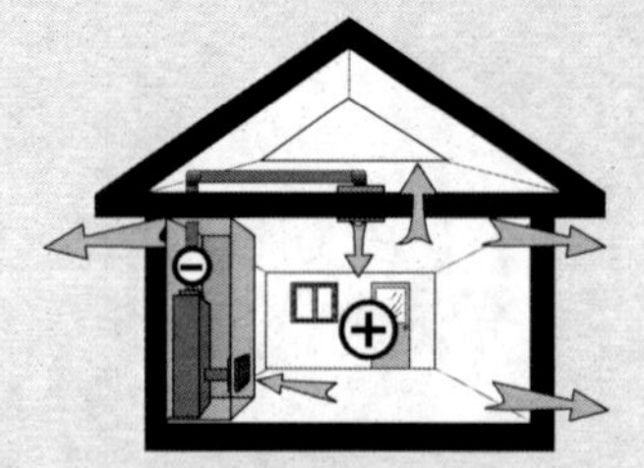

Figure 5-11 Mechanical system driven infiltration

* *A "Pascal" is the metric system unit of measure for stress or force per unit area.*

continued...

continued...

The temperature difference between inside and outside causes warm air inside the building to rise while cooler air falls, creating a driving force known as the *stack effect.* The stack effect is weak but always present. Most buildings have large holes into the attic and crawl space. Because the stack effect is so prevalent and the holes through which it drives air are often so large, it is usually a major contributor to air leakage, moisture, and air quality problems.

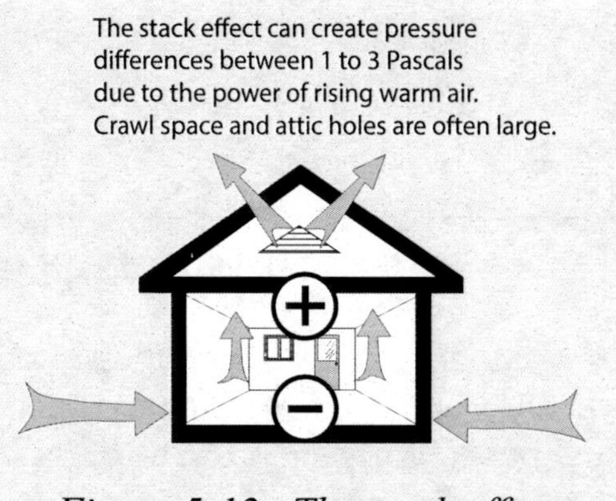

Figure 5-12 The stack effect

Air Sealing and Insulation

Building envelope air tightness and insulation installation shall be demonstrated to comply with the following:

FBC, Energy Conservation, Chapter 4 Residential Energy Efficiency

R402.4.1.2 Testing. The building or dwelling unit shall be tested and verified as having an air leakage rate of not exceeding seven air changes per hour in Climate Zones 1 and 2, and three air changes per hour in Climate Zones 3 through 8. Testing shall be conducted in accordance with ANSI/RESNET/ICC 380 and reported at a pressure of 0.2 inches w.g. (50 Pascals). Where required by the *code official*, testing shall be conducted by an *approved* third party. Testing shall be conducted by either individuals as defined in Section 553.993(5) or (7), *Florida Statutes*, or individuals licensed as set forth in Section 489.105(3)(f),(g), or (i) or an *approved* third party. A written report of the results of the test shall be signed by the party conducting the test and provided to the *code official*. Testing shall be performed at any time after creation of all penetrations of the *building thermal envelope*.

Exception: Testing is not required for additions, alterations, renovations or repairs of the building thermal envelope of existing buildings in which the new construction is less than 85 percent of the building thermal envelope.

During testing:

1. Exterior windows and doors, fireplace and stove doors shall be closed, but not sealed, beyond the intended weatherstripping or other infiltration control measures;
2. Dampers including exhaust, intake, makeup air, backdraft and flue dampers shall be closed, but not sealed beyond intended infiltration control measures;
3. Interior doors, if installed at the time of the test, shall be open;
4. Exterior doors for continuous ventilation systems and heat recovery ventilators shall be closed and sealed;

5. Heating and cooling systems, if installed at the time of the test, shall be turned off; and
6. Supply and return registers, if installed at the time of the test, shall be fully open.

MEASURING AIRTIGHTNESS WITH A BLOWER DOOR

While there are many well known sources of air leakage, virtually all buildings have unexpected air leakage sites called ***bypasses***. These areas can be difficult to find and correct without the use of a ***blower door***. This diagnostic equipment consists of a temporary door covering which is installed in an outside doorway, and a fan which pressurizes (forces air into) or depressurizes (forces air out of) the building. When the fan operates, it is easy to feel air leaking through cracks in the building envelope. Most blower doors have gauges which can measure the relative leakiness of a building (Figure 5-13).

Figure 5-13 Blower door

One measure of a building's leakage rate is air changes per hour (ACH), which estimates how many times in one hour the entire volume of air inside the building leaks to the outside. For example, a home that has 2,000 square feet of living area and 8-foot ceilings has a volume of 16,000 cubic feet. If the blower door measures leakage of 80,000 cubic feet per hour, the home has an infiltration rate of 5 ACH (80,000 ÷ 16,000). The leakier the house, the higher the number of air changes per hour, the higher the heating and cooling costs, and the greater the potential for moisture, comfort, and health problems (Table 5-2).

Table 5-2 Typical infiltration rates for home

	ACH50 (Air changes /hour at 50 Pascals)
New home with special airtight construction and a controlled ventilation system	1.5 – 3.0
Energy efficient home with continuous air barrier system	4.0 – 6.0
Standard new home	7.0 – 15.0
Standard existing home	10.0 – 25.0
Older, leaky home	20.0 – 50.0

continued...

continued…

To determine the number of air changes per hour, many experts use the blower door to create a negative pressure of 50 Pascals. A ***Pascal*** is a small unit of pressure about equal to the pressure that a pat of butter exerts on a piece of toast — about 0.004 inches water gauge. Fifty Pascals is approximately equivalent to a 20 mile-per-hour wind blowing against all surfaces of the building. Energy efficient builders should strive for less than 5 air changes per hour at 50 Pascals pressure (ACH50).

Given ACH50, a natural infiltration rate (resulting from wind and temperature effects) can be estimated. In Florida's climate, ACH50 can be divided by 40 to yield an expected natural infiltration rate (Cummings, Moyer, Tooley, 1990). For example, if ACH50 = 10, then the estimate for natural infiltration for Florida homes would be 10/40 = 0.25 ach. This means that under normal wind and temperature conditions, we would expect about 25% of the house air to be replaced with outdoor air each hour. Note that this is only an estimate of long-term average infiltration. Actual infiltration will vary considerably based on changes in wind, temperature, and time of day.

Airtight Drywall Approach (ADA)

The airtight drywall approach is an air sealing system that connects the interior finish of drywall and other building materials together to form a continuous barrier (Figure 5-14). The airtight drywall approach has been used on hundreds of houses and has proven to be an effective technique to reduce air leakage as well as keep moisture, dust, and insects from entering the building.

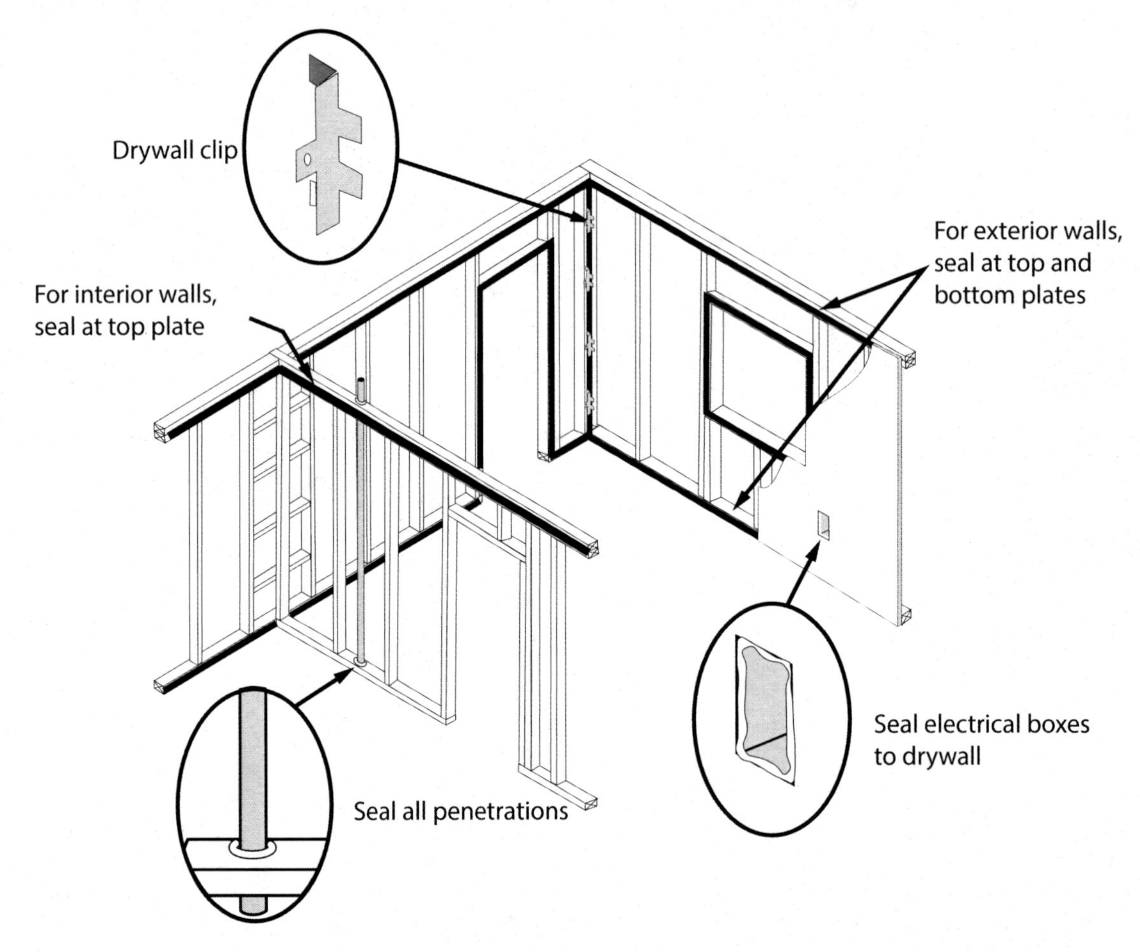

Figure 5-14 Airtight drywall approach air barrier

In a typical drywall installation, most of the seams are sealed by tape and joint compound. However, air can leak in or out of the building in the following locations:

- Between the edges of the drywall and the top and bottom plates of exterior walls.
- From inside the attic down between the framing and drywall of partition walls.
- Between the window and door frames and drywall.
- Through openings in the drywall for utilities and other services.

The airtight drywall approach uses either caulk or gaskets to seal these areas and make the drywall a continuous air barrier system.

Advantages

Effective—the airtight drywall approach has proven to be a reliable air barrier.

Simple—does not require specialized subcontractors or unusual construction techniques. If gasket materials are not available locally, they can be shipped easily.

Does not cover framing—the use of the airtight drywall approach does not prevent the drywall from being glued to the framing.

Scheduling—gaskets can be installed anytime between when the house is "dried-in" and when the drywall is attached to framing.

Adaptable—builders can adapt airtight drywall approach principles to suit any design and varying construction schedules.

Cost—materials and labor for standard designs should only cost a few hundred dollars.

Disadvantages

Unfamiliar—although the airtight drywall approach is a proven technique, many building professionals and code officials are not familiar with its use.

Requires thought—while the airtight drywall approach is simple, new construction techniques require careful planning to ensure that the air barrier remains continuous. However, the airtight drywall approach is often the most error-free and reliable air barrier for unique designs.

Requires care—gaskets and caulking can be damaged or removed by subcontractors when installing the drywall or utilities.

Installation Techniques

Slab floors

- Seal expansion joints and penetrations with a concrete sealant such as one-part urethane caulk.

Exterior framed walls

- Seal between the bottom plate and subflooring with caulk or gaskets.
- Install gaskets or caulk along the face of the bottom plate so that when drywall is installed it compresses the sealant to form an airtight seal against the framing. Some builders also caulk the drywall to the top plate to reduce leakage into the wall.
- Use drywall joint compound or caulk to seal the gap between drywall and electrical boxes. Install foam gaskets behind coverplates and caulk holes in boxes.

- Seal penetrations through the top and bottom plates for plumbing, wiring, and ducts. Code requires firestopping for leaks through top plates.

Partition walls

- Seal the drywall to the top plate of partition walls with unconditioned space above.
- Install gaskets or caulk on the face of the first stud in the partition wall. Sealant should extend from the bottom to the top of the stud to keep air in the outside wall from leaking inside.
- Seal penetrations and shafts (ductwork) where it projects through partition walls. See **FBC, Energy Conservation, Table R402.4.1.1 Air Barrier and Instulation Installation (Mandatory)**.
- Seal penetrations through the top and bottom plates for plumbing, wiring, and ducts. Fire code requirements dictate how this is accomplished.

Windows and doors

- Seal drywall edges to either framing or jambs for windows and doors.
- Fill rough opening with spray foam sealant or suitable substitute.
- Caulk window and door trim to drywall with clear or paintable sealant.

Ceiling

- Follow standard finishing techniques to seal the junction between the ceiling and walls.
- When installing ceiling drywall do not damage gaskets, especially in tight areas such as closets and hallways.
- Seal all penetrations in the ceiling for wiring, plumbing, ducts, attic access openings, and whole house fans (**FBC, Energy Conservation, Table R402.4.1.1 Air Barrier and Instulation Installation (Mandatory)**).
- Seal all openings for chases and dropped soffits above kitchen cabinets and shower/tub enclosures (**FBC, Energy Conservation, Section R402.4.1 Building thermal envelope (Mandatory)**).
- Avoid recessed lights; where used, install airtight, IC-rated and labeled fixtures and seal with gasket or caulk between fixture housings and drywall (see **FBC, Energy Conservation, Section R402.4.4 Recessed Lighting**).

Wood framed floors

- Seal the rim joist to minimize air currents around floor insulation (mandatory; see **FBC, Energy Conservation, Table R402.4.1.1 Air Barrier and Instulation Installation**). Also, seal rim joists for multi-story construction (Figure 5-15).
- For unvented crawl spaces or basements, seal beneath the sill plate.
- Seal the seams between pieces of subflooring with quality adhesive.

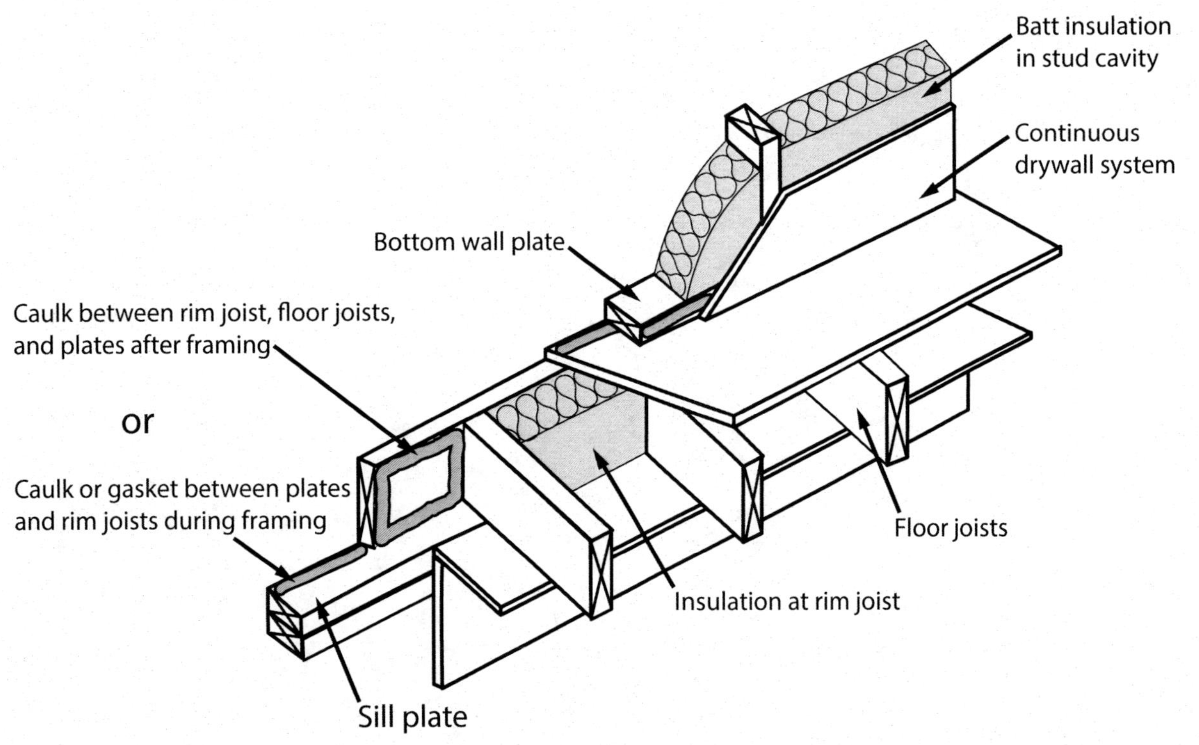

Figure 5-15 Between-floor air barrier

Resources

Note: Web links were current at the time of publication, but can change over time.

Amann, J. T., Wilson, A., & Ackerly, K. (2012). *Consumer Guide to Home Energy Savings* (Tenth Edition). Gabriola Island, BC: New Society Publishers.

An online version of this publication is available at the American Council for an Energy-Efficient Economy (ACEEE) website:
Consumer Guide to Home Energy Savings Online. Retrieved August 12, 2015, from http://aceee.org/consumer-guide-home-energy-savings-online

American Society of Heating, Refrigerating and Air Conditioning Engineers (ASHRAE). *2017 ASHRAE Handbook—Fundamentals*. 2017 Edition. Atlanta, GA: American Society of Heating, Refrigerating, and Air-Conditioning Engineers, Inc., 2017. Retrieved from https://www.techstreet.com/ashrae/lists/ashrae_handbook.tmpl?ashrae_auth_token=.

Carll, C. (2006). *The Ins and Outs of Caulking* (General Technical Report No. FPL-GTR-169). U.S. Department of Agriculture (USDA), Forest Service, Forest Products Laboratory. Retrieved from http://www.fpl.fs.fed.us/documnts/fplgtr/fpl_gtr169.pdf

Cummings, J. B., Tooley, Jr., J. J., & Moyer, N. (1990). *Radon Pressure Differential Project, Phase I* (Final Technical Report No. FSEC-CR-344-90). Cape Canaveral, FL 32920: Florida Solar Energy Center (FSEC). Retrieved from http://www.fsec.ucf.edu/en/publications/pdf/FSEC-CR-344-90.pdf

Cummings, J. B., Moyer, N., & Tooley, Jr., J. J. (1990). *Radon Pressure Differential Project, Phase II* (Final Technical Report: Draft No. FSEC-CR-370-90). Cape Canaveral, FL 32920: Florida Solar Energy Center (FSEC). Retrieved from http://www.fsec.ucf.edu/en/publications/pdf/FSEC-CR-370-90.pdf

Iowa Energy Center. (n.d.). Home Series Energy Guide: Home Tightening, Insulation & Ventilation. Retrieved from http://www.iowaenergycenter.org/wp-content/uploads/2017/02/HomeSeries1.pdf

Lstiburek, J. (2005). *Builder's Guide to Hot-Humid Climates*. Building Science Corporation. Retrieved from https://buildingscience.com/bookstore/ebook/ebook-builders-guide-hot-humid-climates

Lstiburek, J., & Harriman III, L. (2009). *The ASHRAE Guide for Buildings in Hot & Humid Climates* (Second Edition). American Society of Heating, Refrigerating and Air Conditioning Engineers (ASHRAE). Retrieved from https://buildingscience.com/bookstore/books/ashrae-guide-buildings-hot-humid-climates

Miller, C. R., & Porter, W. A. (2015). *Insulation* (Fact Sheet). Gainesville, FL: My Florida Home Energy. Retrieved from http://www.myfloridahomeenergy.com/help/library/weatherization/insulation

My Florida Home Energy. (n.d.). Retrieved July 28, 2015, from http://www.myfloridahomeenergy.com/

A useful resource with a wide array of information on energy and water efficiency, including The Energy Efficient Home series of fact sheets, available at http://www.myfloridahomeenergy.com/help/library

Tooley, J. (1999, April). "The High-Performance House - What does it take?" *Home Energy*, 16(2). Retrieved from http://www.homeenergy.org/show/article/year/1999/magazine/114/id/1458

Trechsel, H. R., & Bomberg, M. (Eds.). (2009). *Moisture Control in Buildings: The Key Factor in Mold Prevention* (Second Edition). West Conshohocken, PA: ASTM International. Retrieved from http://www.astm.org/DIGITAL_LIBRARY/MNL/SOURCE_PAGES/MNL18-2ND.htm

U.S. Department of Energy (DOE) Energy Saver:

(2014). Energy Saver: Tips on Saving Money & Energy at Home. This publication is available in several formats:

As an online multi-page article: http://energy.gov/energysaver/articles/energy-saver-guide-tips-saving-money-and-energy-home

As a downloadable PDF file, a file for use on an Amazon device, or a file for use on an Apple, Android, or Nook device: http://energy.gov/energysaver/downloads/energy-saver-guide

(n.d.). Savings Project: How to Seal Air Leaks with Caulk. Retrieved August 12, 2015, from http://energy.gov/energysaver/projects/savings-project-how-seal-air-leaks-caulk

U.S. Department of Housing and Urban Development. (1999). *The Rehab Guide: Exterior Walls* (Vol. 2). Washington, DC 20410-6000: U.S. Department of Housing and Urban Development. Retrieved from http://www.huduser.org/Publications/pdf/walls.pdf

Periodicals

Energy Design Update. Aspen Publishers, 76 Ninth Avenue, New York, NY 10011. (212) 771-0600. Subscriptions: 1-800-638-8437.

Home Energy Magazine. Retrieved August 11, 2015, from http://www.homeenergy.org/

The Journal of Light Construction (JLC). Retrieved August 12, 2015, from http://www.jlconline.com/

6

Insulation—Materials and Techniques

Insulation is rated in terms of thermal resistance, called R-value, which indicates the resistance to heat flow. Although insulation can slow all types of heat flow—conduction, convection and radiation—its greatest impact is on conduction.

The higher the R-value, the greater the insulation effectiveness. The R-value of thermal insulation depends on the type of material, the thickness and density. When calculating the R-value of a multi-layered installation, the R-values of the individual layers are added.

The effectiveness of an insulated wall or ceiling also depends on how and where the insulation is installed. For example, compressed insulation will not give its full rated R-value. Also, the overall R-value of a wall or ceiling will be somewhat different from the R-value of the insulation itself because some heat flows around the insulation through the studs and joists (thermal bridging). With careful design, this short-circuiting can be reduced.

The key to an **effective insulation** system is proper installation of quality insulation products. A building should have a continuous layer of insulation around the entire building envelope (Figure 6-1). Studies show that improper installation can cut performance by 30 percent or more.

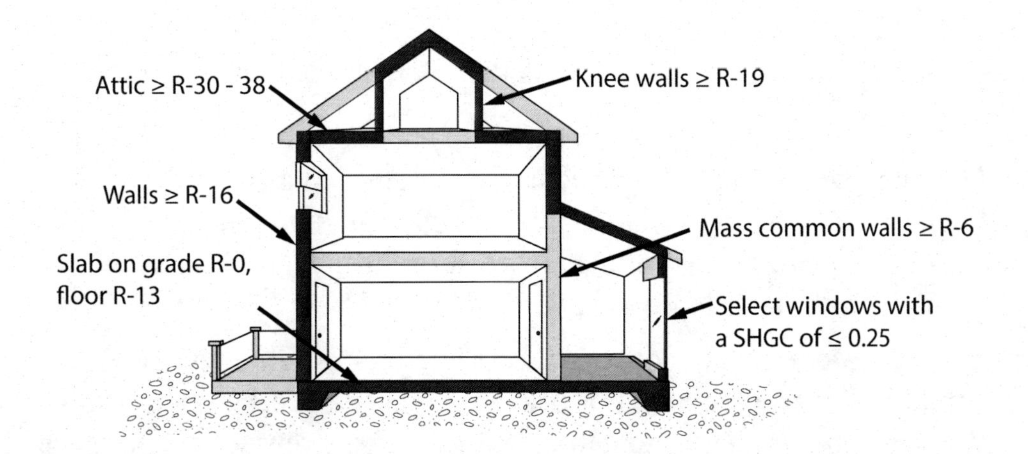

Figure 6-1 Insulating the building envelope—typical new construction R-values

INSULATION MATERIALS

The wide variety of insulation materials makes it difficult to determine which products and techniques are the most cost effective (Table 6-1). **Whichever product is chosen, make sure to install it per the manufacturer's specifications.**

Table 6-1 Comparison of Insulating Materials

Material	Typical R-Value (per inch)
Batts, blankets and loose-fill insulation	
Mineral wool and fiber glass	2.2 – 4.0
Cellulose (loose-fill)	3.0 – 3.7
Cotton (batts)	3.0 – 3.7
Perlite (loose-fill)	2.5 – 3.3
Foam insulation and sheathing	
Polyisocyanurate	6.0 – 6.5
Closed-cell, spray polyurethane	5.8 – 6.8
Open-cell, low-density polyurethane	3.6 – 3.8
Extruded polystyrene	5.0
Molded expanded polystyrene (beadboard)	4.0
Fiberboard sheathing (blackboard)	1.3
Air-Krete	3.9
OSB sheathing (3/8″)	0.5
Foil-faced OSB	Depends on installation
Polyicynene	3.6

Determine actual R-values and costs from manufacturers or local suppliers.

Here are short descriptions of a few of the insulation products available today:

- ***Fiber glass insulation*** products come in batt, roll, and loose-fill form, as well as a semi-rigid board material. Many manufacturers use recycled glass in the production process of fiber glass building insulation, with most using between 20 and 40 percent recycled glass in their product. Fiber glass is used for insulating virtually every building component—from walls to attics to ductwork.
- The term ***mineral wool*** refers to both slag wool and rock wool. Slag wool is manufactured from industrial waste product. It is primarily (~75 percent) produced from iron ore blast furnace slag, a by-product of smelting. Rock wool is fireproof and produced from natural rocks—basalt primarily—under high heat. Mineral wool insulation is available as a loose-fill product, batts, semi-rigid or rigid board. Usage of this product is down as

more and more building codes are requiring active sprinklering of buildings.

- ***Cellulose insulation***, primarily made from post-consumer recycled newsprint with up to 20 percent ammonium sulfate and/or borate flame retardants, is installed in loose-fill, wall-spray (damp), dense-pack and stabilized forms. Because of its high density, cellulose can help reduce air leaks in wall cavities, but air sealing other areas of air infiltration, such as under wall plates and rim joists, must be done to get an effective air barrier. However, given certain conditions and applications, cellulose may hold moisture.
- ***Molded expanded polystyrene (MEPS)***, often known as beadboard, is a foam product made from molded beads of plastic. MEPS is used in serveral alternative building products discussed in this chapter, including insulated concrete forms and structural insulated panels (SIPs).
- ***Extruded polystyrene (XPS)***, also a foam product in rigid board form, is a homogenous polystyrene produced primarily by three manufacturers with characteristic colors of blue, pink, and green.
- ***Polyisocyanurate***, foil-faced rigid board, is an insulating foam with one of the highest available R-values per inch.
- ***Closed-cell, high density spray polyurethane***, used both for cavity insulation and as insulating roofing materials (often referred to as spray polyurethane foam (SPF)). It has structural properties, good adhesive properties and good compressive strength.
- ***Open-cell low-density polyurethane foam*** is used primarily to seal air leaks and provide an insulating layer. Produced primarily from petrochemicals, some of these products are now manufactured in part from soybeans.
- ***Aerated concrete***, including lightweight, autoclaved (processed at high temperature) concrete, can provide a combination of moderate R-values and thermal mass for floors, walls, and ceilings, in addition to structural framing.
- ***Reflective insulation***, often used between furring strips on concrete block walls to reflect the heat. Note that reflective insulation products differ from radiant barriers in that they include a trapped air space as part of the product. These trapped air spaces may be a result of the way the reflective insulation is manufactured or from the way it is installed.

Note: Many new types of insulation are rapidly becoming incorporated into conventional construction. However, always research a material's characteristics and suitability to a particular situation before buying any new product. For instance, many insulation products require covering for fire rating.

INSULATION AND THE ENVIRONMENT

There has been considerable study and debate about potential negative environmental and health impacts of insulation products. These concerns range from detrimental health effects for the installer to depletion of the earth's ozone layer.

Fiber glass and mineral wool—questions about effects on health from breathing in fibers. The International Agency for Research on Cancer, in 2001, changed its classification for fiber glass and mineral wool from "possible human carcinogen" to "not a known human carcinogen."

Cellulose—concerns about inhalation of dust during installation to VOC emissions from printing inks (these are now almost entirely vegetable-based) and limited evidence of toxicity of boric acid flame retardants. Long term fire retardancy is unknown. Limited health and safety research has been done on these products.

Foam products and chlorofluorocarbons—for years, many foam products contained chlorofluorocarbons (CFCs), which are quite detrimental to the earth's ozone layer. The CFCs were the blowing agent which helped create the lightweight foams. Current blowing agents are:

- Expanded polystyrene - pentane, which has no impact on ozone layer, but may increase potential for smog formation.
- Extruded polystyrene, polyisocyanurate and polyurethane - use primarily hydrochlorofluorocarbons (HCFCs) which are 90 percent less harmful to the ozone layer than CFCs. Some companies are moving to non-HCFC blowing agents.
- Open-cell polyurethane including the products made by Icynene, Inc. and Demilec, Inc. as well as the newer soy-based foams - use water, which is much less detrimental than other blowing agents.

See Table 6-2 for a comparison of insulation materials by environmental characteristics and other information.

Table 6-2 Comparison of Insulation Materials (Environmental characteristics and other information)

Fiber Insulation

Type of Insulation	Installation Method(s)	R-value per Inch (RSI/m)*	Raw Materials	Pollution from Manufacture	Indoor Air Quality Impacts	Comments
Cellulose	Loose fill; wall-spray (damp); dense-pack; stabilized	3.0–3.7 (21–26)	Old newspaper, borates, ammonium sulfate	Vehicle energy use and pollution from newspaper recycling	Fibers and chemicals can be irritants. Should be isolated from interior space.	High recycled content; very low embodied energy
Fiber glass	Batts; loose fill; semi-rigid board	2.2–4.0 (15–28)	Silica sand; limestone; boron; recycled glass, phenol formaldehyde resin or acrylic resin	Formaldehyde emissions and energy use during manufacture; some manufactured without formaldehyde	Fibers can be irritants, and should be isolated from interior spaces. Formaldehyde is a carcinogen. Less concern about cancer from respirable fibers.	
Mineral wool	Loose fill; batts; semi-rigid or rigid board	2.8–3.7 (19–26)	Iron-ore blast furnace slag; natural rock; phenol formaldehyde binder	Formaldehyde emissions and energy use during manufacture	Fibers can be irritants, and should be isolated from interior spaces. Formaldehyde is a carcinogen. Less concern about cancer from respirable fibers.	Rigid board (e.g., Roxul) can be an excellent foundation drainage and insulation material.
Cotton	Batts	3.0–3.7 (21–26)	Cotton and polyester mill scraps (especially denim)	Negligible	Considered very safe	Two producers; also used for flexible duct insulation
Perlite	Loose fill	2.5–3.3 (17–23)	Volcanic rock	Negligible	Some nuisance dust	

* RSI/m: The standard unit of measurement in the United States has been the Imperial unit. The country is converting to the International System (SI) unit—or metric standard—which predominates internationally. To differentiate like terms, you may find "SI" added to the term symbol. For example, RSI refers to the R-value in International System (metric) units.

Conversion Factors:

R-value conversions	To Get	Multiply	By
Thermal Resistance (R)	RSI (m^2C/w)	R (ft^2hF/Btu)	0.1761
Insulation R/unit thickness	RSI/mm	R/in	0.00693

In the chart, the heading is R-value per inch (RSI/m); to obtain this number the RSI/mm is divided by 1000.

Table 6-2 (cont'd) Comparison of Insulation Materials Environmental characteristics and other information)

Foam Insulation

Type of Insulation	Installation Method(s)	R-value per Inch (RSI/m)*	Raw Materials	Pollution from Manufacture	Indoor Air Quality Impacts	Comments
Poly-isocy-anurate	Foil-faced rigid boards; nail-base with OSB sheath-ing	6.0–6.5 (42–45)	Fossil fuels; some recy-cled PET; pentane blowing agent; TCPP flame retardant; aluminum facing	Energy use during manufac-ture	Potential health concerns during manufacture. Negligible emissions after installation.	Phaseout of HCFC ozone-deplet-ing blowing agents com-pleted
Extruded polystyrene (XPS)	Rigid board	5.0 (35)	Fossil fuels; HCFC-142b blowing agent; HBCD flame retar-dant	Energy use during manu-facture. Ozone depletion.	Potential release of residual sty-rene monomer (a carcinogen) and HBCD flame retardant.	Last remain-ing insulation material with ozone-deplet-ing blowing agents.
Expanded polystyrene (EPS)	Rigid board	3.6–4.4 (25–31)	Fossil fuels; pentane blowing agent; HBCD flame retar-dant.	Energy use during manufac-ture.	Potential release of residual sty-rene monomer (a carcinogen) and HBCD flame retardant.	
Closed-cell spray po-ly-urethane	Spray-in cavity-fill or spray-on roofing	5.8–6.8 (40–47)	Fossil fuels, HCFC-141b (through early 2005) or HFC-245fa blowing agent; nonbromi-nated flame retardant	Energy use during manu-facture, glob-al-warming potential from HFC blowing agent	Quite toxic during installa-tion (respirators or supplied air required). Allow several days of airing out prior to occupancy.	
Open-cell, low-density poly-ure-thane	Spray-in cavity-fill	3.6–3.8 (25–27)	Fossil fuels and soy-beans; water as a blowing agent; nonbromi-nated flame retardant	Energy use during manufac-ture.	Quite toxic during installa-tion (respirators or supplied air required). Allow several days of airing out prior to occupancy.	
Air-Krete	Spray-in cavity-fill	3.9 (27)	Magnesium oxide from seawater; ceramic talc	Negligible	Considered very safe	Highly fire-re-sistant; inert; remains friable

Table 6-2 (cont'd) Comparison of Insulation Materials Environmental characteristics and other information)

Radiant Barrier

Type of Insulation	Installation Method(s)	R-value per Inch (RSI/m)*	Raw Materials	Pollution from Manufacture	Indoor Air Quality Impacts	Comments
Bubble back	Stapled to framing	Depends on installation	Aluminum; fossil fuels	Energy use during manufacture	Minimal offgassing from plastic	Exaggerated R-value claims have been common.
Foil-faced polyethylene foam	Stapled to framing; requires air space for radiant benefit	Depends on installation	Aluminum; fossil fuels; recycled polyethylene	Energy use during manufacture	Minimal offgassing from polyethylene	Exaggerated R-value claims have been common. Recycled content in some.
Foil-faced paperboard sheathing	Stapled to framing; requires air space for radiant benefit	Depends on installation	Aluminum; fossil fuels; recycled paper	Energy use during manufacture	Considered very safe	High recycled content. Structural sheathing available (e.g., Thermo-Ply®)
Foil-faced OSB	Most common as attic sheathing	Depends on installation	Wood fiber; formaldehyde binder in OSB; aluminum	Energy use and VOC emissions during manufacture	Formaldehyde emissions	Primary benefit is reduced heat gain.

Source: *Adapted with permission from Environmental Building News. www.BuildingGreen.com*

INSULATION STRATEGIES

In general, commonly used insulation products are the most economical. Prices can vary according to installer and location. Look at all of the choices as they offer different R-values, suggested uses, and environmental and health considerations.

Critical guidelines

When installing any insulating material, the following guidelines are critical for optimum performance:

- Seal all air leaks between conditioned and unconditioned areas
- Obtain complete coverage of the insulation, especially around doors and windows
- Minimize air leakage through the material with air sealing measures
- Avoid compressing insulation
- Avoid lofting (installing too much air) in loose-fill products

Foam insulation strategies

Foam products are primarily economical when they can be applied in thin layers as part of a structural system or to help seal air leaks. Examples include:

- Exterior sheathing over wall framing
- Forms in which concrete can be poured
- As part of a structural insulated panel for building walls
- Spray-applied foam insulation

FOUNDATIONS

Slab-on-Grade

Most of Florida's buildings are built concrete slab-on-grade, meaning that a slab situated near ground level serves as the floor itself. Florida has no minimum requirement for slab-on-grade insulation. Raised floor systems (wood and concrete) have specific requirements depending on climate zones.

WALLS

Walls are the most complex component of the building envelope to provide adequate thermal insulation, air sealing, and moisture control.

Concrete Wall Insulation

Foundation walls and other masonry walls are usually built of concrete block or poured concrete. Insulating concrete block walls is more difficult than framed walls.

Insulating concrete block cores

Builders can insulate the interior cores of concrete block walls with insulation such as:

- Vermiculite (<1% asbestos).... R-2.1 per inch (Figure 6-2)
- Polystyrene inserts or beads.... R-4.0 to 5.0 per inch
- Polyurethane foam.................. R-5.8 to 6.8 per inch

Unfortunately, as shown in Figure 6-2, the substantial thermal bridging in the concrete connections between cores continues to depreciate the overall R-value. This approach is only a partial solution to providing a quality, well-insulating wall. Other techniques, as explained in the next few pages, provide a more cost-effective solution.

6: INSULATION

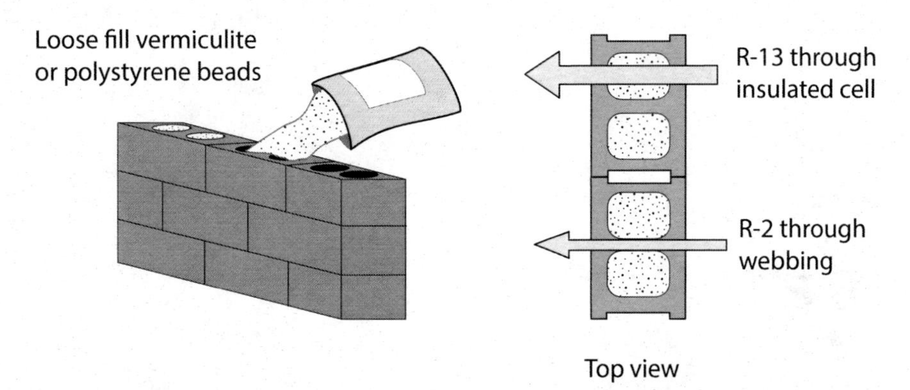

Figure 6-2 Insulating concrete block cores (R-4 to R-6 overall)

Exterior rigid fiber glass or foam insulation

Rigid insulation is generally more expensive per R-value than mineral wool or cellulose, but its rigidity is a major advantage (Figures 6-3 and 6-4). However, it is difficult and expensive to obtain R-values as high as in framed walls.

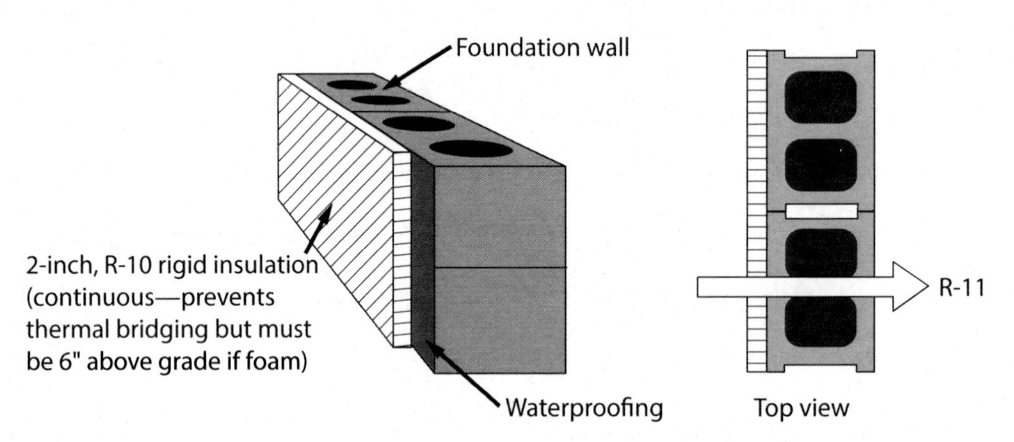

Figure 6-3 Exterior foam insulation (R-11 to R-12 overall)

Interior foam wall insulation

Foam insulation can be installed on the interior of concrete block walls (Figure 6-4); however, it must be covered with a material that resists damage and meets local fire code requirements. Half-inch drywall will typically comply, but furring strips will need to be installed as nailing surfaces. Furring strips are usually installed between sheets of foam insulation; however, to avoid the direct, uninsulated thermal bridge between the concrete wall and the furring strips, a continuous layer of foam should be installed underneath or on top of the furring strips.

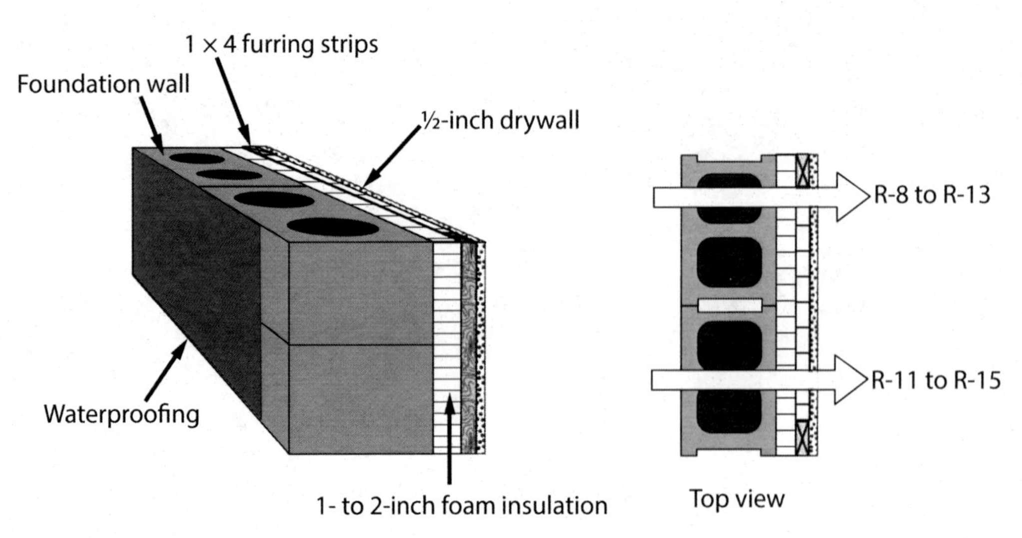

Figure 6-4 Interior foam wall insulation (R-10 to R-14 overall)

Interior framed wall

In some cases, designers will specify a framed wall on the interior of a masonry wall (Figure 6-5). Standard framed wall insulation and air sealing practice can then be applied.

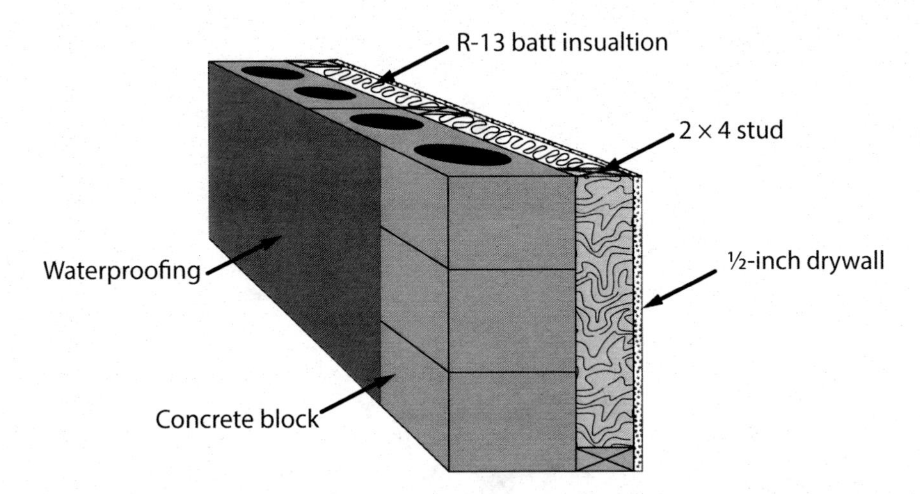

Figure 6-5 Interior framed wall (R-11 to R-13 overall)

Insulated concrete form systems

Insulated concrete forms (ICFs) are permanent rigid plastic foam forms that are filled with reinforced concrete to create structural walls with significant thermal insulation (Figure 6-6). The foam is typically either expanded polystyrene (EPS) or extruded polystyrene (XPS) and occasionally polyurethane, but may also be made from a composite of cement and foam insulation or a composite of cement and processed wood.

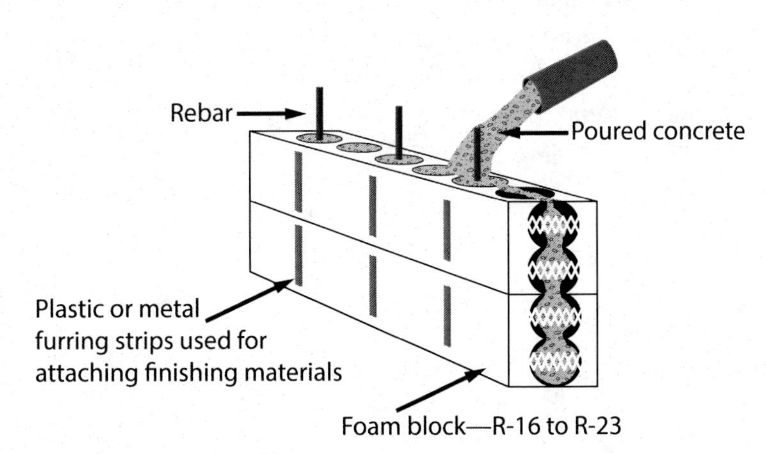

Figure 6-6 Insulated concrete form system (R-17 to R-24 overall)

INSULATED CONCRETE FORM SYSTEMS

Foam insulation systems that serve as formwork for concrete foundation walls can save on materials and cut heat flow. Among these types of products are:

Foam blocks (Figure 6-7) - Several companies manufacture foam blocks that can be installed quickly on the footings of a building. Once stacked, reinforced with rebar, and braced, they can be filled with concrete. Key considerations are:

Figure 6-7 Foam blocks

- Bracing requirements - bracing the foam blocks before construction may outweigh any labor savings from the system. However, some products require little bracing.
- Stepped foundations - make sure of the recommendations for stepping foundations—some systems have 12" high blocks or foam sections, while others are 16" high.
- Reinforcing - follow the manufacturer's recommendations for placement of rebar and other reinforcing materials.
- Concrete fill - make sure that the concrete ordered to fill the foam foundation system has sufficient slump to meet the manufacturer's requirements. These systems have been subject to "blow-outs" when the installer did not fully comply with the manufacturer's specifications. A blow-out occurs when the foam or its support structure breaks and concrete pours out of the form.
- Termites - be sure to follow the Florida Building Code and consult a reputable termite control expert.

Spray-on systems - Concrete can be applied to foam panels covered by a metal reinforcing grid, part of which is exposed. Structural concrete mixture is sprayed onto the exposed reinforcing metal. As with foam block systems, installers must follow manufacturer's recommendations carefully for a successful system.

Foam panel/snap tie systems (Figure 6-8) - Some companies produce systems in which insulation panels are locked together with plastic snap ties. A space, typically eight inches, is created between the foam panels that is filled with concrete. As with foam block systems, installers must follow manufacturer's recommendations carefully for a successful system.

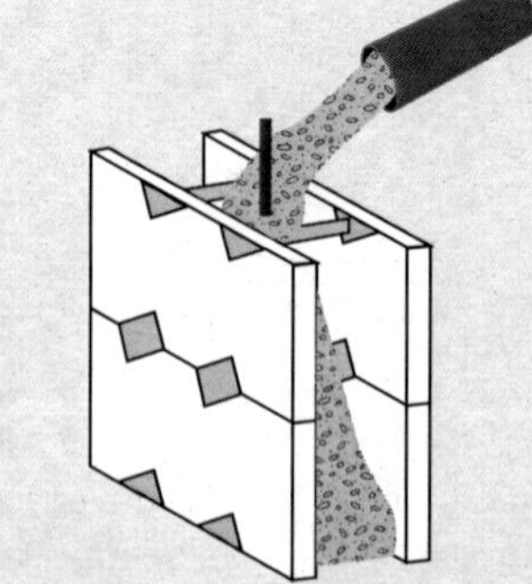

Figure 6-8 Foam panel

The concrete will be one of several shapes: flat, waffle- or screen-grid, or post-and-beam, depending on the specific form design. ICF construction was first used primarily for single-family residential construction, and has grown to about five percent of new construction. In recent years, ICF construction in multifamily residences and commercial buildings, such as retail stores, schools and hotels, has increased at a rate of about 25 to 30 percent per year.

Above-grade ICF walls cost more to build than typical wood-framed walls. As wood-framed walls approach the thermal insulation value of ICFs, cost differential decreases. In most cases, materials' costs (concrete and forms) are primarily responsible for increased costs, while labor costs are often similar to wood framing. Cost premium depends on relative material prices, labor efficiency for each system, necessity for engineering, and effect on other practices or trades, among other factors.

The cost premium for ICF houses is smaller in areas such as high-wind regions that require additional labor, time, and materials for special construction of wood-framed houses. According to an NAHB Research Center study, costs are estimated to increase by 1 to 8 percent of total house cost over a wood-framed house.

Lightweight concrete products

Lightweight, air entrained concrete is an alternative wall system (Figure 6-9). Autoclaved aerated concrete (AAC), sometimes referred to as precast autoclaved aerated concrete (PAAC), which can be shipped either as blocks or panels, combines elevated R-values (compared to standard concrete) with thermal mass.

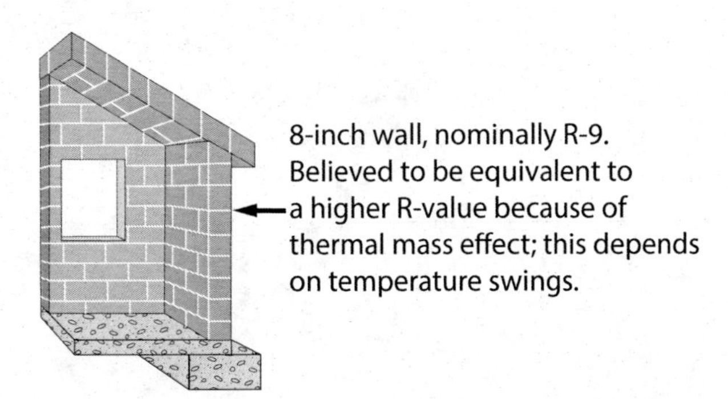

Figure 6-9 Lightweight concrete products (R-1.1 per inch plus mass effect)

2 × 4 Wall Insulation

Throughout the United States, debates continue on optimal wall construction. Table 6-3 summarizes typical problems and solutions in walls framed with 2 × 4 studs. To solve some of the energy and moisture problems in standard wall construction, builders should follow the steps shown in Chapter 1, "Step-by-Step Energy Efficient Construction." Some of these steps involve preplanning, especially the first time these procedures are used. In addition to standard framing lumber and fasteners, the following materials will also be required during construction:

- Foam sheathing for insulating headers
- 1 × 4 or metal T-bracing for corner bracing (Figure 6-10)
- R-13 batts for insulating areas during framing behind shower/tub enclosures and other hidden areas
- From the ***Florida Building Code, Residential***: **R307.2 Bathtub and shower spaces.** Bathtub and shower floors and walls above bathtubs with installed shower heads and in shower compartments shall be finished with a nonabsorbent surface. Such wall surfaces shall extend to a height of not less than **6 feet** above the floor.
- Caulking or foam sealant for sealing areas that may be more difficult to seal later

Table 6-3 2 × 4 Framed Wall Problems and Solutions

Problem	Solution
Small space available for insulation.	Install continuous exterior foam sheathing and medium (R-13) to high (R-15) density cavity insulation.
Enclosed cavities are more prone to cause condensation, particularly when sheathing materials with low R-values are used.	Install a continuous air barrier system. Use continuous foam sheathing.
Presence of wiring, plumbing, ductwork, and framing members lessens potential R-value and provides pathways for air leakage.	Locate mechanical systems in interior walls; avoid horizontal wiring runs through exterior walls; use air sealing insulation system.

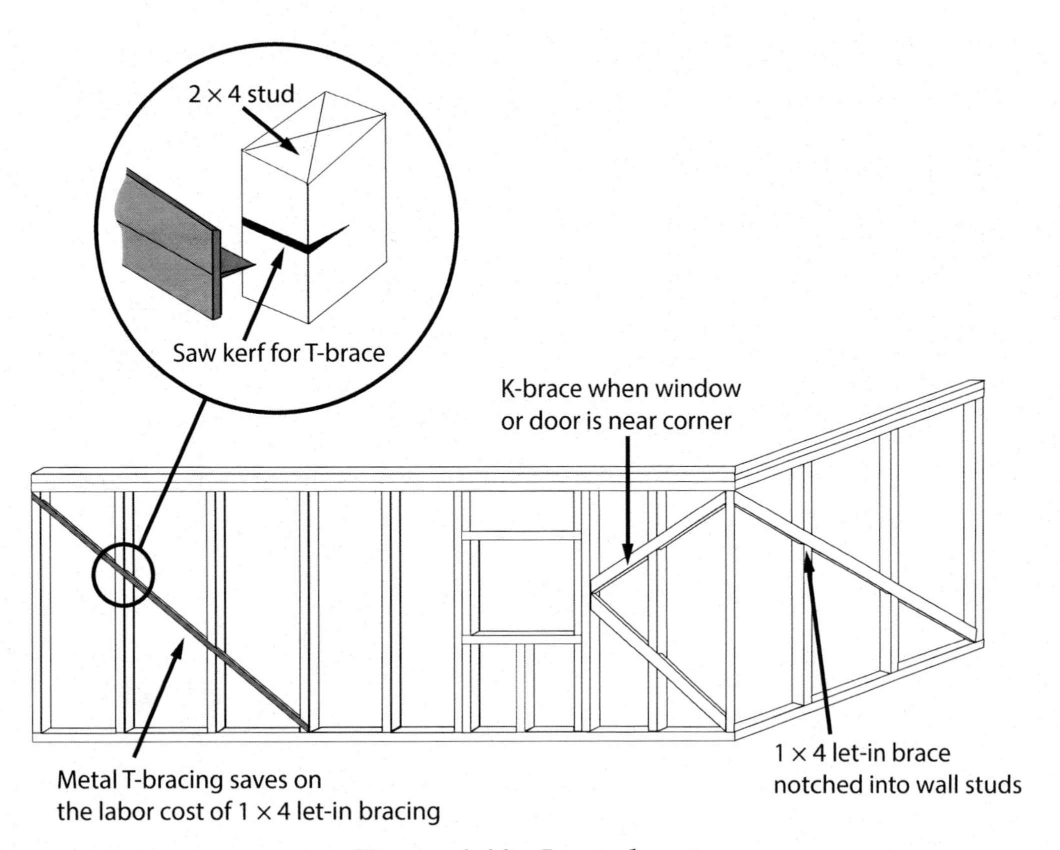

Figure 6-10 Let-in bracing

Avoid side stapling

Walls are usually insulated with batts having an attached vapor retarder facing. Many builders question whether it is best to side staple or face staple batt insulation. The common arguments illustrate that face stapling results in less compression, while side stapling interferes less with drywall installation.

The ideal solution should focus on where the kraft paper (vapor retarder) is, rather than how it is installed.

The face stapling question is an appropriate question in northern or "heating dominated" climates. In northern areas, vapor retarders should be installed on the "warm" side of the wall cavity. In southern or "cooling dominated" climates the vapor retarder, if used, should be on the outside surface of the wall cavity. Because of this, the **use of unfaced batts is recommended** in Florida (Figure 6-11).

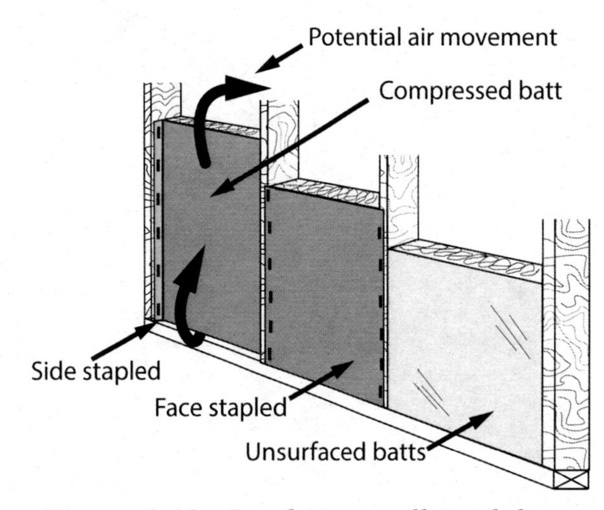

Figure 6-11 Insulating walls with batts

Unfaced batts are slightly larger than the standard 16- or 24-inch stud spacing and rely on a friction-fit for support. Since unfaced batts are not stapled, they can often be installed in less time. In addition, it is easier to cut unfaced batts to fit around wiring, plumbing, and other obstructions in the walls.

Blown loose-fill insulation

Loose-fill cellulose, fiber glass, and rock wool insulation can also be used to insulate walls. It is installed with a blowing machine and held in place with a glue binder or netting (Figure 6-12). This technique can provide good insulation coverage in the stud cavities; however, it is very important that excess moisture in the binder be allowed to evaporate before the wall cavities are enclosed by an interior finish. Keep in mind that insulation products are not replacements for proper air sealing technique.

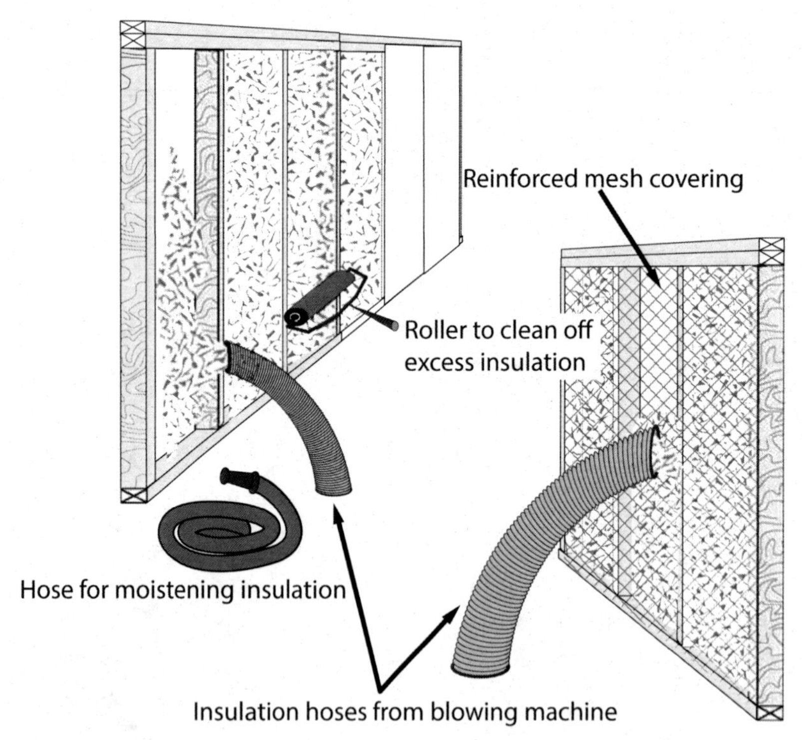

Figure 6-12 Blown sidewall insulation options

Blown foam insulation

Some insulation contractors are now blowing polyurethane or polyicynene insulation into walls and ceilings of new buildings (Figure 6-13). This technique provides high R-values in relatively thin spaces and seals air leaks effectively. The economics of foam insulation should be examined carefully before deciding on its use.

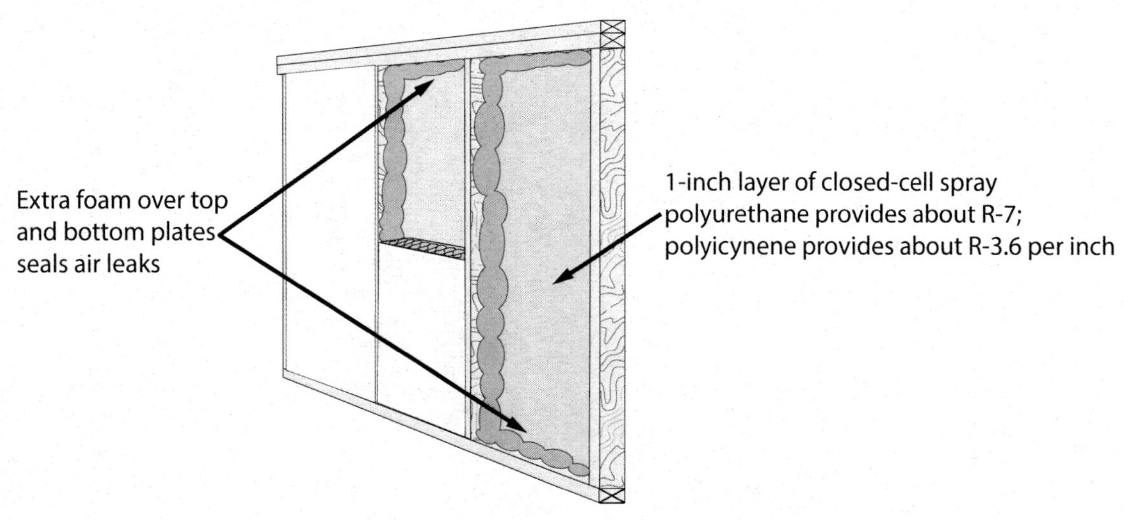

Figure 6-13 Blown foam insulation

Structural insulated panels

Another approach to wall construction is the use of structural insulated panels (SIP), also known as stress-skin panels (Figures 6-14 and 6-15). They consist of 4-inch or 6-inch thick foam panels onto which structural sheathing such as oriented strand board (OSB), cement fiber board, or various types of metal have been attached. They reduce labor costs, and because of the reduced framing in the wall, have higher R-values and less air leakage than standard walls.

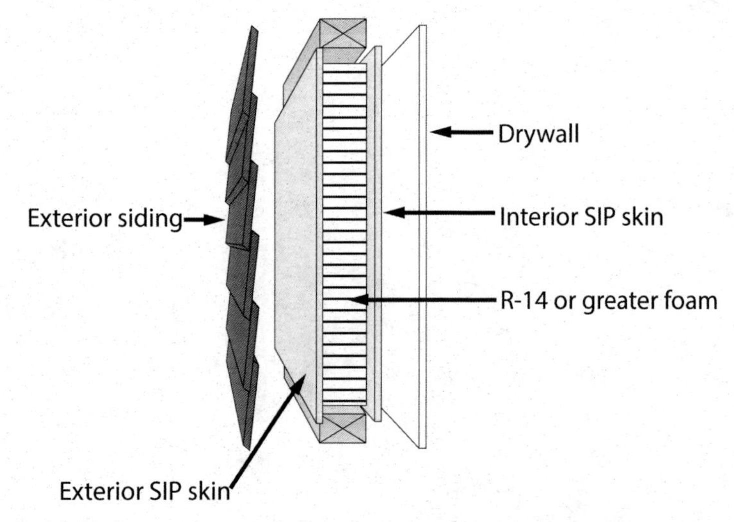

Figure 6-14 Structural insulated panels (SIP)

6: INSULATION

STRUCTURAL INSULATED PANELS CONSTRUCTION

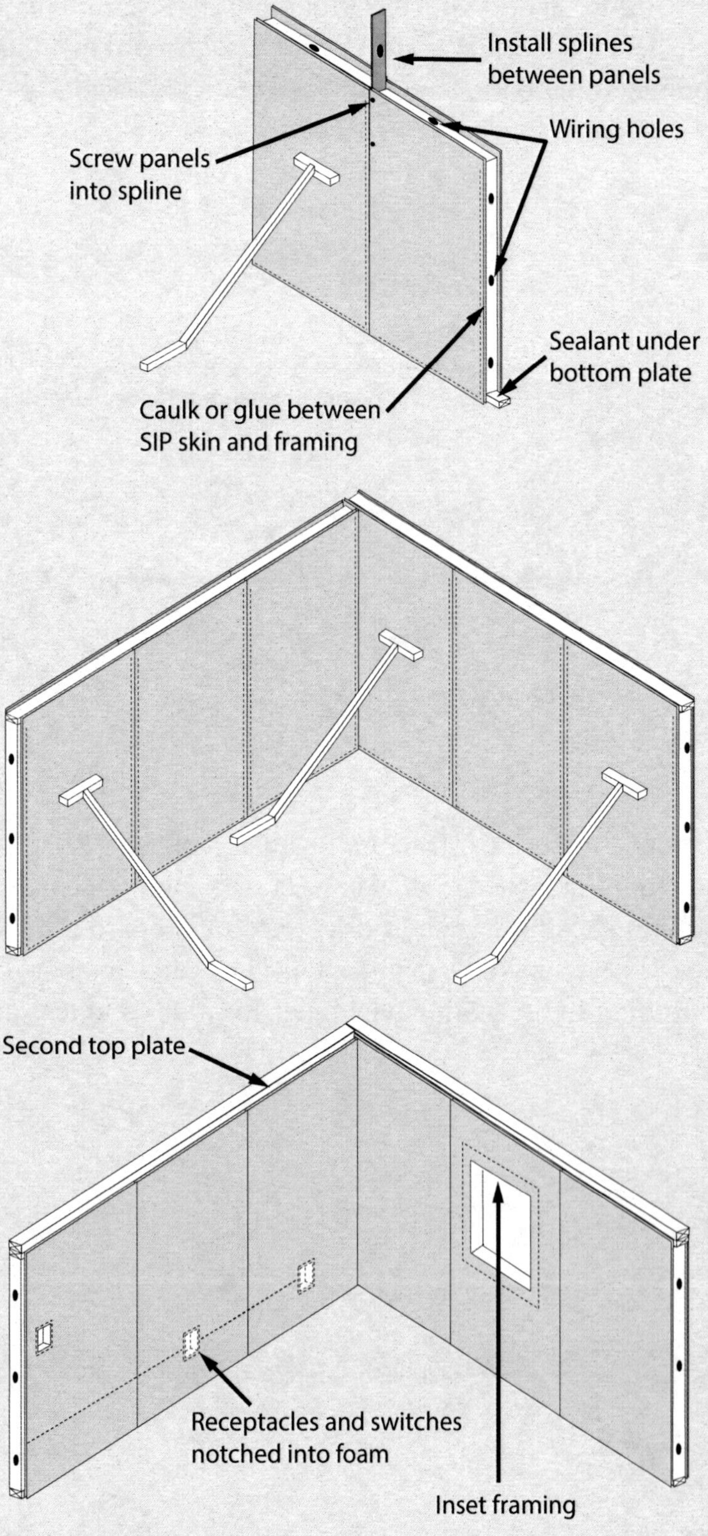

- Install first panel on top of bottom plate—caulk in place.
- Be careful to install panel plumb and level.

- Continue installing panels and splines, caulking all seams, and checking for plumb and level.
- Install continuous top plate.

- Install second top plate.
- Run wiring.
- Cut holes for windows if necessary.
- Notch into foam using a special tool for inset framing around rough openings.
- Install and caulk framing.

Figure 6-15 Structural insulated panels

SIPs are generally 4 feet wide and 8 to 12 feet long. There are a wide variety of manufacturers, each with its own method of attaching panels together. Procedures for installing windows, doors, wiring, and plumbing have been worked out by each manufacturer. Some SIPs come from the factory with pre-installed windows. In addition to their use as wall framing, SIPs can also be used in ceilings, floors, and roofs.

While buildings built with SIPs may be more expensive than those with standard framed and insulated walls, research studies have shown SIP-built buildings have higher average insulating values per inch than most commonly-used insulation materials. Due to their typical modular style of construction, infiltration losses are also reduced. Thus, they can provide substantial energy savings. Be sure to follow the Florida Building Code with regard to termites, including leaving a 6" inspection zone above final earth grade (**FBC Residential, Section R318.7, Inspection for termites**.).

Variables Affecting Performance of Structural Insulated Panels

The performance of any structural insulated panel (SIP) wall depends on its component materials and installation processes. There are a few important variables to take into consideration when building with SIP systems:

1. Panel fabrication (proper panel gluing, pressing, and curing) is critical to prevent delamination.
2. Panels must be flat, plumb, and have well-designed connections to ensure tightness of construction.
3. Though SIPs offer ease of construction, installers may need training in installation of the system being used.
4. Fire-rating of SIP materials and air-tightness of SIP installation affect the system's fire safety.
5. There may be potential insect and rodent mitigation issues, depending on SIP materials and construction.
6. Proper HVAC design and installation must take the SIP system being used into account.

Steel framing

Builders and designers are well aware of the increasing cost and decreasing quality of framing lumber. As a consequence, interest in alternative framing materials, such as steel framing, has grown. While steel framing offers advantages over wood, such as consistency of dimensions, lack of warping, and resistance to moisture and insect problems, it has distinct disadvantages from an energy perspective.

Steel framing is an excellent conductor of heat. Buildings framed with steel studs and plates usually have steel ceiling joists and rafters as well. Thus, the entire structure serves as a highly conductive thermal grid. Insulation placed between steel studs and joists is much less effective due to the extreme thermal bridging that occurs across the framing members.

The American Iron and Steel Institute is well aware of the challenges involved in building an energy efficient steel structure. In their publication *Thermal Design and Code Compliance for Cold-Formed Steel Walls 2015 Edition*, the Institute provides information on the thermal performance of steel-framed buildings. Table 6-4 summarizes some of their findings.

Table 6-4 Whole-wall Test Results of Cold-formed Steel Assemblies with 22 - 25% Framing

Stud Size	Stud Spacing	Cavity Insulation	Continuous Exterior Insulation	Assembly R-Value	Assembly U-Factor
1.5" × 3.5"	24" o.c.	R-13	None	9.25	0.108
1.5" × 3.5"	24" o.c.	R-13	R-4	12.8	0.078
1.5" × 3.5"	24" o.c.	R-13	R-5	14.08	0.071
1.5" × 3.5"	16" o.c.	R-13	None	8.06	0.124
1.5" × 3.5"	16" o.c.	R-13	R-4	11.76	0.085
1.5" × 3.5"	16" o.c.	R-13	R-5	12.98	0.077

Notes:

All assemblies have ½-inch gypsum board on the interior and ½-inch OSB sheathing on the exterior.
For Extruded Polystyrene Insulation (XPS), ¾ inches equals R-4 and 1 incwh equals R-5.
Other insulation materials are permissible with the same R-Values.
Assembly values include interior and exterior air films.

Source: Oak Ridge National Laboratory Hot Box Test database available at http://www.ornl.gov/sci/roofs+walls/AWT/Ref/steel.htm (except U-factor which were calculated as the inverse of the R-Values).

Moisture-related problems have been reported in steel frame buildings that do *not* use sufficient insulated sheathing on exterior walls. Steel studs cooled by the air conditioning system can cause outdoor air to condense, leading to mildew streaks (or ghosting) where one can see the framing members on the inside and outside of a home. In winter, studs covered by cold outside air can also cause streaking. Attention to proper insulation techniques can alleviate this problem.

The **FBC, Energy Conservation, Section R402.2.6 Steel-frame ceilings, walls and floors**, states "Steel-frame ceilings, walls and floors shall meet the insulation requirements of Table R402.2.6 or shall meet the equivalent U-factor requirements in Table R402.1.4."

2 × 6 Wall Construction

There has been interest in Florida in the use of 2 × 6s for construction. The advantages of using wider wall framing are:

- More space provides room for R-19 or R-21 wall insulation.
- Thermal bridging across the studs is less of a penalty due to the higher R-value of 2 × 6s.
- Less framing reduces labor and material costs.
- There is more space for insulating around piping, wiring, and ductwork.

Disadvantages of 2 × 6 framing include:

- Wider spacing may cause the interior finish or exterior siding to bow slightly between studs.
- Window and door jambs must be deeper, resulting in additional cost.
- Walls with substantial window and door area may require almost as much framing as 2 × 4 walls and leave relatively little area for actual insulation.

The economics of 2 × 6 wall insulation are affected by the number of windows in the wall, since each window opening adds extra studs and may require the purchase of a jamb extender. Walls built with 2 × 6s having few windows provide a positive economic payback. However, in walls in which windows make up over 10 percent of the total area, the economics become questionable.

CEILINGS AND ROOFS

Attics over flat ceilings are usually the easiest part of a building's exterior envelope to insulate. They are accessible and have ample room for insulation. However, many homes use cathedral ceilings that provide little space for insulation. It is important to insulate both types of ceilings properly.

Attic Ventilation

In summer, properly-designed ventilation reduces roof and ceiling temperatures, thus potentially saving on cooling costs and lengthening the life of the roof. In winter, roof vents expel moisture which could otherwise accumulate and deteriorate insulation or other building materials.

At present, several research studies are investigating whether attic ventilation is beneficial. For years, researchers have believed the cooling benefits of ventilating a well-insulated attic are negligible. However, some experts are now questioning whether ventilation is even effective at moisture removal. The *Florida Building Code Sixth Edition (2017), Residential* provides provisions for "unvented attic assemblies" (**FBC, Residential, Section R806.5, Unvented attic and unvented enclosed rafter assemblies**) as long as certain conditions are met. When attic ventilation is provided, ventilation openings shall have a least dimension of 1/16" and maximum ¼". If greater than ¼" then ventilation openings shall be provided with corrosion-resistant wire mesh with 1/16" minimum to ¼" maximum openings (**FBC, Residential, Section R806.1, Ventilation required**). Total net free ventilation area shall not be less than 1-to-150 of the area of the space ventilated (**FBC, Residential, Section R806.2, Minimum vent area**). An exception for 1-to-300 is provided in the code.

Vent selection

The amount of attic ventilation needed is based on requirements found in the FBC Residentail, **FBC, Residential, R806.2, Minimum vent area**. If ventilating the roof, locate vents high along the roof ridge and low along the eave or soffit. Vents should provide air movement across the entire roof area (Figure 6-16). There is a wide variety of products available including ridge, gable, soffit, and mushroom vents.

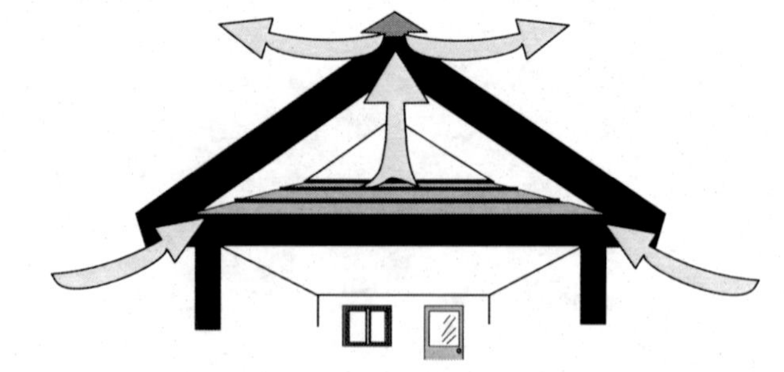

Figure 6-16 Ridge and soffit vents

The combination of continuous ridge vents along the peak of the roof and continuous soffit vents at the eave provides the most effective ventilation. Ridge vents come in a variety of colors to match any roof. Some brands are made of plastic covered by cap shingles to hide the vent from view.

Manufacturer or product testing is being performed by a variety of organizations to verify leak-free operation of continuous ridge vents in high wind situations. Care should be taken to ensure that the vents chosen are appropriate for hurricane-prone areas of Florida.

Powered attic ventilator problems

Electrically powered roof ventilators can consume more electricity to operate than they save on air conditioning costs and are **not recommended** for most designs. Power vents can create negative pressures in the home which may have detrimental effects such as (Figure 6-17):

- Drawing outside air into the home
- Removing conditioned air from the home through ceiling leaks and bypasses
- Pulling pollutants such as radon and sewer gases into the home
- Backdrafting fireplaces and fuel-burning appliances

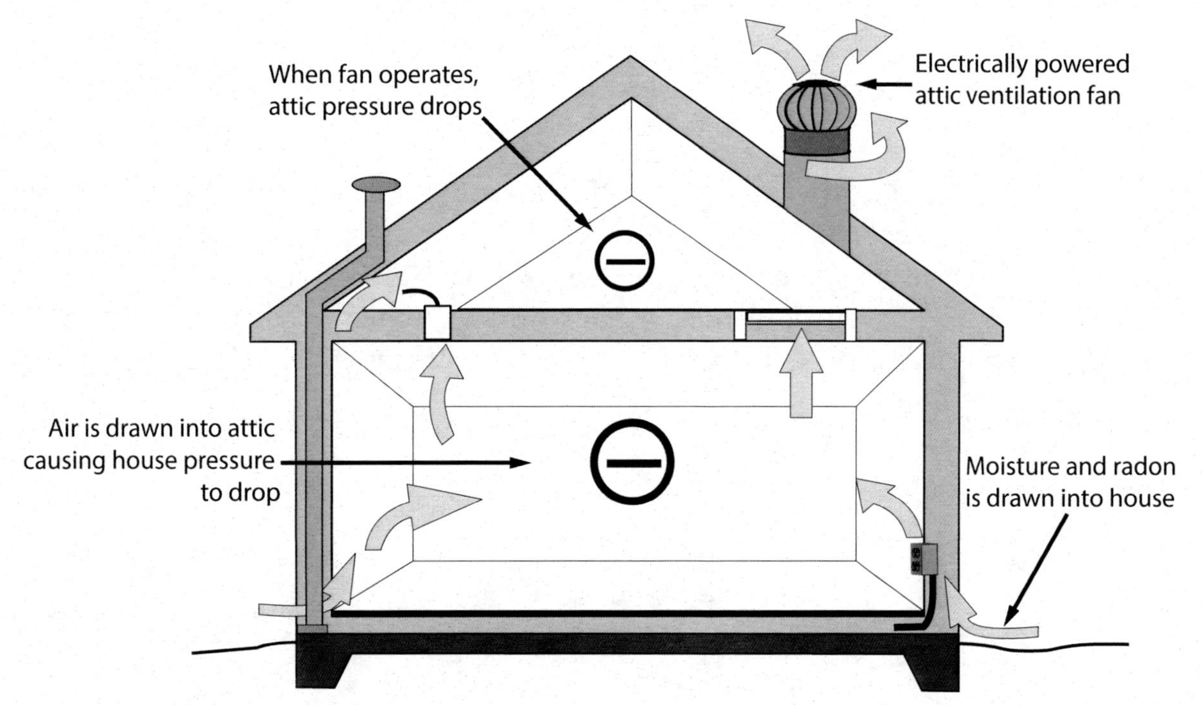

Figure 6-17 Pressure problems due to powered attic ventilators

Attic Floor Insulation Techniques

Either loose-fill or batt insulation can be installed on an attic floor. Generally, blowing loose-fill attic insulation is usually less expensive than installing batts or rolls. Many attics have either blown fiber glass, rock wool or cellulose. According to the **FBC, Energy Conservation, Section R303.2.1 Insulation installation,** ceilings with a slope of more than three in twelve should not be insulated with blown-in insulation. See **FBC, Energy Conservation, Section R303.1.1 Building thermal envelope insulation** and **Section R303.1.1.2.1 Blown or sprayed roof/ceiling insulation** for specific restrictions for blown-in insulation.

Steps for installing loose-fill attic insulation:

1. Seal attic air leaks, as prescribed by fire and energy codes.
2. Follow manufacturer's and FBC clearance requirements for heat-producing equipment found in an attic, such as flues or exhaust fans. One example of attic blocking is shown later in this chapter.
3. Use baffles to preserve ventilation space at eave of roof for soffit vents (**FBC, Energy Conservation, Section R303.2.1 Insulation installation**).
4. Insulate the attic hatch or attic stair to a level equivalent to the insulation on the surrounding surfaces. Foam boxes are available for providing a degree of insulation over a pull-down attic stairway. A baffle or retainer is required when using loose-fill insulation (**FBC, Energy Conservation, Section R402.2.4 Access hatches and doors**).
5. Determine the attic insulation area; based on the spacing and size of the joists, use the chart on the insulation bag to determine the number of bags to install. Table 6-5 shows a sample chart for cellulose insulation. Cellulose is heavier than fiber glass for the same R value. Closer spacing of roof joists and thicker drywall is required for larger R values. Check this detail with the insulation contractor. Weight limits and other factors at R-38 insulation levels are shown in Table 6-6 for the three primary types of loose fills.

Table 6-5 Blown Cellulose in Attics

R-value at 75° F	Minimum thickness (in)	Minimum weight (lb/ft²)	2 × 6 joists spaced 24" on center		2 × 6 joists spaced 16" on center	
			Coverage per 25-lb bag (ft²)	Bags per 1,000 ft²	Coverage per 25-lb bag (ft²)	Bags per 1,000 ft²
R-40	10.8	2.10	12	83	13	77
R-32	8.6	1.60	16	63	18	56
R-24	6.5	.98	21	48	23	43
R-19	5.1	.67	37	27	41	24

6. Avoid "fluffing" the insulation—blowing with too much air—by using the proper air-to-insulation mixture in the blowing machine. A few insulation contractors have fluffed loose-fill insulation to give the impression of a high R-value. The insulation may be the proper depth, but if too few bags are installed, the R-values will be less than claimed.

7. Obtain complete coverage of the blown insulation at relatively even insulation depths. **(FBC, Energy Conservation, Section R303.1.1.2.1 Blown or sprayed roof/ceiling insulation)**. Use attic rulers—obtainable from insulation contractors—affixed to trusses or joists to ensure uniform depth of insulation. Each marker shall face the attic access opening.

Steps for installing batt insulation

1. Seal attic air leaks, as prescribed by fire and energy codes.

2. Block around heat-producing devices, as described in Step 2 for Loose-fill Insulation.

3. Insulate the attic hatch or attic stair as described in Step 4 for Loose-fill Insulation.

4. Determine the attic insulation area; based on the spacing and size of the joists, order sufficient R-30 insulation for the flat attic floor. Choose batts that are tapered—cut wider on top—so that they cover the top of the ceiling joists. (See Figure 6-18.)

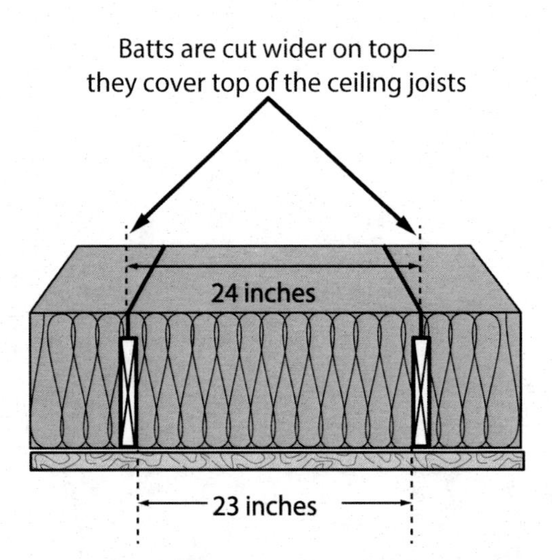

Figure 6-18 Full width batts

Table 6-6 Recommended Specifications by Insulation Type

	Cellulose	Fiber glass	Rock Wool
R-value/inch	3.2–3.8	2.2–2.7	3.0–3.3
Inches (cm) needed for R-38	10–12 (25–30)	14–17 (35–43)	11.5–13 (29–33)
Density in lb/ft^3 (kg/m^3)	1.5–2.0 (24–36)	0.5–1.0 (10–14)	1.7 (27)
Weight at R-38 in lb/ft^2 (kg/m^2)	1.25–2.0 (6–10)	0.5–1.2 (3–6)	1.6–1.8 (8–9)
OK for ½" drywall, 24" on center?	No	Yes	No
OK for ½" drywall, 16" on center?	Yes	Yes	Yes
OK for 5/8" drywall, 24" on center?	Yes	Yes	Yes

5. When installing the batts, make certain they completely fill the joist cavities. Shake batts to ensure proper loft. If the joist spacing is uneven, patch gaps in the insulation with scrap pieces. Try not to compress the insulation with wiring, plumbing or ductwork. In general, obtain complete coverage of full-thickness, non-compressed insulation.

6. Attic storage areas can pose a problem. If the ceiling joists are shallower than the depth of the insulation (generally less than 2 × 10s), raise the finished floor using 2 × 4s or other spacing lumber. Install the batts before nailing the storage floor in place. (See Figure 6-19.)

   Note: Often attic framing is not designed for storage. Check engineered loads of framing before increasing loads and piggy-backing ceiling joists.

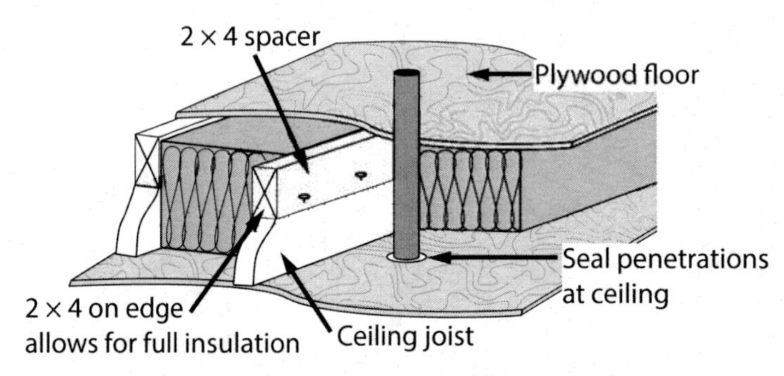

Figure 6-19 Insulating under attic floors

Increasing the roof height at the eave

One problem area in many standard roof designs is at the eave, where there is not enough room for full R-30 to R-38 insulation without preventing air flow from the soffit vent or compressing the insulation, which reduces its R-value. Figures 6-20 and 6-21 show several solutions to this problem. If using a truss roof, purchase *raised heel trusses* that form horizontal overhangs. They should provide adequate clearance for both ventilation and insulation.

In stick-built roofs, where rafters and ceiling joists are cut and installed on the construction site, an additional top plate that lies across the top of the ceiling joists at the eave will prevent compression of the attic insulation. Note: This needs to be a double plate for bearing unless rafters sit directly above joists. The rafters sitting on this *raised top plate* allow for both insulation and ventilation.

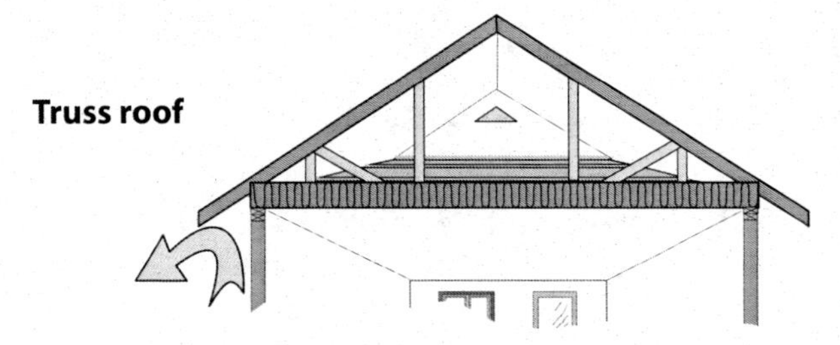

Problem: Roof deck compresses insulation and blocks air flow from soffit vent

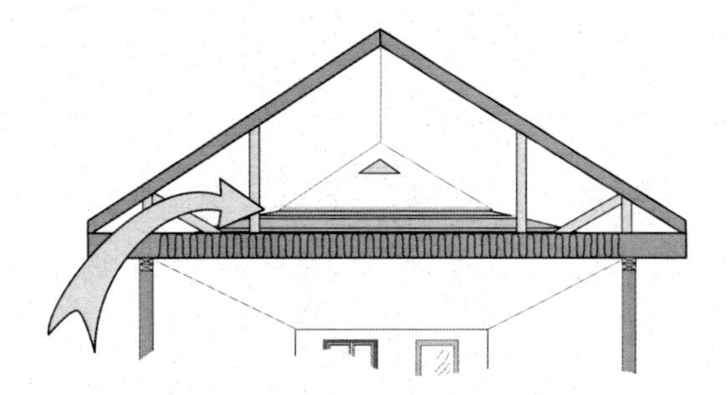

Solution: Raised heel trusses—insulation not compressed; air flow path is open

Figure 6-20 Insulation options for eaves - truss roof

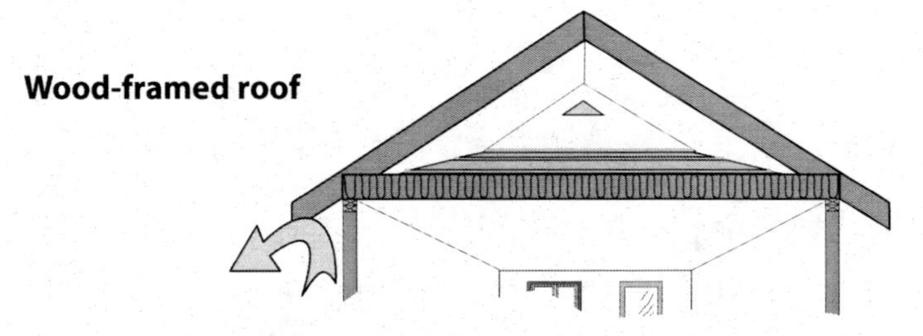

Problem: Roof compresses insulation at eave and blocks air flow from soffit vent

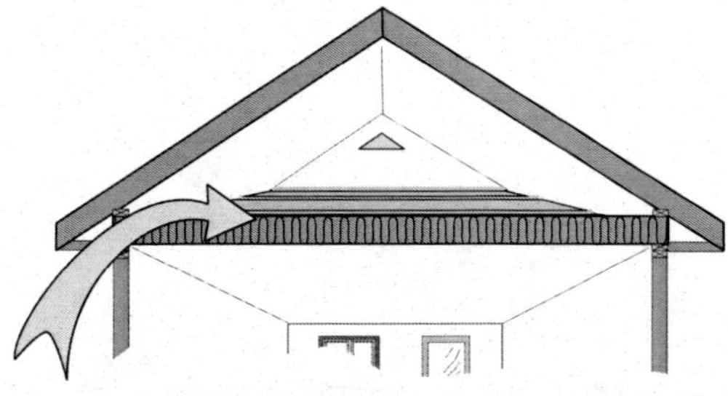

Solution: Raised top plate—insulation not compressed; air flow path is open

Figure 6-21 Insulation options for eaves - wood-framed roof

Cathedral Ceiling Insulation Techniques

Cathedral ceilings are a special case because of the limited space for insulation and ventilation within the depth of the rafter. Fitting in a 10-inch batt (R-30) and still providing ventilation is impossible with a 2 × 6 or 2 × 8 rafter (R-19 or R-25 respectively).

Building R-30 cathedral ceilings

Cathedral ceilings built with 2 × 12 rafters can be insulated with standard R-30 batts and still have plenty of space for ventilation. Some builders use a *vent baffle* between the insulation and roof decking to ensure that the ventilation channel is maintained (Figure 6-22).

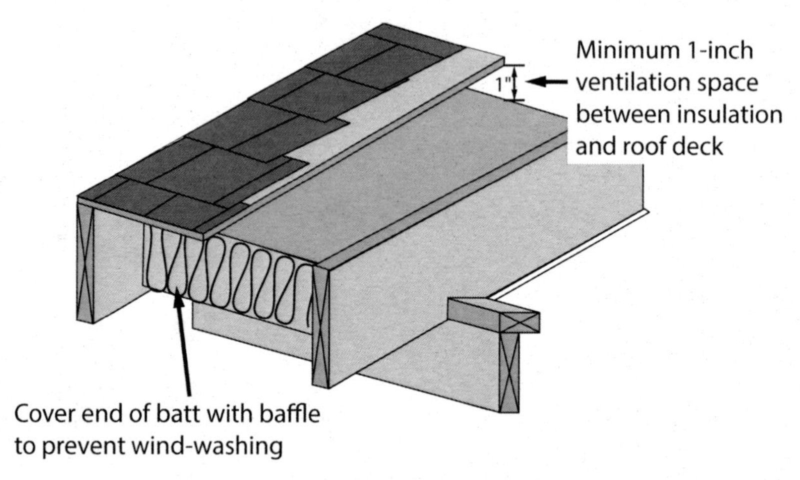

Figure 6-22 Insulating cathedral ceilings

If 2 × 12s are not required structurally, most builders find it cheaper to construct cathedral ceilings with 2 × 10 rafters and high-density R-30 batts, which are 8¼ inches thick (Table 6-7).

Some contractors wish to avoid the higher cost of 2 × 10 lumber and use 2 × 8 rafters. These roofs are usually insulated with R-19 batts.

In framing with 2 × 6 and 2 × 8 rafters, higher insulating values can be obtained by installing rigid foam insulation under the rafters. Note that the rigid foam insulation must be covered with a fire-rated material when used on the interior of the building. Drywall usually meets the requirement.

Table 6-7 Cathedral Ceiling Insulation Options

Rafter	Batt
2 × 8	R-19
2 × 10	R-25
2 × 10	moderate density R-30
2 × 12	standard density R-30

Any sized rafter; blown-in cellulose, fiber glass or rock wool held in place; provide 1" ventilation space above

Scissor trusses

Scissor trusses are another cathedral ceiling framing option. Make certain they provide adequate room for both R-30 insulation and ventilation, especially at their ends, which form the eave section of the roof.

Difficulties with exposed rafters

A cathedral ceiling with exposed rafters or roof decking is difficult and expensive to insulate well. Often, foam insulation panels are used over the attic deck as shown in Figure 6-23. However, to achieve R-30, 4 to 7 inches of foam insulation are needed. Ventilation is also a problem.

In homes where exposed rafters are desired, it may be more economical to build a standard, energy efficient cathedral ceiling, and then add exposed decorative beams underneath. Note that homes having tongue-and-groove ceilings can experience substantially more air leakage than solid, drywall ceilings. Install a continuous air barrier, sealed to the walls, above the tongue-and-groove roof deck.

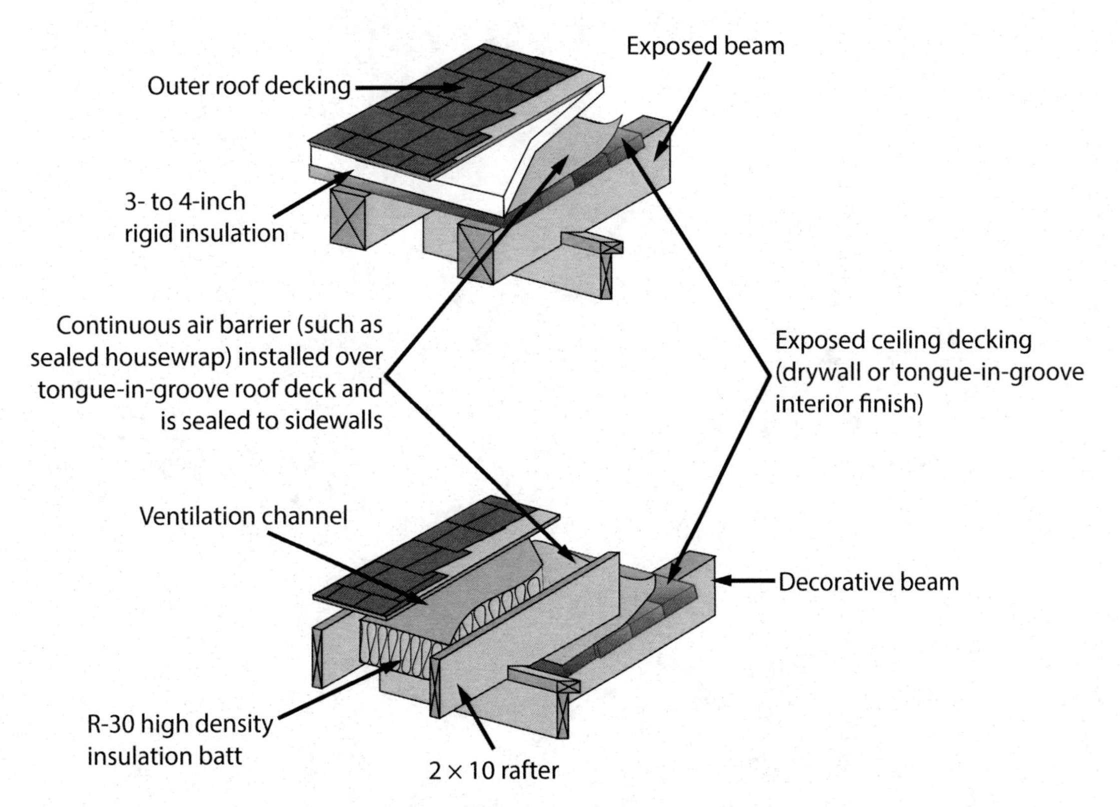

Figure 6-23 Cathedral ceiling - exterior roof insulation

ATTIC BLOCKING REQUIREMENTS

Object	Recommended Action*
Recessed light	Shall be sealed and IC rated, otherwise 3-inch clearance on all sides.
Doorbell transformer	Do not cover; no clearance on sides required
Masonry chimney	See **FBC, Building, Section 2113, Masonry Chimneys** and/or follow manufacturer's recommendations
Metal chimney	See **FBC, Mechanical, Section 806, Metal Chimneys (NFPA 211)** and/or follow manufacturer's recommendations
Vent pipes from fuel-burning equipment	Follow manufacturer's recommendations
Kitchen/bath exhaust	Duct to the outside
Heat/light/ ventilation	3-inch clearance on all sides unless IC-rated
Uncovered electric junction boxes	Shall have are barrier extending behind box or air-sealed boxes, otherwise, cover the box and insulate over it. If it is left uncovered, leave a 3-inch clearance.
Whole house fan	Install blocking up to the fan housing; leave 3-inch clearance around fan motor
Attic access door	Block around the door if blowing in loose-fill insulation (Mandatory; see **FBC, Energy Conservation, Section R402.2.4, Access hatches and doors**.)

* *These are general guidelines for existing blocking. Follow specific manufacturer's and FBC recommendations for new installations.*

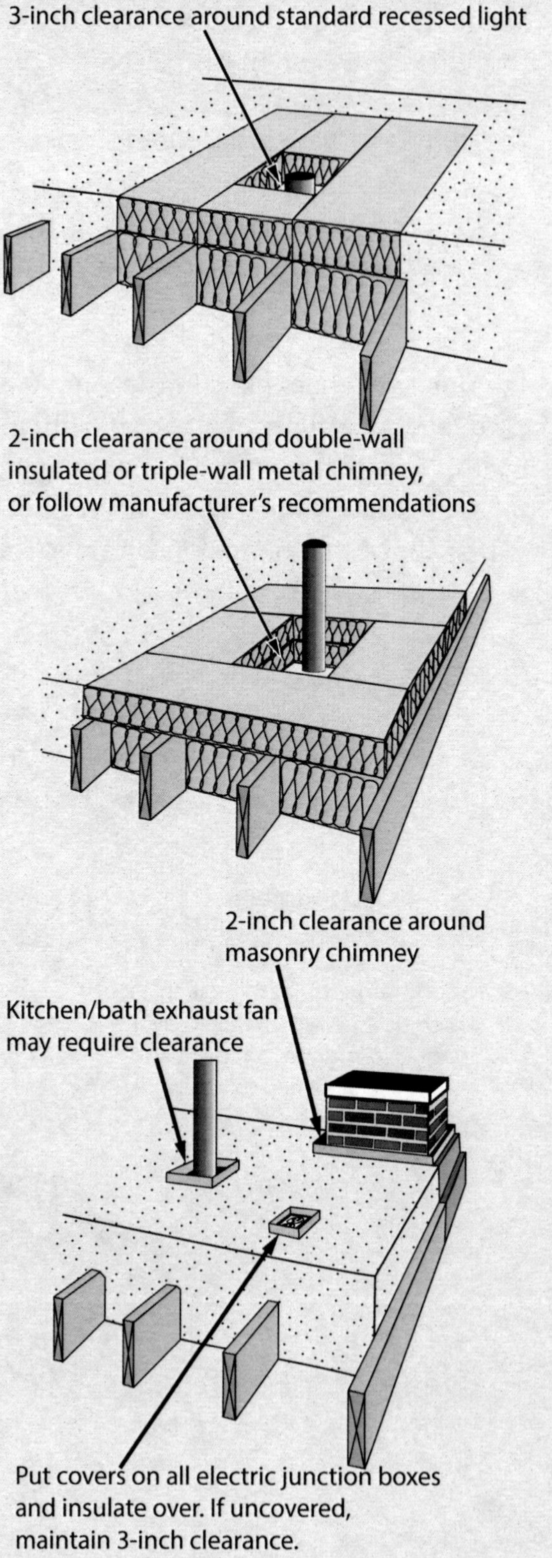

3-inch clearance around standard recessed light

2-inch clearance around double-wall insulated or triple-wall metal chimney, or follow manufacturer's recommendations

2-inch clearance around masonry chimney

Kitchen/bath exhaust fan may require clearance

Put covers on all electric junction boxes and insulate over. If uncovered, maintain 3-inch clearance.

Figure 6-24 Attic blocking requirements

Radiant Heat Barriers

Radiant heat barriers (RHB) are reflective materials that can reduce summer heat gain by the insulation and building materials in attics and walls. RHBs work two ways: first, they *reflect* thermal radiation well and second, they *emit* (give off) very little heat. RHBs should always face a vented airspace and be installed to prevent dust build-up. They are usually attached to the underside of the rafter or truss top chord or to the underside of the roof decking. Acceptable attic radiant barrier configurations can be found in Figure 6-25. If using the performance method, see the **FBC, Energy Conservation, Section R405.7.1 Installation criteria for homes claiming the radiant barrier option** for specific details.

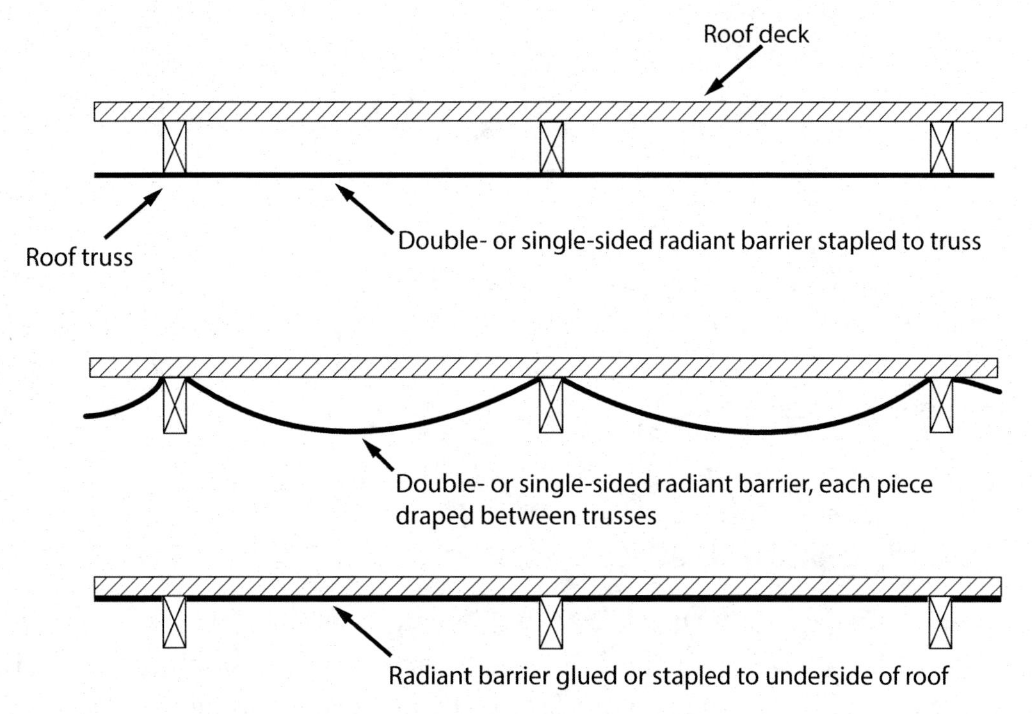

Figure 6-25 Radiant barriers

How Radiant Barrier Systems Work

A radiant barrier reduces heat transfer. Thermal radiation, or radiant heat, travels in a straight line away from a hot surface and heats any object in its path.

When sunshine heats a roof, most of the heat conducts through the exterior roofing materials to the inside surface of the roof sheathing. Heat then transfers by radiation across the attic space to the next material, either the top of the attic insulation or the attic floor. A radiant barrier, properly installed in one of many locations between the roof surface and the attic floor, will reduce radiant heat flow. Thermal insulation on the attic floor resists the flow of heat through the ceiling into the living space below. The rate at which insulation resists this flow determines the insulation's R-value. *The amount of thermal insulation affects the potential radiant barrier energy savings*. For example, installing a radiant

6: Insulation

barrier in an attic that already has high levels of insulation (R-30 or above) would result in much lower energy savings than an attic insulated at a low level (R-11 or less).

All radiant barriers use reflective foil that blocks radiant heat transfer. In an attic, a radiant barrier that faces an air space can block up to 95 percent of the heat radiating down from a hot roof. Only a single, thin, reflective surface is necessary to produce this reduction in radiant heat transfer. Additional layers of foil do little more to reduce the remaining radiant heat flow.

Conventional types of insulation consist of fibers or cells that trap air or contain a gas to retard heat conduction. These types of insulation reduce conductive and radiant heat transfer at a rate determined by their R-value. Radiant barriers reduce only radiant heat transfer. There is no current method for assigning an R-value to radiant barriers. The reduction in heat flow achieved by the installation of a radiant barrier depends on a number of factors, such as ventilation rates, roof reflectivity, ambient air temperatures, geographical location, amount of roof solar gains, and the amount of conventional insulation present.

Several factors effect the cost-effectiveness of installing a radiant barrier. You should examine the performance and cost savings of at least three potential insulation options: adding additional conventional insulation, installing a radiant barrier, and adding both conventional insulation and a radiant barrier.

Because radiant barriers redirect radiant heat back through the roofing materials, shingle temperatures may increase between 1° to 10°F (17.2° to 12.2°C). This increase does not appear to exceed the roof shingle design criteria.

Remember, radiant barriers are most effective in blocking summer radiant heat gain and saving air-conditioning costs. Although the radiant barrier may be somewhat effective in retaining heat within a cold-climate home, it may also block any winter radiant solar heat gain in the attic.

Problems with recessed lights

Standard recessed fixtures require a clearance of several inches between the sides of the lamp's housing and the attic insulation. In addition, insulation cannot be placed over the fixture. Even worse, recessed lights leak considerable air between attics and the home.

Insulated ceiling (IC) rated fixtures have a heat sensor switch which allows the fixture to be covered—except for the top—with insulation. (See Figure 6-26 for proper insulation methods for these fixtures.) (However, these units also leak air. Specific air tightness requirements can be found in the **FBC, Energy Conservation, Section R402.4.5 Recessed lighting.**) If you have to use recessed lights, install airtight IC-rated fixtures. There are alternatives to recessed lights, including surface-mounted ceiling fixtures and track lighting, which typically contribute less air leakage to the home.

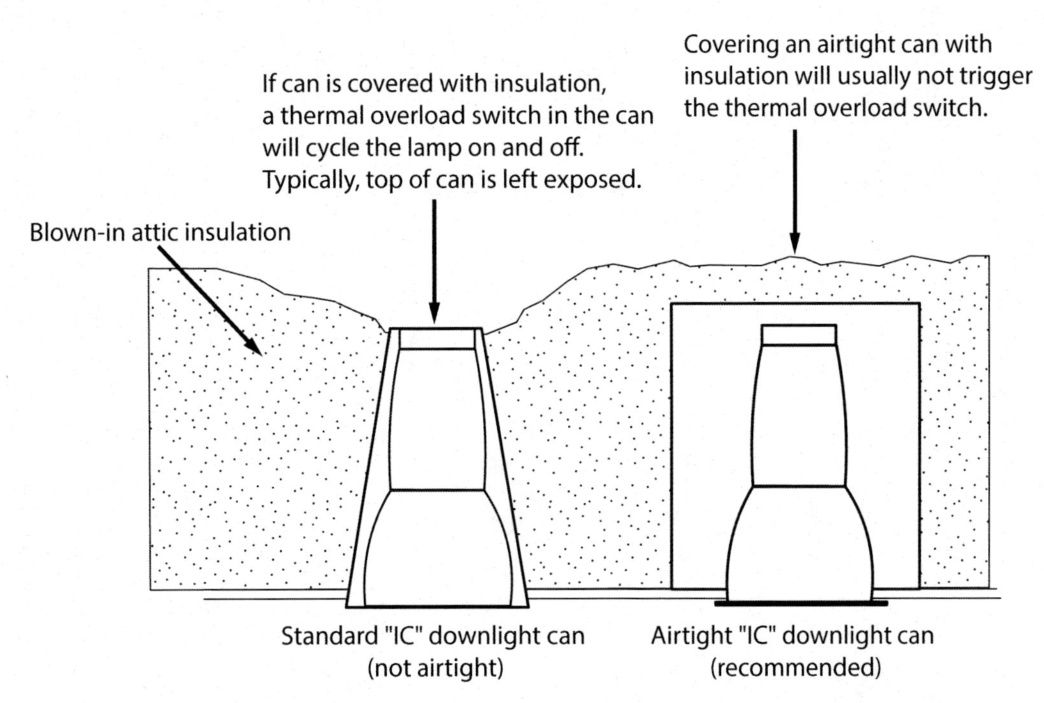

Figure 6-26 Choose quality recessed fixtures

Special Note:

The Federal Trade Commission (FTC) authorized the publication of final amendments to the FTC's Trade Regulation Rule Concerning the *Labeling and Advertising of Home Insulation (16 CFR Part 460; File No. R811001)*, commonly known as the R-value Rule. The Rule, issued in 1979, requires insulation industry members to provide consumers with important information about the thermal performance of home insulation products. This information, which must be based on uniform test procedures, helps consumers make decisions about heating and cooling their homes to maximize energy efficiency and reduce costs.

The amendments, effective November 28, 2005, mark the completion of a broad review of the Rule. After analyzing the comments received, the Rule has been amended to increase its benefits for consumers and sellers, minimize its costs, and respond to new technical developments in the industry. For example, the amendments require disclosures that will make it easier for installers, consumers, and inspectors to ensure that the correct amount of loose-fill insulation is installed in homes. In addition, the amendments streamline advertising disclosure requirements; update the required tests for some insulation products; delete disclosures for insulation products no longer sold; and eliminate duplicative disclosure requirements for sellers of do-it-yourself home insulation. You can be fined heavily, i.e., up to $11,000 plus an adjustment for inflation, each time you break the Rule. The *Federal Register* notice explains these changes in detail (see Resources for website).

Resources

Note: Web links were current at the time of publication, but can change over time.

Air Conditioning Contractors of America Association, Inc. (ACCA). (2011). *Manual J - Residential Load Calculation* (Eighth Edition). Air Conditioning Contractors of America, Inc. (ACCA). Retrieved from http://www.acca.org/standards/technical-manuals

Davidson, Sr., J. E., Heinberg, J., & Williamson, B. (2009). *Guide to Energy Efficient Homes in Louisiana*. Louisiana Department of Natural Resources. Retrieved from http://dnr.louisiana.gov/assets/docs/energy/buildersguide/Builder's%20Guide.pdf

Desjarlais, A. O., Petrie, T. W., & Stovall, T. (2004). "Comparison of Cathedralized Attics to Conventional Attics: Where and When do Cathedralized Attics Save Energy and Operating Costs?" In *Performance of Exterior Envelopes of Whole Buildings, IX International Conference, Conference Proceedings*. ASHRAE. Retrieved November 21, 2017, from http://web.ornl.gov/sci/buildings/conf-archive/2004%20B9%20papers/062_Desjarlais.pdf

Federal Trade Commission. Labeling and Advertising of Home Insulation (2005). 16 C.F.R. Part 460. Retrieved from https://www.ftc.gov/sites/default/files/documents/federal_register_notices/labeling-and-advertising-home-insulation-trade-regulation-rule-16-cfr-part-460/050531labelingadvertisingofhomeinsulation.pdf

ICF Builders Network. (n.d.). ICF Homes. Retrieved August 19, 2015, from http://www.icfhomes.com/index.html

Knowles, Hal S. (2005). *Structural Insulated Panel (SIP) Walls* (Fact Sheet). Gainesville, FL: University of Florida, School of Natural Resources and Environment, Program for Resource Efficient Communities. Retrieved from http://www.flash.org/resources/files/SIP2005-05-04.pdf

My Florida Home Energy. (n.d.). Retrieved December 21, 2017, from http://www.myfloridahomeenergy.com/

A useful resource with a wide array of information on energy and water efficiency, including The Energy Efficient Home series of fact sheets, available at http://www.myfloridahomeenergy.com/help/library

Miller, C. R., & Porter, W. A. (2015). *Insulation* (Fact Sheet). Retrieved from http://www.myfloridahomeenergy.com/help/library/weatherization/insulation

North Carolina Clean Energy Technology Center at N.C. State University. (n.d.). Database of State Incentives for Renewables & Efficiency®. Retrieved August 3, 2015, from http://www.dsireusa.org/

Nowak, M. (2015). *Thermal Design and Code Compliance for Cold-Formed Steel Walls 2015 Edition*. Washington, DC: Steel Framing Alliance and American Iron and Steel Institute. Retrieved from http://www.steelframing.org/PDF/energy/thermal_design_guide_2015_edition.pdf

Oak Ridge National Laboratory. (n.d.) "Radiation Control: Effect of Solar Radiation Control on Energy Costs." Accessed November 28, 2017. http://web.ornl.gov/sci/buildings/tools/radiation-control/.

Oak Ridge National Laboratory. (n.d.). Radiant Barrier Fact Sheet. Retrieved November 21, 2017, from http://web.ornl.gov/sci/buildings/tools/radiant/

Southface Energy Institute, & Oak Ridge National Laboratory:

(2000a). *Attic Access: Provide adequate insulation coverage and air sealing for the acceess between living space and the unconditioned attic* (Technology Fact Sheet No. DOE/GO10099-768). U.S. Department of Energy, Office of Building Technology, State and Community Programs, Energy Efficiency and Renewable Energy. Retrieved from http://www.southface.org/factsheets/AA-Atticaccess%2000-768.pdf

(2000b). *Ceilings and Attics: Install Insulation and Provide Ventilation* (Technology Fact Sheet No. DOE/GO10099-771). U.S. Department of Energy, Office of Building Technology, State and Community Programs, Energy Efficiency and Renewable Energy. Retrieved from http://www.southface.org/factsheets/CA-Ceilings-attics%2000-771.pdf

(2000c). *Wall Insulation: Provide Moisture Control and Insualtion in Wall Systems* (Technology Fact Sheet No. DOE/GO-102000-0772). U.S. Department of Energy, Office of Building Technology, State and Community Programs, Energy Efficiency and Renewable Energy. Retrieved from http://www.southface.org/factsheets/WI-Wallinsulation%2000-772.pdf

U.S. Department of Energy (DOE) Energy Saver:

(2012, May 30). Radiant Barriers. Retrieved August 13, 2015, from http://energy.gov/energysaver/articles/radiant-barriers

(2015a, April 27). Adding Insulation to an Existing Home. Retrieved August 13, 2015, from http://energy.gov/energysaver/articles/adding-insulation-existing-home

(2015b, April 27). Insulation Materials. Retrieved August 13, 2015, from http://energy.gov/energysaver/articles/insulation-materials

(2015c, April 27). Types of Insulation. Retrieved August 13, 2015, from http://energy.gov/energysaver/articles/types-insulation

(n.d.). Energy Saver. Retrieved August 13, 2015, from http://energy.gov/energysaver/energy-saver

Office of Energy Efficiency & Renewable Energy (EERE), Department of Energy (DOE). (n.d.). Retrieved August 13, 2015, from http://energy.gov/eere/office-energy-efficiency-renewable-energy

U.S. Department of Housing and Urban Development:

(1999a). *Exterior Walls* (The Rehab Guide Vol. 2). Washington, DC 20410-6000: U.S. Department of Housing and Urban Development. Retrieved from http://www.huduser.org/Publications/pdf/walls.pdf

(1999b). *Roofs* (The Rehab Guide Vol. 3). Washington, DC, 20401: U.S. Department of Housing and Urban Development. Retrieved from http://www.huduser.org/Publications/pdf/roofs.pdf

U.S. Environmental Protection Agency (EPA) ENERGY STAR:

(2007). *A Do-It-Yourself Guide to Sealing and Insulating with ENERGY STAR®: Sealing Air Leaks and Adding Attic Insulation* (No. EPA-430-F-04_024) (p. 13). Retrieved from http://www.energystar.gov/ia/partners/publications/pubdocs/DIY_Guide_May_2008.pdf

(2008). ENERGY START *Qualified Homes Thermal Bypass Checklist Guide* (Version 2.1). Retrieved from http://www.energystar.gov/ia/partners/bldrs_lenders_raters/downloads/TBC_Guide_062507.pdf

Wilson, A. (2005, January). "Insulation - Thermal Performance is Just the Beginning." *Environmental Building News*, 14(1). Retrieved from https://www2.buildinggreen.com/article/insulation-thermal-performance-just-beginning

Organizations and associations that provide professional information on building insulation:

American Society of Heating, Refrigeration and Air Conditioning Engineers, Inc. (ASHRAE). Retrieved August 13, 2015, from https://www.ashrae.org/home

Cellulose Insulation Manufacturers Association (CIMA). Retrieved August 13, 2015, from http://www.cellulose.org/

Center for the Polyurethanes Industry (CPI). Retrieved August 13, 2015, from http://polyurethane.americanchemistry.com/

EIFS Industry Members Association (EIMA). Retrieved August 13, 2015, from http://w-ww.eima.com/
(EIFS = Exterior Insulation and Finish Systems)

EPS Industry Alliance. Retrieved August 13, 2015, from http://www.epsindustry.org/
(EPS = Expanded Polystyrene)

EPS Industry Alliance - Insulated Concrete Forms. Retrieved August 13, 2015, from http://www.forms.org/

High Performance Insulation Professionals (HPIP). Retrieved August 13, 2015, from http://www.hpipros.org/

Insulation Contractors Association of America (ICAA). Retrieved August 13, 2015, from http://www.insulate.org/

National Insulation Association (NIA). Retrieved August 13, 2015, from http://www.insulation.org/index.cfm

North American Insulation Manufacturers Association (NAIMA). Retrieved August 13, 2015, from http://www.naima.org/index.php

Perlite Institute. Retrieved August 13, 2015, from https://perlite.org//

Polyisocyanurate Insulation Manufacturers Association (PIMA). Retrieved August 13, 2015, from http://www.polyiso.org/

Reflective Insulation Manufacturers Association International (RIMA International). Retrieved August 13, 2015, from http://www.rimainternational.org/

Society of the Plastics Industry (SPI). Retrieved August 13, 2015, from http://www.plasticsindustry.org/

Structural Insulated Panel Association (SIPA). Retrieved August 13, 2015, from http://www.sips.org/

Analysis Tools:

Oak Ridge National Laboratory:

Building Technologies Research & Integration Center (BTRIC). Retrieved August 13, 2015, from http://web.ornl.gov/sci/buildings/

The mission of the Building Technologies Research & Integration Center (BTRIC) is to identify, develop, and deploy sustainable and energy-efficient building system technologies by forming partnerships between the public sector and private industry for analysis, well-characterized experiments, technology development, and market outreach.

The following web-based interactive calculators allow you to enter values specific to your construction type and location to determine the energy efficiency of your new or existing building.

DOE Steep Slope Calculator. (2005). Estimate annual energy cost savings from the use of solar radiation control on steep-slope roofs. Retrieved November 21, 2017, from http://web.ornl.gov/sci/buildings/tools/SteepSlopeCalc/

Cool Roof Calculator. (n.d.) Estimates cooling and heating savings for flat roofs with non-black surfaces. Accessed December 1, 2017a. http://web.ornl.gov/sci/buildings/tools/cool-roof/.

Modified Zone Method Calculator. (n.d.). Calculates thermal resistance (R-values) for metal stud walls with insulated cavities. Retrieved August 13, 2015, from http://web.

ornl.gov/sci/roofs+walls/calculators/modzone/index.html

Zip Code Insulation Program. (n.d.). Find the most economic insulation level and cost value for new or existing construction based on the climate in your area. (Requires Java-enabled browser). Retrieved November 21, 2017, from http://web.ornl.gov/sci/buildings/tools/zip/

Oak Ridge National Laboratory, and Fraunhofer IBP. (n.d.) WUFI Software. A menu-driven PC program which allows realistic calculation of the transient coupled one-dimensional heat and moisture transport in multi-layer building components exposed to natural weather. Accessed November 28, 2017. http://web.ornl.gov/sci/buildings/tools/wufi/.

Oak Ridge National Laboratory, and Lawrence Berkeley National Laboratory Roof Savings Calculator (RSC). (n.d.) The Roof Savings Calculator was developed as an industry-consensus roof savings calculator for commercial and residential buildings using whole-building energy simulations. Accessed December 1, 2017c. http://rsc.ornl.gov/.

7
Windows and Doors

Windows connect the interior of a house to the outdoors, provide ventilation and daylighting, and are key structural elements (Figure 7-1). Windows and doors are often the architectural focal point of residential designs, yet they provide the lowest insulating value in the building envelope. Poorly chosen windows can increase the costs of keeping a house cool, cause glare, fading of fabrics, and reduce residents' comfort. Well designed buildings carefully consider location, size and performance.

Although recent developments in energy efficient products have markedly improved the efficiency of windows, poorly engineered windows still represent a major energy liability in new construction. The type, size, and location of windows greatly affect cooling and heating costs. Select good quality windows, but shop wisely for the best combination of price and performance. Make certain that windows are chosen for hot climates. Many suppliers do not understand that premium windows designed for northern and central climates are not suitable for Florida; they may increase energy consumption for cooling. Many housebuilding budgets have been exceeded by spending thousands of additional dollars on premium windows suitable for northern and central climates with marginal, if any, energy savings for southern climates. In general, if the windows are well-built, reduce solar heat gain and have good weather-stripping, they will serve you well.

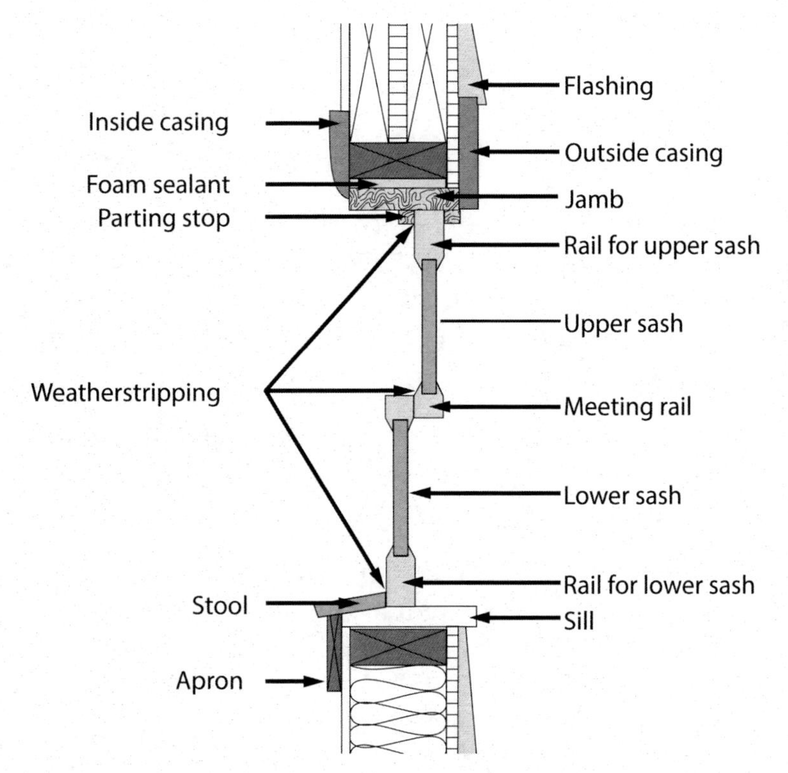

Figure 7-1 Cross-section of single-hung window

In passive design buildings, windows can provide a significant amount of heat, ventilation and light for a house. Consider orientation, overhang and shading in window selection, even if the house is not a passive design building. Details are covered in Chapter 3, "Siting and Passive Design Features", and Chapter 8, "Heating, Ventilation, Air Conditioning (HVAC)".

WINDOWS

Windows should be considered and specified as a system composed of the frame, glazing, spacers separating panes and glass coatings. The number of options has grown rapidly with new glazing and coatings and improved energy efficiency. Choices for window frame materials have expanded beyond wood, aluminum and polyvinyl chloride (PVC) to include hybrid (mixed) materials, composite wood, or aluminum with non-conducting thermal breaks. A thorough understanding of the system components and their purposes can ensure significant energy savings without compromising functionality and comfort for building occupants.

High performance windows certainly are more expensive than basic counterparts, but they can often reduce heat loads enough to allow smaller HVAC units, reducing the effective cost of the windows.

Heat transfers through windows and skylights in four different ways:

- **Solar radiant energy** passes through glass–both visible light and invisible wavelengths (ultraviolet and infrared). Most radiant heat is caused by infrared wavelengths, and is usually the most significant source of heat transfer in windows.
- **Conduction of heat** (energy transferred by direct contact of molecules) through the glass and window frame. Energy is conducted quickly through metals, but more slowly through wood or vinyl. Conduction occurs much more slowly through gases, which is the main reason that double-paned windows reduce heat transfer.
- **Convection of air** carries heat as the air circulates. As warm gases rise and cooler gases sink, a circular movement is created that distributes heat. This occurs between the panes of double-glazed windows, or within a room, mixing the air that has been either heated or cooled adjacent to a window throughout the room.
- **Air leakage**. Air leaking around window frames and sashes transfers heat as it moves in or out of the building. Poorly constructed or installed windows that allow significant air leakage can negate any benefits gained by energy efficient design and materials.

Radiant energy comes from the sun in many wavelengths, composed of visible light as well as ultraviolet light (UV) and infrared light that we cannot see. Objects, such as windows, absorb the energy and re-radiate or emit much of it, transferring the energy indoors (see Figure 7-2). More than half of the energy in solar radiation is from infrared wavelengths, not visible light. Of course the purpose of windows is to allow light into a building. If windows can permit in

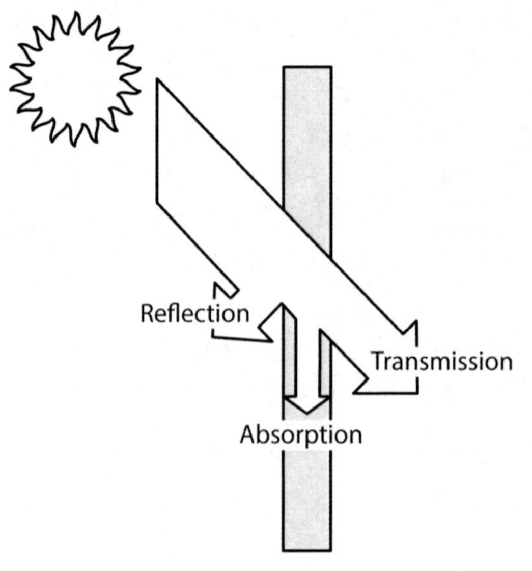

Figure 7-2 Solar radiation

only the visible light and reflect infrared wavelengths, heat transfer is reduced. This is what spectrally selective low emissivity (low-e) windows do. A metal oxide coating on one pane reflects radiant energy back, rather than absorbing and re-emitting it. To prevent damage to the coating it is located on an interior surface of double paned glazing. For more, see Figures 7-3 and 7-4, and further discussion on low-e windows later in this chapter.

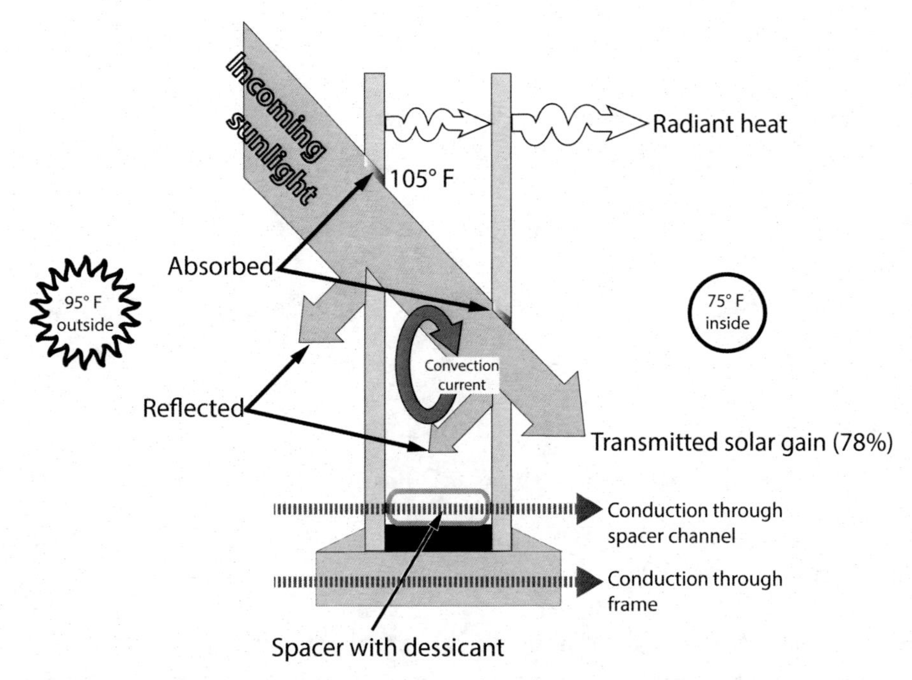

Figure 7-3 Summer heat gain through a double-glazed window

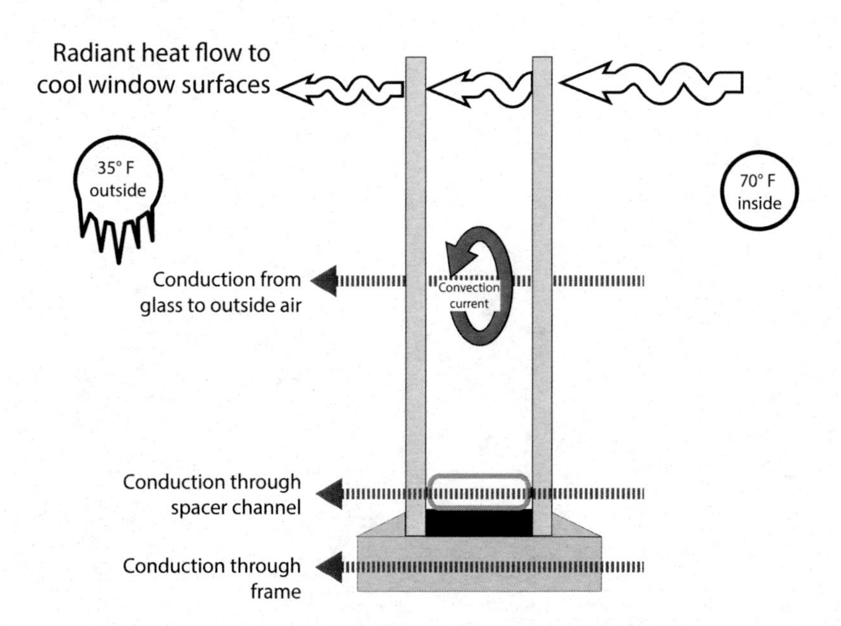

Figure 7-4 Winter heat loss in a double-glazed window

Conduction of heat through window frames can also be a significant component of energy transmission. Metals conduct heat much more quickly than materials such as wood or vinyl. A strip of non-conducting material between the inner and outer metal frames, called a thermal break, will reduce heat movement through the window. Thermal breaks can make aluminum frames as energy efficient as vinyl or wood frames. (See Figure 7-5.) Another option for increasing energy efficiency in double or triple glazed windows is to fill the space between the panes with argon, krypton or carbon dioxide. These gases are better insulators, as they conduct heat more slowly than air.

THERMAL BREAKS AND WINDOW SPACERS

Thermal breaks in metal window frames are of particular importance. Metal is a very poor insulator—in fact, it is a good **conductor** of heat. A thermal break separates inside and outside pieces of the window frame with an insulating material, thus improving U-factors.

When shopping for windows note:

- the U-factor for the entire window, not just for the glass
- whether NFRC 100 was used to derive it, as required by code
- gas fill will improve the window's U-factor (insulation), but is less important than preventing solar gain

It makes no sense to pay top dollar for a window that looks great on paper, but performs poorly in the real world. Depend on certification like National Fenestration Rating Council (NFRC) for energy features, and AAMA (American Architectural Manufacturers Association), WDMA (Window and Door Manufacturers Association) or FMA/Keystone (Fenestration Manufacturers Association) for structural features, to get the quality and performance you purchased.

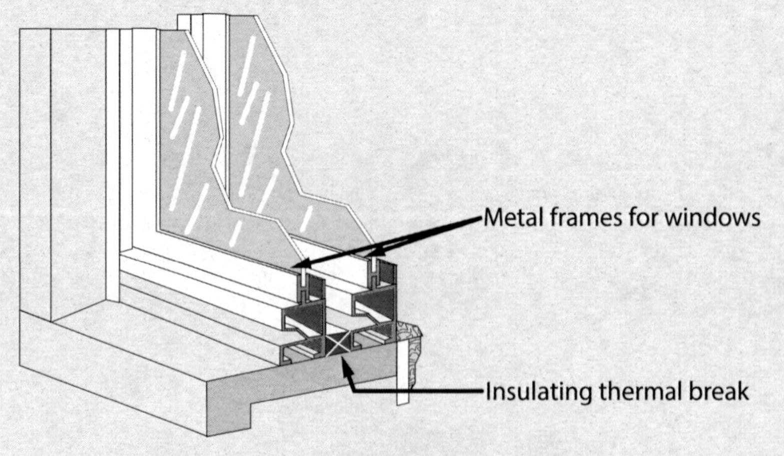

Figure 7-5 Metal window frame with thermal break

Convection within a room can be reduced by anything that reduces the temperature difference between the interior and exterior, such as multiple glazing or shading of the window. It also can be lessened by window coverings on the inside that block the circular flow of air, such as floor length curtains, shades or shutters that extend to the window sill and pelmets across the top of the window.

Air leakage around fenestration is obviously influenced by the tightness of the window assembly as well as its installation. Most air leaks between the sash and frame or between sliding sashes. Horizontal sliders are the most difficult to seal; casement and awning windows have the least air leakage. Look for manufacturers that have performed standard testing of their window assemblies and report leakage data. Rates no more than 0.3 cfm/square foot are considered acceptable. Caulking and sealing around windows and other building envelope penetrations is covered in Chapter 5, "Air Leakage—Materials and Techniques".

Window Performance

The following terms are commonly used to compare window performance:

- **Solar Heat Gain Coefficient (SHGC),** is the most important in Florida for the reasons described above. It is expressed as a decimal between 0 and 1, and smaller values are desirable.
- **U-factor** expresses the effective insulation value of the window—its ability to prevent heat loss through conduction and convection and long wave infra-red radiation (not included in light wavelengths). It is the inverse of R values generally used with wall and ceiling insulation. For example, a window with a U-factor of 0.4 has a R value of 2.5. U-factors are usually in the range of 0.20 to 1.2 and are expressed in the units of Btu/h ft^2 °F. When comparing u-factors, be certain they represent full assemblies, not just glass ratings.
- **Visible Transmittance (VT)** is the decimal percent of light that passes through the window, so a VT of 1.0 allows all visible light in and 0.50 allows half to pass through. Human eyes do not perceive the difference of slight to moderate decreases. A $VT \geq 0.70$ is considered good, and usually should not be < 0.50.
- **Air Leakage (AL)** is usually quoted as a measurement of the air in cubic feet per minute passing through each square foot (cfm/ft^2) of window assembly. Lower values are better, and a typical range is 0.1 to 0.3, although casement windows can achieve as low as 0.01 cfm/ft^2.
- **Condensation Resistance (CR),** a number between 0 and 100, provides a relative comparison of the window's resistance to condensation forming on its interior. Bigger numbers are better.

Types of Low-E Windows

The low-e windows designed for **hot** climates are called *low-solar-gain* or *spectrally selective*. Their coatings select mainly infrared energy and reflect it back outdoors. These windows aim to allow a high transmission of visible light; to achieve significant savings in solar heat gain without appearing dark inside. In Florida's hot climate, reducing solar heat gain is best way to improve energy efficiency of windows.

It is important to know that low-e coatings reflect radiant energy in only one direction—either to the outside or the inside of the building, depending on the placement of the coating layer. Low-e windows were first designed for **cold** climates, and are *high solar gain*. Their coatings reflect radiant energy back indoors, similar to the way a greenhouse functions. They should not be used in Florida, as summer heat gain would be magnified indoors, increasing demand for air conditioning.

Warm climate low-E glass: low-solar-gain, double-glazed window with argon or krypton gas fill

An example of low solar gain low-e, double glazed window with argon or krypton gas is illustrated in Figure 7-6. The low SHGC value indicates it will reduce summer heat gain, and the low U-factor indicates it will reduce heat loss in the winter. This is suitable for areas that require both summer cooling and winter heating. A window with a high VT factor will transmit most of the visible light indoors. A window with triple glazing or double glazing with an interior suspended film and argon or krypton gas fill can achieve a U-factor as low as 0.13 with a SHGC of about 0.33.

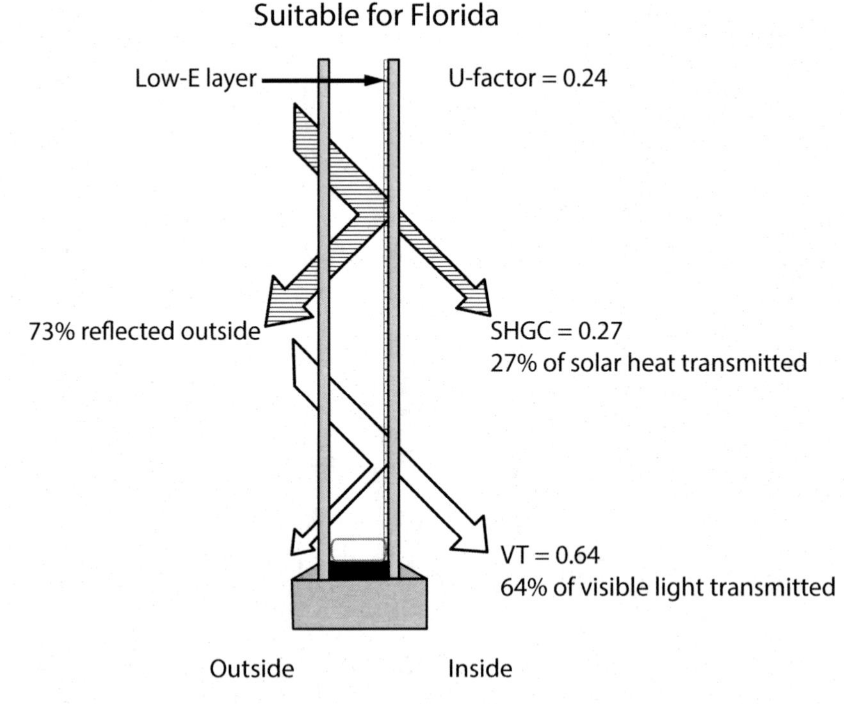

Figure 7-6 Warm climate low-e glass

These values represent the glass only; remember the performance of the window will also depend on its type of framing and installation.

Cold climate low-E glass:
high-solar-gain, double-glazed window with argon or krypton gas fill

Figure 7-7 is an illustration of a typical window designed for climates where heating governs the choice of performance characteristics. It will slow heat loss, and its low-e coating allows more solar radiation into the interior, increasing heat gain year-round. It is not suitable for Florida climates.

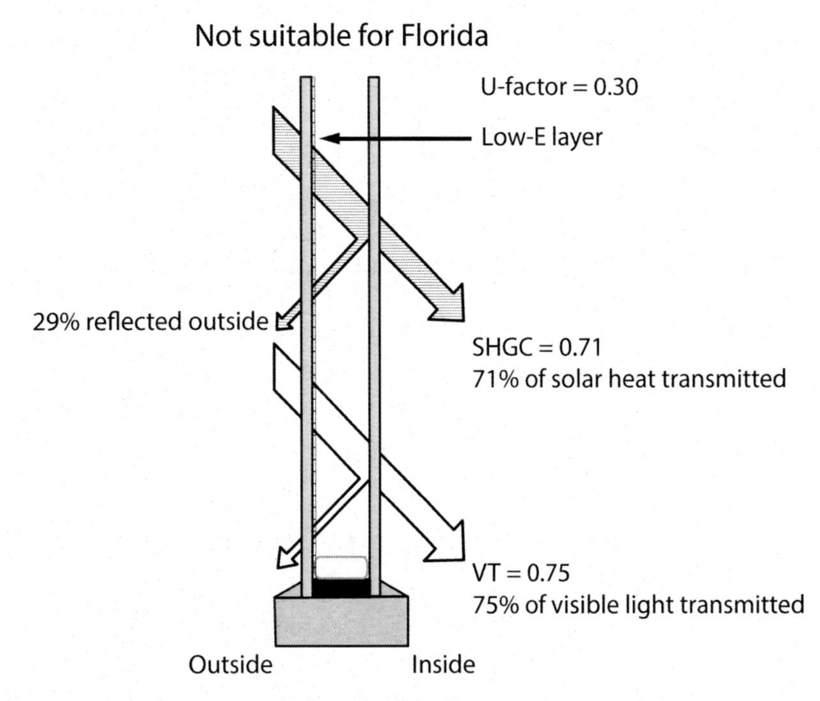

Figure 7-7 Cold climate low-e glass

NFRC Ratings

The National Fenestration Rating Council (NFRC) is a nonprofit organization representing groups related to all facets of fenestration, from designers and architects to manufacturers, suppliers and government regulators. They offer a voluntary testing and certification program for thermal performance of windows, skylights and doors. They do not certify that windows, doors or skylights are energy efficient or meet any particular standards; rather they provide a standard method for rating and labeling fenestration.

If a product is certified by NFRC, it will have a label showing the test results (an example is shown in Figure 7-8). Note the entire unit, including glazing, frame and spacers is rated. Manufacturers sometimes report center-of-glass U-factors to describe the performance of the glazing alone without the effects of the frame. For most energy efficient windows, the whole window U-factor is higher than the center-of-glass U-factor.

U-factors for all fenestration products must be determined by an independent, accredited laboratory as described in NFRC 100. The manufacturer must label and certify the rating. SHGC values also must be determined by an accredited, independent laboratory in accordance with NFRC 200, and labeled and certified by the manufacturer. *If no ratings are provided for fenestration products, the following default values must be assumed.* (See Table 7-1 and Table 7-2). Notice that they are consistent with very inefficient windows, so it is highly advisable to avoid purchasing unlabeled fenestration.

Table 7-1 Default Fenestration Ratings

From FBC Table R303.1.3 (1)

Rating Type	Frame	Single Glazed		Double Glazed		Skylight		Glazed Block
		Clear	Tinted	Clear	Tinted	Single	Double	
U-Factor	Metal	1.2		0.8		2	1.3	
	Metal with Thermal Break	1.1		0.65		1.9	1.1	0.6
	Nonmetal or Metal Clad	0.95		0.55		1.75	1.05	
SHGC	All Frames	0.8	0.7	0.7	0.6			0.6

Table 7-2 Typical Properties for Complete Windows[1]

Frame	Metal Frame	Metal Frame with Thermal Break	Non-metal Frame	Non-metal Frame, Thermally Improved
Single-Glazed, Clear Glass				
U-Factor	≥1.00		0.71-0.99	
SHGC	≥0.60		≥0.61	
VT	≥0.60		≥0.60	
Double-glazed, Clear Glass				
U-Factor	0.71-0.99	0.56-0.70	0.41-0.55	
SHGC	>0.60	>0.60	0.41-0.60	
VT	>0.60	>0.60	0.51-0.60	
Double-glazed, Medium-solar-gain Low-E Glass, Argon/Krypton Gas				
U-Factor	0.56-0.70	0.41-0.55	0.31-0.40	0.26-0.30
SHGC	0.26-0.40	0.26-0.40	0.26-0.40	0.26-0.40
VT	0.51-0.60	0.51-0.60	0.51-0.60	0.51-0.60
Double-glazed, Low-solar-gain Low-E Glass, Argon/Krypton Gas				
U-Factor	0.56-0.70	0.41-0.55	0.31-0.40	0.26-0.30
SHGC	≤0.25	≤0.25	≤0.25	≤0.25
VT	0.51-0.60	0.51-0.60	0.41-0.50	0.41-0.50

1 Values for specific windows will vary. Use NFRC ratings for individual products.

Data from Efficient Windows Collaborative, http://www.efficientwindows.org. (Nov 2012)

The **Product Description** provides information about the characteristics of the product.

The **Manufacturer Name** displays the name of the window manufacturer.

The **U-Factor** measures the amount of heat that escapes through the product. It refers to the whole window U-factor. The lower the rating, the better the window is at preventing heat loss. NFRC certified products require U-factor ratings. In Florida, select windows with a U-factor at least as low as 0.65.

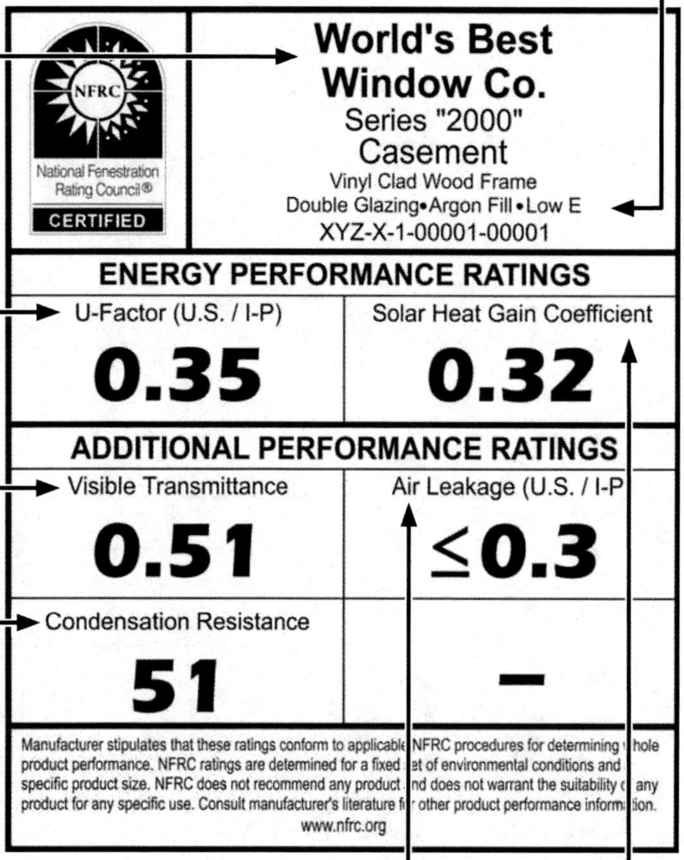

Visible Transmittance (VT) measures how much light comes through a product. The visible transmittance is an optical property that indicates the amount of visible light transmitted. VT is expressed as a number between 0 and 1. The higher the VT, the more light is transmitted.

Condensation Resistance (CR) is an optional rating that a manufacturer may or may not include on the label. It measures the ability of a product to resist the formation of condensation on the interior surface of that product. The higher the CR rating, the better that product is at resisting condensation formation. While this rating cannot predict condensation, it can provide a credible method of comparing the potential of various products for condensation formation. CR is expressed as a number between 0 and 100.

Air Leakage (AL) is indicated by an air leakage rating, expressed as the equivalent cubic feet of air passing through a square foot of window area (cfm/sq ft). Heat loss and gain occur by infiltration through cracks in the window assembly. The lower the AL, the less air will pass through cracks in the window assembly. In Florida, look for an AL value of no more than 0.3 cfm/sq ft.

Solar Heat Gain Coefficient (SHGC) measures how well a product blocks heat caused by sunlight. The SHGC is the fraction of incident solar radiation admitted through a window, both directly transmitted, and absorbed and subsequently released inward. SHGC is expressed as a number between 0 and 1. The lower a window's solar heat gain coefficient, the less solar heat it transmits. In Florida, select windows with a SHGC of 0.30 or less.

Figure 7-8 Components of the NFRC label for windows

See **FBC, Energy Conservation, Table R402.1.2 Insulation and Fenestration Requirements by Component** and **Section R402.4.3 Fenestration air leakages** for detailed information.

Table 7-3 provides a summary of minimum efficiency requirements.

Table 7-3 Florida Fenestration Energy Efficiency Requirements

Summarized from FBC, Table R402.1.2

Construction Type	Other Criteria	Fenestration U-Factor	Skylight U-Factor	Fenestration SHGC	Skylight SHGC
Residential[1]	ENERGY STAR[1]	≤ 0.40	≤ 0.60	≤ 0.25	≤ 0.28
	Standard	≤ NR[2] / 0.40[3]	≤ 0.75[2] / 0.65[3]	≤ 0.25	≤ 0.30
	Impact Resistant[4]	≤ 0.75[2] / 0.65[3]	≤ 0.75	≤ 0.30	≤ 0.30
New Commercial and Multi-family	0-40% WW Ratio 40-50% WW Ratio > 50% WW Ratio	≤ 0.45	≤ 1.36	≤ 0.25 ≤ 0.19	≤ 0.19
Commercial Renovations and Alterations	0-40% WW Ratio > 40% WW Ratio	≤ 0.45	≤ 1.36	≤ 0.25 ≤ 0.25	≤ 0.19

1 *ENERGY STAR Southern Climate Zone*

2 *Florida Building Code Climate Zone 1*

3 *Florida Building Code Climate Zone 2*

4 *Impact Resistant fenestration must comply with* ***FBC, Residential, Section R301.2.1.2 Protection of openings*** *or* ***FBC, Building, Section 1609.1.2, Protection of openings****.*

An independent verification program began in late 2011 to verify the advertised performance of ENERGY STAR and NFRC certified residential windows, skylights and doors. This ensures that the stated ratings accurately reflect performance. NFRC/AAMA/WDMA/CSA 101/I.S.2/A440 testing, which tests structural characteristics, including air and water leakage, may be used.

In addition to its certification program for residential products, NFRC offers a Site-Built certification and rating program that can be used for site-assembled fenestration in non-residential buildings. Called *Component Modeling Approach* (CMA), the software contains a database of tested and approved fenestration component parts. They can be combined to produce ratings for assembled fenestration. A "Label Certificate Form" to be posted on the side of the building replaces the labels used on residential products.

> ***How is the NFRC label different than the ENERGY STAR label?***
>
> The ENERGY STAR label indicates that a product is energy-efficient. The NFRC label allows comparisons between energy-efficient products by giving independent ratings in several energy performance categories.

ENERGY STAR® windows and skylights

ENERGY STAR windows and skylights must meet criteria suitable for the climate in various parts of the country. All of Florida is in the "Southern Zone", and fenestration installed here should display a label shaded as shown here (Figure 7-9). Windows that meet Southern Zone ENERGY STAR criteria will have a U-factor of ≤ 0.40 and SHGC ≤ 0.25 or lower; skylights will have a U-factor ≤ 0.60 and SHGC ≤ 0.28.

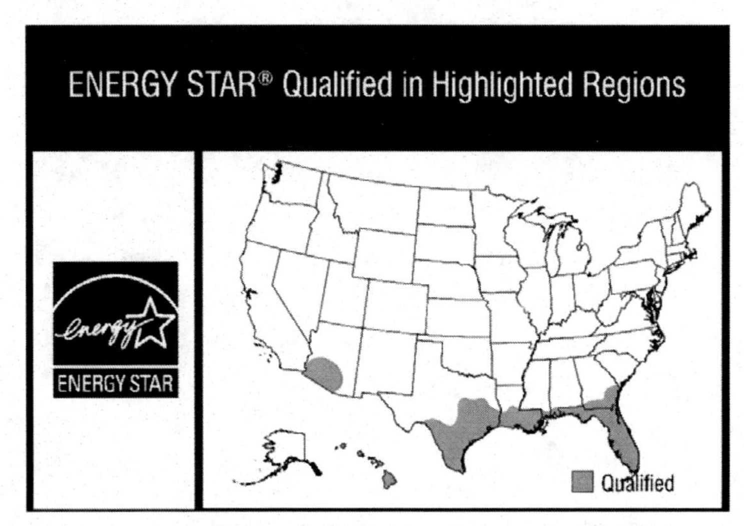

Figure 7-9 ENERGY STAR Southern Climate Zone label for windows

Visible transmittance is not included in the criteria. However, it should be considered, given current technologies. Again, high efficiency does not mean reduced visibility. Glazing with low SHGC rates and high visible transmittance (0.50 and above) may be more expensive, but the investment may be worth it for energy saving, unobstructed views.

The majority of the fenestration market is now choosing ENERGY STAR products. A Ducker Research study published in 2010 found that 81% of windows, 99% of skylights and 71% of swinging doors purchased were rated ENERGY STAR. The EPA is continuing to assess research and coordinate with industry stakeholders to regularly update ENERGY STAR criteria. Check their website for updates (listed in the Resources section at the end of the chapter.)

Difficulties in Reporting Window Values

Insulating Values

Window insulating values are typically reported in U-factors—the inverse of R-values. Double-glazed products can have R-values as high as 3.3, or U-factors of about 0.30 (1/R = 1/3.3 = 0.30). A single pane of glass generally has R-values of 1.0 and thus has U-factors of 1.0 for glass only. The R-value is no longer commonly used by certifying agencies to evaluate window performance, but window manufacturers may include it in their window ratings. It can be confused with the R-classification of structural design pressure values, so ask for the U-factor to avoid confusion.

Component vs. Whole Window Performance

Window U-factors and Solar Heat Gain Coefficients may be reported as the value through the glass surface alone. However, windows are made of more than just glass. They have a frame or sash, spacer bars that hold the sections of glass in a double-glazed window apart, and a jamb. The claimed value should reflect the overall insulating value of *all* of the components. Manufacturers are encouraged to report window U-factors consistently and accurately. Moreover, code compliance calculations reference whole window values tested to NFRC procedures as well for specific product performance. (See **FBC, Energy Conservation, Section R303.1.3 Fenestration product rating.**)

For an example on insulating value, there are two companies that produce extremely efficient windows. Both have two outer glass panes and two inner layers of low-e coated film. In one case, all air spaces are filled with argon, providing an U-factor of 0.13 for the glass. However, losses through the edges and frames changes the overall window U-factor to 0.25. (Remember, *lower* numbers indicate *better* ratings.) Another window is filled with air rather than argon yielding an U-factor of 0.15 for the glass. Due to its unique edge system, which has thermal breaks made of nylon spacers with insulation in between, the ***overall*** window U-factor is 0.16—better than the window with the higher glass U-factor.

Solar Heat Gain and Shading Coefficients

In warm climates such as Florida, a low Solar Heat Gain Coefficient or SHGC is the most important criteria in window selection. SHGC measures the amount of solar radiation that enters a building through the window and causes heat gain. A poor performance window that has a SHGC of 0.80 allows 80% of the solar heat to pass through it. Windows with low SHGCs are good at preventing heat gain caused by solar radiation. Select windows with a SHGC of less than 0.25 in warm climates.

Construction professionals may be more familiar with the *shading coefficient* or SC, which measures how the solar gain of a glazing system compares to that transmitted through clear, single-pane glass. The more layers of glass, coatings, or tints that a window has, the more sunlight it impedes and hence, the lower the shading coefficient.

SC was formerly widely used to measure solar control properties, but has now been phased out in favor of the more accurate SHGC. (References to shading coefficients have been removed from the Florida Building Code.) However, you can still find SC rating listed in window and window treatment performance literature (Table 7-4). To make an approximate conversion from SC to SHGC, multiply the SC value by 0.87. Divide by 0.87 to convert from SHGC to SC. To claim accurate energy performance on code calculations, obtain SHGC values from accredited laboratories that test to NFRC procedures. (See **FBC, Energy Conservation, Section R303.1.3 Fenestration product rating.**)

Table 7-4 Window Treatment Shading Coefficients

Treatment	Window Type	Approximate Solar Heat Gain Coefficient	Shading Coefficient*
Single-glazed window	1/8-inch glass	0.87	1.00
	1/4-inch glass	0.82	0.94
	tinted (1/4-inch)	0.68	0.78
Double-glazed window	1/8-inch glass	0.76	0.88
	1/4-inch glass	0.70	0.81
	tinted (1/4-inch)	0.58	0.67
Venetian blinds	1/4-inch single glass	0.52	0.60
	1/4-inch double glass	0.47	0.54
Roller blinds (white)	1/4-inch single glass	0.22	0.25
	1/4-inch double glass	0.22	0.25
Light, airy drapes	1/4-inch single glass	0.61	0.70
	1/4-inch double glass	0.50	0.58
Heavy drapes	1/4-inch single glass	0.39	0.45
	1/4-inch double glass	0.37	0.42
Exterior shade screen	1/4-inch single glass	0.33	0.38
Louvered sun screen	1/4-inch double glass	0.37	0.42

* *Fraction of sunlight that passes through glass and window treatment. Assumes sunlight strikes perpendicular to glass.*

New Window Technologies

Two types of dynamic glass have entered the market and are expected to grow significantly over the next 5-10 years. Dynamic glazing has the ability to change its performance properties and reverse the process. Thermochromic glass temporarily darkens in response to direct sunlight, similar to photo-grey lenses in glasses. The tint of electrochromic glass is controlled electronically, either on a programed schedule or as desired by the occupant. A darker tint decreases heat gains and reduces interior cooling demand. The average cost per square meter was about $400 for thermochromic and $700 for electrochromic glass in 2011, but the cost of each is expected to be cut by half by about 2016. If energy costs continue to rise, they may become cost effective options.

Skylights

The same features that improve energy efficiency in windows apply to skylights, but since they are often in direct sunlight, it is even more important that low SHGC be selected. Tubular daylighting devices have domed glass lenses that transmit sunlight through roofs to interior areas. They can contribute to energy efficiency by increasing natural daylight in areas where windows are not possible and reducing the need for electric lighting.

In commercial buildings, skylights may not exceed 3% of the total roof area.

Financial Incentives for Energy Efficiency

Beyond the money saved on energy consumption, many utilities in Florida offer incentives or rebates for installation of energy efficient windows. These range from various loan programs to rebates for replacement windows or installation of window films meeting specified criteria. A summary of available options is found in the document *Incentives and Rebates for Energy-Efficient Windows Offered through Utility and State Programs* at the Efficient Windows Collaborative website (see Resources, at the end of this chapter), or check with your local utility.

Tools for selecting fenestration

A Condensation Resistance Factor (CRF) Tool has been developed by the AAMA to calculate a suggested minimum CRF rating based on the indoor and outdoor temperature, and indoor relative humidity. The calculations are described in AAMA 1503-09, *Voluntary Test Method for Thermal Transmittance and Condensation Resistance of Windows, Doors and Glazed Wall Sections*. The tool and a user's manual can be accessed from the AAMA website (see Resources).

Table 7-5 contains some options for design tools for residential and commercial windows:

Table 7-5 Software for Window Design

Software	Description	Web address	Fee for Use	Required Expertise
RESFEN	Calculates peak & annual energy use and cost based on window characteristics, HVAC system data, floor and window areas, orientation, and shading.	https://windows.lbl.gov/software/resfen	no cost	low
WINDOW	Performs heat transfer analysis for NFRC rating criteria, such as U-value SHGC, shading coefficient, VT, and more. Often used for design.	https://windows.lbl.gov/software/window	no cost	moderate
THERM	Calculates heat conduction in windows frames and other building components. Works with WINDOW software.	https://windows.lbl.gov/software/therm	no cost	moderate
Home Energy Saver (HES)	US DOE program. Homeowners describe their home, estimate energy use and potential savings from structural improvements, new appliances, etc.	http://hes.lbl.gov/consumer/	no cost	general public
EnergyPlus	Whole building simulation for optimizing energy and water use, airflow within the building and photovoltaic systems.	https://www.energyplus.net/	no cost	high
Building Energy Software: Tools Directory	Directory of many models and tools for analysis of materials and systems in building envelope and design. It includes both free and purchased software packages.	http://www.buildingenergysoftwaretools.com/	N/A	varies

Proper Window Installation

As referenced in the Florida Building Code, windows should be installed according to manufacturer's instructions. If there are no instructions included, consult ASTM E2112-07(2016). The American Architectural Manufacturers Association (AAMA) provides installer certification for those who successfully pass their training course based on this standard.

Options for Reducing Solar Gain

When choosing windows, consider:

- Orientation
- Desired size
- Shading potential
- Energy performance

To maximize energy savings from window orientation, locate as many windows as possible on the north side, or the south side if large overhangs are provided to shield the interior from summer afternoon sun. To the east and west, the sun's rays enter at lower angles and cannot be effectively shielded by overhangs. Therefore, it is more important to invest in low SHGC windows for those with east and west exposure, and of lesser importance on the north and south facing walls.

Up to 40% of unwanted heat gain is due to windows, according to the Energy Efficient and Renewable Energy Network at the U.S. Department of Energy. Construction professionals in Florida have a wide choice of options in addition to high efficiency windows to reduce heat gain:

- Window tints and films
- Overhangs
- External shades and shutters
- Interior shades and shutters
- Landscaping and trees (see Chapter 3, "Siting and Passive Design Features," for detailed information)

All of these measures reduce energy use, lower utility costs and increase homeowner comfort. Overhangs can be used for credit in energy code calculations and ENERGY STAR Homes qualification, because permanent overhangs have a consistent, measurable effect on energy consumption. External screening and internal shading are included in energy calculations; however, they have negative impact on the calculations.

Reflective Films and Tints

Reflective or low-emittance films, which adhere to glass and are found often in commercial buildings, can block up to 85% of incoming sunlight. If visible transmittance is similarly reduced, this can lead to higher interior electric lighting use, and the same or higher electricity bills. In fact, many of the same films are inserted between glazings during assembly of low-emittance windows by manufacturers. However, exposure to external elements can cause degradation when applied post-installation, requiring replacement. It is not recommended for windows that experience partial shading because as the film absorbs sunlight it may heat the glass unevenly. The uneven heating of windows may break the glass or ruin the seal between double-glazed units. Many double-paned window manufacturers void their warranties if unapproved window films are applied to their products.

In cooler parts of Florida, where solar heat gain is desired in the winter, choose films with moderate SHGC ratings. The installed cost of reflective films varies. Price should not be the sole criterion when selecting a film installer—quality is a vital consideration affecting the appearance of the house and the beauty of the view to the outside. NFRC has a searchable product listing of certified window films (as well as windows, doors and skylights) on its website; go to http://search.nfrc.org. When shopping, look for an NFRC label that is oval, rather than the rectangular label for windows.

Most window manufacturers also offer tinted glass—a color added to the glazing mixture. Some have a reflective finish to block additional sunlight. Traditionally, bronze and gray tints have been popular in Florida, but new high-performance tints can be as effective as low-e coatings. They are more durable in single-pane applications, but the ones that perform as well as low-e coatings may not appeal to consumers as much, since they are often blue or green. Some homeowners have a perception that their entire home will have a blue hue, although it is hardly noticeable once installed. The benefit from tinted glass will be reflected in the SHGC rating.

Shading

The effectiveness of different window shading options depends on the composition of the incoming sunlight. Sunlight reaches the building in three forms: *direct, diffuse*, and *ground reflected*. On a clear day, most sunlight is direct, traveling as a beam without obstruction from the sun to a building's windows. In winter, most of the direct sunlight striking a window is transmitted; however, in summer, the sun strikes south windows at a much steeper angle, and much of the direct sunlight is reflected.

The majority of the sunlight entering south-facing windows in the summer is either diffuse—bounced between the particles in the sky until it arrives as a bright haze—or is reflected off the ground.

In developing a strategy for effectively shading windows, both direct and indirect sources of sunlight must be considered. Overhangs, long thought to be totally effective for shading south-facing windows, are best at blocking direct sunlight and are therefore only a partial solution. Excessive overhangs make the home building structurally vulnerable to hurricane winds. A better approach would be to use exterior shutters, such as Bahama shutters, for shading the window, allowing the view and protecting the window from damage. Only devices with NFRC-tested SHGC and U-factor values can be included in code compliance.

Overhangs

Overhangs shade direct sunlight on windows facing within about 30 degrees of south. Overhangs on east and west windows are ineffective unless they are as long as the window is high. Keep in mind long overhangs should be adequately designed for typical hurricane wind loads.

Overhangs should be sized to account for differences in sun angles, elevation, window height and width, and wall height above the window. Free and low-cost computer programs and tools are available to help. Some examples (web addresses are listed in Resources):

- SunAngle, a program to determine the angle of the sun for any point in the country, and Overhang Design, a program to calculate overhang dimensions, are both available from Sustainable By Design.
- Latitude, longitude, and elevation data is available from Weather Underground.
- A listing of free and for-purchase energy models, including solar design tools can be found at the Building Energy Codes Program, which is part of the U.S. Department of Energy's Building Technology Office.
- A low-cost sun angle calculator is available from the Society of Building Science Educators.

Retractable awnings allow full winter sunlight, yet provide effective summer shading where they do not conflict with hurricane codes. They should have open sides or vents to prevent accumulation of hot air underneath. Awnings may be more expensive than other shading options, but they serve as an attractive design feature.

While other overhangs and awnings provide considerable energy savings, they should be carefully considered with regard to hurricane preparedness. Poorly designed overhangs can be a serious detriment in windstorms. Look for awnings or other external operable shutters that are wind resistant. The Florida Building Code overhang parameters are dictated by the structural design.

External Shades and Shutters

Exterior window shading treatments are effective cooling measures because they block both direct and indirect sunlight before it enters windows. Solar shade screens are an excellent exterior shading product with a thick weave that blocks up to 70 percent of all incoming sunlight before it enters the windows. The screens absorb sunlight so they should be used on the outside of the windows. From the outside, they look slightly darker than regular screening, and provide greater privacy. From the inside many people do not detect a difference. Most products also serve as insect screening and come in several colors. They can be removed in winter in north Florida where full sunlight is desired for the colder season.

Hinged decorative exterior shutters which close over the windows are also excellent shading options. However, they obscure the view, block daylight, can be expensive, are subject to wear and tear, and can be difficult to operate on a daily basis. Again, refer to the *Florida Building Code, Building* volume before investing in these devices. There are edge-encapsulated exterior roller shades that may not be as susceptible to these problems.

Interior Shades and Shutters

Many types of interior shades, blinds and curtains can reduce solar radiation as well as their decorative function. They are most effective if they have a white exterior backing that reflects light back through the window. Those that isolate the window from the room by closing gaps at the top and bottom of the window can also slow convective heat transfer within the room. Shades that fill the window frame, floor length curtains and cornice boards across the top of the window are examples to slow convection.

Shutters and shades located inside the building include curtains, roll-down shades, and Venetian blinds. More sophisticated devices, such as shutters that slide over the windows on a track and interior movable insulation, are also available.

Interior shutters and shades are generally the least effective shading measures because they try to block sunlight that has already entered the room (Table 7-6, next page). However, it is the least expensive choice to the builder because interior shading is usually provided by the homebuyer. The most effective interior treatments are solid shades with a reflective surface facing outside. In fact, simple white roller blinds keep the building cooler than more expensive louvered blinds, which do not provide a solid surface and allow trapped heat to migrate between the blinds into the building.

Table 7-6 Shading Coefficients for Window Coverings

Type of Covering	Shading Coefficient*
None	0.88
Medium-colored venetian blinds	0.57
Opaque dark shades	0.60
Opaque white shades	0.25
Translucent light shades	0.37
Open weave dark draperies	0.62
Close weave light draperies	0.45

* *Lower numbers shade better. The table assumes windows are double-glazed. Source: ASHRAE Handbook of Fundamentals, 2009.*

Doors

Energy efficient exterior doors can be fabricated of various cladding materials such as wood, fiberglass or steel around a polyurethane or polystyrene foam core (Figure 7-10). Or, fiberglass insulation and steel stiffeners can be sandwiched between steel layers. These can have U-factors below 0.2 (0.18–0.09), compared to solid wood doors with U-factors about 0.5. Steel and fiberglass skins can be molded to look like wood, and some incorporate recycled materials. These doors have long lifetimes, will not warp and offer increased security. Doors without glass are rated by U-factors only, and those with glass include SHGC ratings as well.

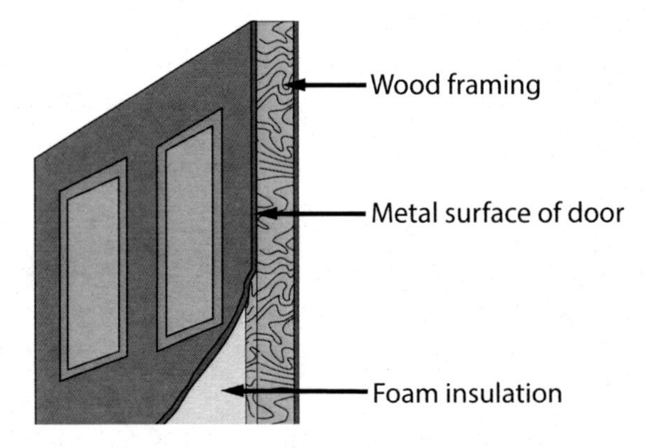

Figure 7-10 Insulated metal door

Other options for core materials include honeycomb core with steel facing. These are strong structurally, but have poor thermal performance.

Pre-hung doors are preferred to those with custom, on-site constructed frames. They generally have better weather stripping and a tighter fit. If patio doors are being used, choose swinging doors rather than sliding glass doors. The seals are much tighter, reducing air flow around the doors.

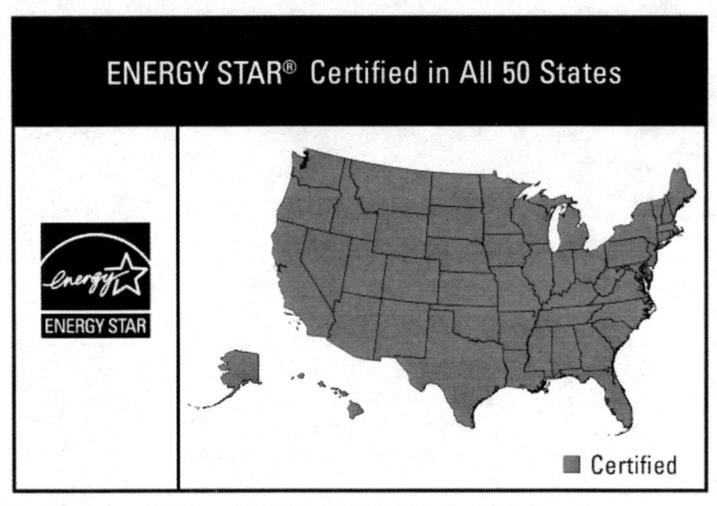

Figure 7-11 ENERGY STAR label for doors

Look for a label indicating the door meets the energy star criteria (Figure 7-11). Doors in all climate zones have the same requirements, as shown in Table 7-7 below. Make sure the row selected matches the glazing level of the door. Notice that ½ Lite doors are less than 1/3 glass (not 50% as you might expect). The NFRC has also developed a label for doors, similar to its window label. (See Figure 7-12, next page.)

Table 7-7 Criteria for ENERGY STAR® Doors

<table>
<tr><th>Glazing Level</th><th>U-Factor[1]</th><th colspan="2">SHGC[2]</th></tr>
<tr><td>Opaque</td><td>≤ 0.17</td><td colspan="2">No Rating</td></tr>
<tr><td>≤ ½-Lite</td><td>≤ 0.25</td><td colspan="2">≤ 0.25</td></tr>
<tr><td rowspan="2">> ½-Lite</td><td rowspan="2">≤ 0.30</td><td>Northern
North-Central</td><td>≤ 0.40</td></tr>
<tr><td>Southern
South-Central</td><td>≤ 0.25</td></tr>
</table>

Air Leakage for Sliding Doors ≤ 0.3 cfm/ft^2
Air Leakage for Swinging Doors ≤ 0.5 cfm/ft^2

1 *Btu/h ft² °F*
2 *Solar Heat Gain Coefficient*

The **Manufacturer Name** displays the name of the door manufacturer.

The **Product Description** provides information about the characteristics of the product.

Glass Area - Amount of glass in door by area.

Door-specific Rating
The circled value shows the rating a door has received. Each rating has 2 values: ***Solar Heat Gain*** and ***U-Factor***.

Solar Heat Gain Coefficient measures how well a product can resist unwanted direct or indirect solar radiation. This radiation can cause your home to heat regardless of outside temperature, which may be favorable or unfavorable depending on whether you're heating or cooling your home. In summer months, a low solar heat gain coefficient helps to keep your home cool. In winter months, a higher solar hear gain coefficient can help to keep your home warm.

- **Range**: 0-1
- **Look for**: Low numbers in cooling conditions; high numbers in heating conditions.

U-Factor measures how well a product can keep heat from escaping from the inside of a room. The lower the number, the better a product is at keeping heat in.

- **Range**: 0.00-2.00
- **Look for**: Low numbers

NFRC

National Fenestration Rating Council®

CERTIFIED

Door Company Ltd.

Entrance Door

Insulated Steel Edge Door
XYZ-X-1*

ENERGY PERFORMANCE RATINGS

Product Description** Default Frame*** Wood	**U-Factor[1] /Solar Heat Gain Coefficient (SHGC)** Individual Option Number			
	1/4 Lite ≤ 0.265 m² (410 in²)	1/2 Lite ≤ 0.581 m² (900 in²)	3/4 Lite ≤ 0.710 m² (1100 in²)	Full Lite ≤ 0.710 m² (1100 in²)
2/A1/na/AIR/0.250	1.65 (0.29) /-	1.70 (0.30) /-	2.04 (0.36) / 0.33	2.27 (0.40) / 0.40
	00001-00001	**00001-00002**	**00001-00003**	**00001-00004**
2/A1/.020(3)/ARG/0.750	1.19 (0.21) /-	1.36 (0.24) /-	1.48 (0.26) / 0.31	1.59 (0.28) / 0.36
	00002-00001	**00002-00002**	**00002-00003**	**00002-00004**
2/A1/na/AIR/0.675	[illegible]	[illegible]	1.87 (0.33) / 0.34	1.93 (0.34) / 0.40
	00003-00001	**00003-00002**	**00003-00003**	**00003-00004**
3/S5/na/AIR/0.250	1.19 (0.21) /-	1.42 (0.25) /-	1.53 (0.27) / 0.35	1.65 (0.29) / 0.40
	00004-00001	**00004-00002**	**00004-00003**	**00004-00004**
Flush/Embossed 00005-00001	U-Factor[1] 1.10 (0.19) SHGC 0.04			
Air Leakage	≤**2.5** l/s•m² / ≤**0.5** cfm/ft²			

Manufacturer stipulates that these ratings conform to applicable NFRC procedures for determining whole product performance. NFRC ratings are determined for a fixed set of environmental conditions and a specific product size. NFRC does not recommend any product and does not warrant the suitability of any product for any specific use. Consult manufacturer's literature for other product performance information.

* Number below the performance ratings are referenced in the NFRC Certified Products Directory (e.g., XYZ-X-1-00001-00001 or 860-X-1-00001-00001)

** #glazing layers / spacer type / low-e emissivity (surface) / gap fill / gap width (na=not applicable)

*** per NFRC 100 1 W/m²•K (btu/hr•ft²•F)

www.nfrc.org

Figure 7-12 Components of the NFRC label for doors

As with windows, it is highly advisable to select doors with NFRC ratings. If none is provided, the following default values (Table 7-8) must be assumed for Climate Zone 2 (Climate Zone 1 exempt).

Table 7-8 Default Door Ratings - U-Factor

Nominal Door Thickness	Description	Storm Door		
		None	Wood	Metal
Wood doors				
1⅜"	Panel door with 7⁄16" panels	0.57	0.33	0.37
1⅜"	Hollow core flush door	0.47	0.30	0.32
1⅜"	Solid core flush door	0.39	0.26	0.28
1¾"	Panel door with 7⁄16" panels	0.54	0.32	0.36
1¾"	Hollow core flush door	0.46	0.29	0.32
1¾"	Panel door with 1⅛" panels	0.39	0.26	0.28
1¾"	Solid core flush door	0.40	—	0.26
2¼"	Solid core flush door	0.27	0.20	0.21
Steel doors				
1¾"	Fiberglass or mineral wool core with steel stiffeners, no thermal break	0.60	—	—
1¾"	Paper honeycomb core without thermal break	0.56	—	—
1¾"	Sold urethane foam core without thermal break	0.40	—	—
1¾"	Sold fire-rated mineral fiberboard core without thermal break	0.38	—	—
1¾"	Polystyrene core without thermal break (18-ga commercial steel)	0.35	—	—
1¾"	Polyurethane core without thermal break (18-ga commercial steel)	0.29	—	—
1¾"	Polyurethane core without thermal break (24-ga residential steel)	0.29	—	—
1¾"	Polyurethane core with thermal break and wood perimeter (24-ga residential steel)	0.20	—	—
1¾"	Solid urethane foam core with thermal break	0.20	—	0.16

All U-factors for exterior doors in this table are for doors with no glazing (glass), except for the storm doors which are in addition to the main exterior door. If any glazing area in an exterior door—not a storm door—is greater than ⅓ of the door area, the door should be treated as a "window" for code compliance. Source: 2001 ASHRAE Handbook, Fundamentals, *IP Edition.*

Insulated core doors can be difficult to trim, so careful installation is required. As with windows, it is important to seal the rough openings. A gap as small as 1/8 inch around an exterior door lets in as much air as a 5-½ inch diameter hole in the wall. Thresholds should seal tightly against the bottom of the door and be caulked underneath. If weather stripping is needed, consider the effectiveness of the seal and the durability of the material. Felts are the least energy efficient and tapes made of rubber or foam may need replacing every few years, so should be avoided. Some of the available types are illustrated in Figure 7-13.

Felt and adhesive-backed foam weather stripping are inexpensive, but are not as durable as vinyl or metal types. Tubular rubber and vinyl barriers, magnetic or reinforced silicone gaskets, and interlocking metal channels can all be effective air barriers, but must be carefully installed.

Tubular gasket

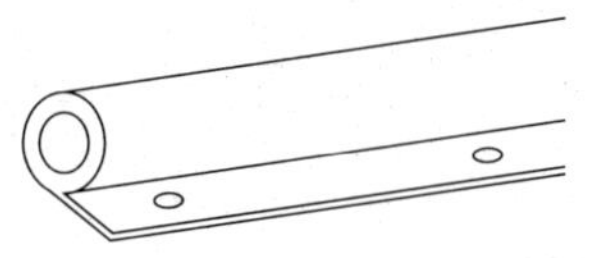

Tubular gaskets are made of vinyl or rubber, with or without a foam filling. They are durable and effective even when gaps around window or door are uneven. Applied from outside, tubular gaskets take sub-zero temperatures well.

Spring metal strip

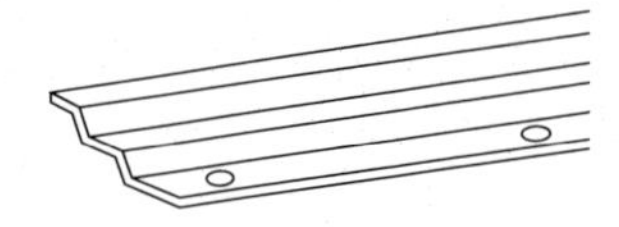

Made of bronze, stainless steel, or aluminum, these long-lasting strips fit unobtrusively in window or door channels and use tension to create a seal. They may make a tight-fitting door or window hard to open.

V-strip

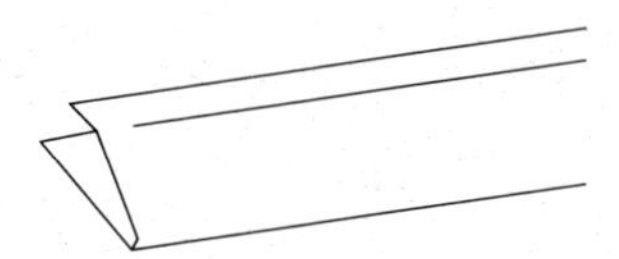

V-strips, made of metal or vinyl, also use tension to create a tight seal. They are installed in window and door channels. Vinyl strips often come with adhesive backing. However, metal V-strips, applied with nails, last longer.

Adhesive-backed foam

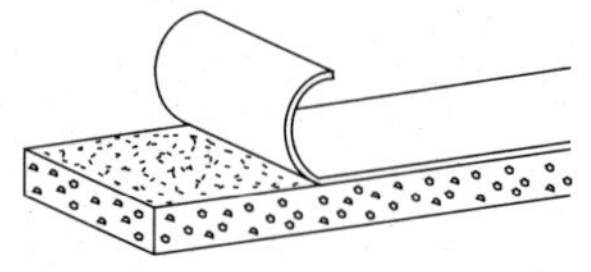

Adhesive-backed foam provides an inexpensive quick fix for a filtration problem. Very easy to apply, the foam may lose its resiliency and effectiveness during a single season.

Foam-edged wood strip

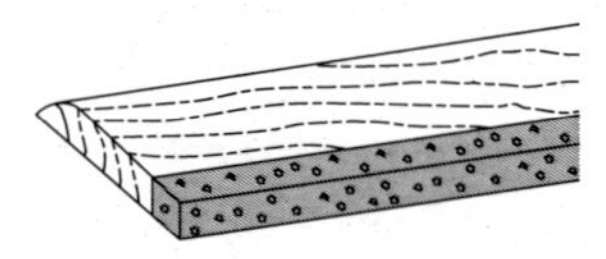

Foam-edged wood lasts longer (and costs more) than plain adhesive-backed foam. Self-sticking, it is easy to install on even surfaces but wears out in several seasons.

Grooved gasket

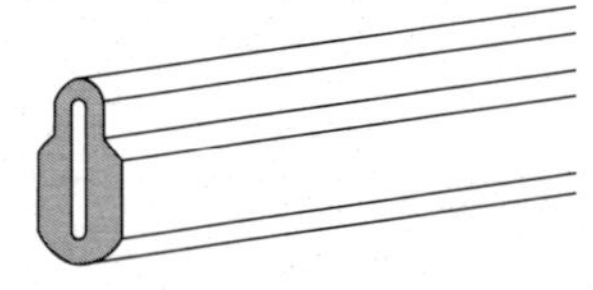

Grooved gaskets, made of various plastics, fit metal casement windows or jalousie windows. Compression makes them effective, and they last 10 years or more.

Astragal

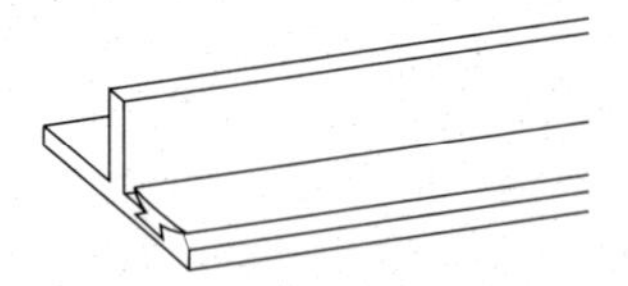

Astragal weather stripping, vinyl or aluminum, is used on double doors (French doors). A T-shaped type consists of a single piece that attaches to the less-used door. Another design interlocks two separate strips, one for each door.

Magnetic

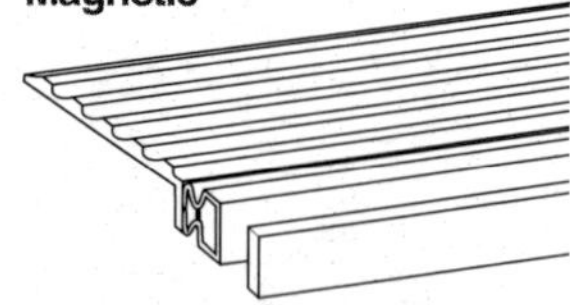

Magnetic seal for gliding doors works like the seal on a refrigerator door. One part, attached to door trim, holds a magnet and a gasket. Other part, attached to door, is metal. The magnet holds door against gasket in a tight seal.

Figure 7-13 Types of weatherstripping
From Virginia Energy Savers Handbook, *Virginia Department of Mines, Minerals and Energy, 2005*

Accessible design

Almost one out of ten people will suffer from physical disabilities during their lifetime. Designing buildings to ensure accessibility for the physically impaired adds little to the cost of a building. One important feature is to ensure that both exterior and interior door openings and hallways are 3'–0" wide to allow passage of a wheelchair or walker. Ensuring that baths and kitchens have adequate room for wheelchairs is another feature that adds little to construction costs but is expensive to retrofit.

Storm doors

Storm doors primarily increase energy efficiency by reducing infiltration of outside air. ENERGY STAR approved storm doors have a tight fit, good weather-stripping, double paned glazing. Storm doors can save energy if used with existing, poorly insulated doors; but are not necessary with new, well-insulated doors.

Garage doors

Garage doors can be constructed of wood, steel, aluminum, vinyl or fiberglass. Doors with good thermal performance typically have an insulating core with cladding material.

The insulation properties of garage doors can be published as either U-factor or R-value of the "calculated door section" or the "tested installed door". These are not the same. Be aware the "calculated door section" is similar to a "center of glass" value for windows; it does not reflect losses around the sides of the door. A performance characteristic reported for the "tested installed door" does account for the full door assembly.

The design of garage doors should be considered along with all other elements of the building envelope. It is particularly important that they have structural strength to withstand likely hurricane wind loads. Often wind damage to residences originates with inadequate garage doors. Doors must meet the wind load requirements of **FBC, Residential, Section R609.4 Garage doors**.

Resources

Note: Web links were current at the time of publication, but can change over time.

American Architectural Manufacturers Association (AAMA), & Window and Door Manufacturers Association (WDMA). (1997). *Voluntary Specifications for Aluminum, Vinyl (PVC) and Wood Windows and Glass Doors (AAMA/NWWDA 101/I.S. 2-97)*. American Architectural Manufacturers Association (AAMA). Retrieved from http://pubstore.aamanet.org/pubstore/ProductResults.asp?cat=0&src=I.S.2-97

American Society of Heating, Refrigerating and Air Conditioning Engineers (ASHRAE). (2013). *ASHREA Handbook—Fundamentals*. American Society of Heating, Refrigerating and Air Conditioning Engineers (ASHRAE). Retrieved from https://www.ashrae.org/resources--publications/handbook/description-of-the-2013-ashrae-handbook--fundamentals

Carl Wagus. (2002, April 1). Personal communication. American Architectural Manufacturers Association (AAMA).

Carmody, J. (Ed.). (2007). *Residential Windows: A Guide to New Technologies and Energy Performance (Third Edition)*. New York: W.W. Norton.

Fisette, P. (n.d.). Windows: Understanding Energy Efficient Performance. University of Massachusetts Amherst, Building and Construction Technology Program. Retrieved from http://bct.eco.umass.edu/publications/by-title/windows-understanding-energy-efficient-performance/?q=bmatwt/publications/articles/windows_understanding_energy_efficient_performance.html

Florida Solar Energy Center (FSEC). (n.d.). Windows. Retrieved August 14, 2015, from http://www.fsec.ucf.edu/en/consumer/buildings/basics/windows/

Lighting Research Center, Rensselaer Polytechnic Institute. (n.d.). Daylighting. Retrieved August 14, 2015, from http://www.lrc.rpi.edu/researchareas/daylighting.asp

My Florida Home Energy. (n.d.). Retrieved July 28, 2015, from http://www.myfloridahomeenergy.com/

A useful resource with a wide array of information on energy and water efficiency, including The Energy Efficient Home series of fact sheets, available at http://www.myfloridahomeenergy.com/help/library

Ruppert, K. C., Porter, W. A., Cantrell, R. A., & Lee, H.-J. (2015). *Windows and Skylights* (Fact Sheet). Retrieved from http://www.myfloridahomeenergy.com/help/library/weatherization/windows

U.S. Department of Energy (DOE) Energy Saver:

Daylighting. (2012, July 29). Retrieved August 14, 2015, from http://www.energy.gov/energysaver/articles/daylighting

Energy-Efficient Windows. (n.d.). Retrieved August 14, 2015, from http://energy.gov/energysaver/articles/energy-efficient-windows

U.S. Environmental Protection Agency (EPA) ENERGY STAR:

Residential Windows, Doors and Skylights for Consumers. (n.d.). Retrieved August 7, 2015, from http://www.energystar.gov/products/certified-products/detail/residential-windows-doors-and-skylights

Residential Windows, Doors and Skylights for Partners. (n.d.). Retrieved August 7, 2015, from http://www.energystar.gov/products/certified-products/detail/7578/partners

Virginia Department of Mines, Minerals and Energy. (2005). *Virginia Energy Savers Handbook: A Guide to Saving Energy, Money, and the Environment (Third Edition)*. Retrieved from https://www.dmme.virginia.gov/DE/EnergySaversHandbook.shtml

Software tools and resources

American Architectural Manufacturers Association (AAMA). (n.d.). Condensation Resistance Factor Tool. Retrieved August 7, 2015, from http://www.aamanet.org/crfcalculator/1/334/crf-tool

Building Energy Codes Program, U.S. Department of Energy (DOE), Energy Efficiency & Renewable Energy, Building Technologies Office. (n.d.). Building Energy Codes Program (BECP) Resource Center. Retrieved August 7, 2015, from https://www.energycodes.gov/resource-center

Building Energy Software Tools (BEST). (n.d.). BEST Directory. Retrieved November 21, 2017, from http://www.buildingenergysoftwaretools.com/

Efficient Windows Collaborative. (2017). Incentives and Rebates for Energy-Efficient Windows Offered through Utility and State Programs (Updated October 13, 2017). Retrieved November 21, 2017, from http://www.efficientwindows.org/downloads/UtilityIncentivesWindows.pdf

Lawrence Berkeley National Laboratory (LBNL):

LBNL Window & Daylighting Software -- RESFEN. (n.d.-a). Retrieved November 21, 2017, from https://windows.lbl.gov/software/resfen

LBNL Windows & Daylighting Software -- THERM. (n.d.-b). Retrieved November 21, 2017, from https://windows.lbl.gov/software/therm

LBNL Windows & Daylighting Software -- WINDOW. (n.d.-c). Retrieved November 21, 2017, from https://windows.lbl.gov/software/window

Lawrence Berkeley National Laboratory (LBNL), Environmental Energy Technologies Division, & U.S. Department of Energy (DOE). (n.d.). Home Energy Saver. Retrieved November 21, 2017, from http://hes.lbl.gov/consumer/

My Florida Home Energy. (n.d.). Retrieved July 28, 2015, from http://www.myfloridahomeenergy.com/

Society of Building Science. (n.d.). Pilkington Sun Angle Calculator (SAC). Retrieved August 7, 2015, from http://www.sbse.org/resources/sac/index.htm

Sustainable By Design:

Overhang Design. (n.d.-a). Retrieved August 7, 2015, from http://susdesign.com/overhang/

SunAngle. (n.d.-b). Retrieved August 7, 2015, from http://www.susdesign.com/sunangle/

U.S. Department of Energy (DOE). (n.d.). EnergyPlus. Retrieved November 21, 2017, from https://www.energyplus.net/

Weather Underground (Wunderground). (n.d.). Weather Forecast & Reports - Long Range & Local. Retrieved August 7, 2015, from http://www.wunderground.com/

Organizations and associations that provide information on windows and doors:

American Architectural Manufacturers Association (AAMA). (n.d.). Retrieved August 14, 2015, from http://www.aamanet.org/

Efficient Windows Collaborative™. (n.d.). Retrieved August 14, 2015, from http://www.efficientwindows.org/

Fenestration Manufacturers Association (FMA). (n.d.). Retrieved August 14, 2015, from http://www.fmausaonline.org/

National Fenestration Rating Council (NFRC). (n.d.). Retrieved August 14, 2015, from http://www.nfrc.org/

Window and Door Manufacturers Association (WDMA). (n.d.). Retrieved August 14, 2015, from http://www.wdma.com/

8
Heating, Ventilation, Air Conditioning (HVAC)

One of the most important decisions regarding a new building is the type of heating, ventilation, and air conditioning (HVAC) system to install. Equally critical is the heating and cooling contractor selected. The operating efficiency of a system relies on proper installation to achieve the performance rating. Keys to obtaining the design efficiency of a system in the field include:

- Sizing the system for the specific heating and cooling load of the building being built or remodeled. Remember, one size does *not* fit all! A proper load analysis must be performed for the HVAC system to operate in an efficient and predictable manner.
- Selection and proper installation of thermostats or controls.
- A efficient ductwork system designed to deliver the correct amount of conditioned air to each space within the building.
- Insulating and proper sealing of all ductwork.

Improper installation of components, in particular the ductwork, has a negative impact on comfort, increases energy bills, and can dramatically degrade the quality of air in a building. Poorly designed and installed ducts can create conditions that may reduce comfort, degrade indoor air quality, or even threaten the lives of the occupants.

As of December 31, 2017, the state of Florida implemented the Florida Building Code. The Florida Energy Code was updated and is found in *Florida Building Code Sixth Edition, (2017), Energy Conservation* (existing buildings).

Air conditioners and heat pumps (Figures 8-1 and 8-2) provide the majority of the HVAC systems installed in Florida residences. Other systems, including gas furnaces are also used. This chapter will start with a short description of the most common HVAC systems.

AIR CONDITIONERS AND HEAT PUMPS

Air Conditioners use the vapor compression cycle, a 4-step process:

1. The compressor (in the outside unit) pressurizes a gaseous refrigerant. The refrigerant heats up during this process.
2. Fans in the outdoor unit blow air across the heated, pressurized gas in the condensing coil; the refrigerant gas cools and condenses into a liquid.
3. The pressurized liquid is piped inside to the air-handling unit. It enters a throttling or expansion valve, where it expands and cools.
4. The cold gas circulates through evaporator coils. Inside air is blown across the coils and cooled while the refrigerant warms and evaporates. The cooled air is blown through the ductwork. The refrigerant, now a gas, returns to the outdoor unit where the process starts over.

continued...

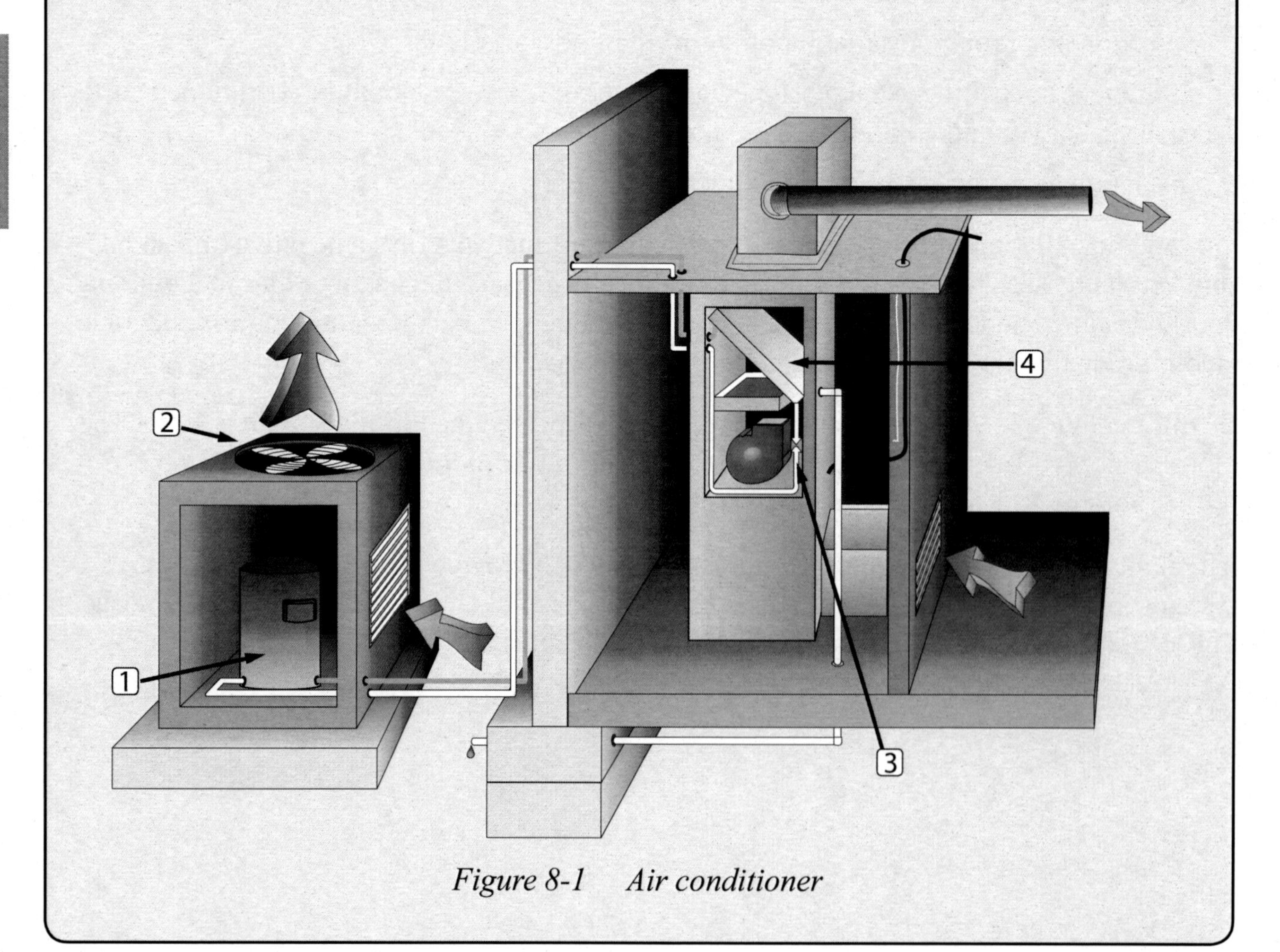

Figure 8-1 Air conditioner

continued...

Heat pumps use a reversed version of the same cycle for heating. A reversing valve allows the heat pump to work automatically in either heating or cooling mode. The steps for heating are:

1. The compressor (in the outside unit) pressurizes the refrigerant, which is piped inside.
2. The hot gas enters the inside condensing coil. Room air passes over the coil and is heated. The refrigerant cools and condenses.
3. The refrigerant, now a pressurized liquid, flows outside to a throttling valve where it expands to become a cool, low pressure liquid.
4. The outdoor evaporator coil, which serves as the condenser in the cooling process, uses outside air to boil the cold, liquid refrigerant into a gas. This step completes the cycle.
5. If the outdoor air is so cold that the heat pump cannot adequately heat the building, electric resistance strip heaters usually provide supplemental heating.
6. Periodically in winter, the heat pump must switch to a "defrost cycle," which melts any ice that has formed on the outdoor coil.

Packaged systems and room units use smaller versions of these components in a single box.

8: HVAC

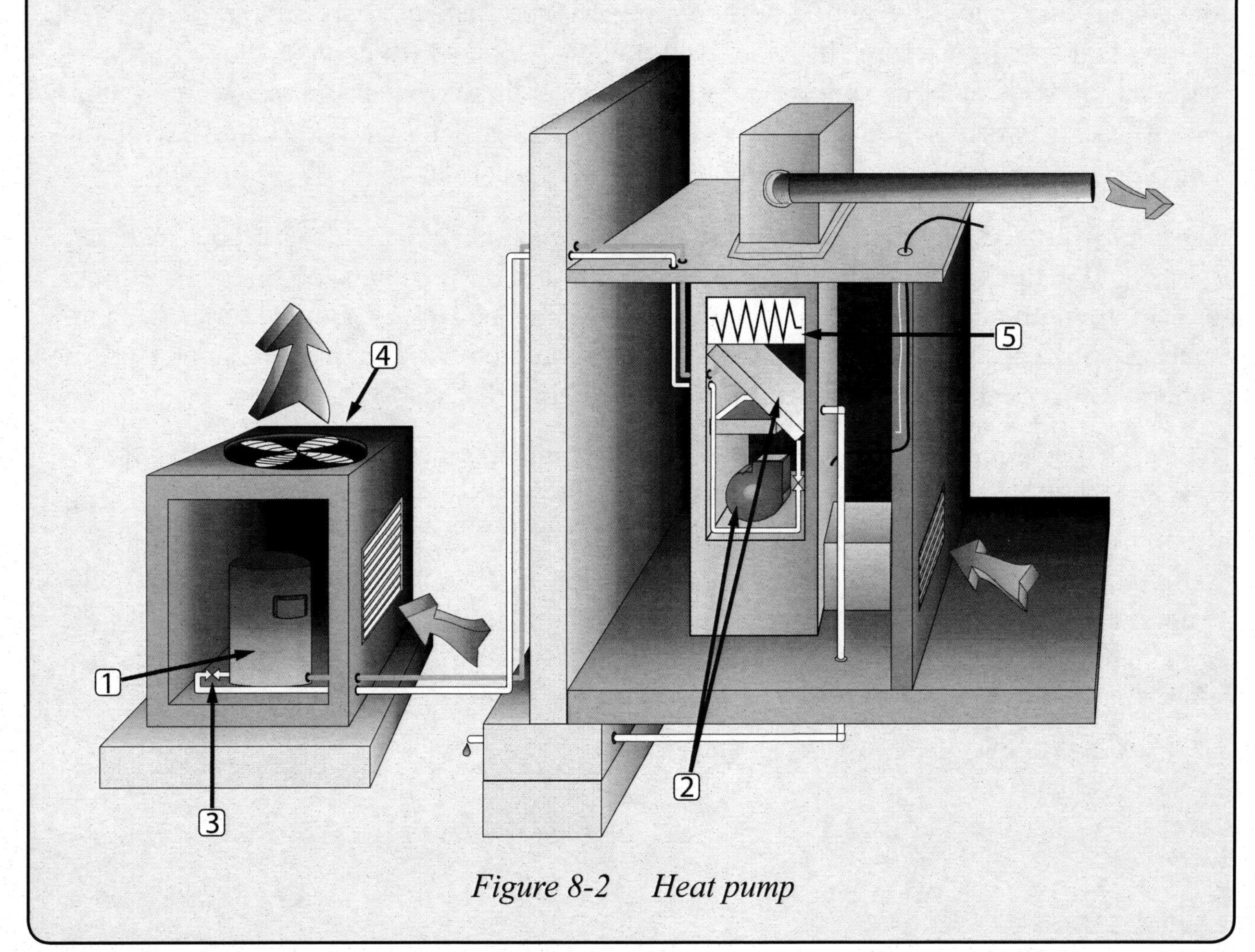

Figure 8-2 Heat pump

Furnace Equipment

Furnaces burn fuels such as natural gas, propane, and fuel oil to produce heat and provide warm, comfortable indoor air during cold weather in winter. They come in a variety of efficiencies. The comparative economics between heat pumps and furnaces depend on the type of fuel burned, its price, the building's design, and the outdoor climate. In general, moderately efficient natural gas furnaces (paired with central air conditioning) and air-source heat pumps have similar installation and operating costs.

Furnace operation

Furnaces require oxygen for combustion and extra air to vent exhaust gases. Most furnaces are *non-direct vent* units—they use the surrounding air for combustion. Others, known as *direct vent* or *uncoupled* furnaces, bring combustion air into the burner area via sealed inlets that extend to outside air. All buildings equipped with a fuel burning furnace of any type or style should be equipped with a carbon monoxide alarm.

Direct vent furnaces can be installed within the conditioned area of a building since they do not rely on inside air for safe operation. Non-direct vent furnaces must receive adequate outside air for combustion and exhaust venting. The primary concern with non-direct vent units is that a malfunctioning heater may allow flue gases, which could contain poisonous carbon monoxide, into the area around the furnace. If there are leaks in the return system, or air leaks between the furnace area and living space, carbon monoxide could enter habitable areas and cause potentially severe health problems or death.

Most new furnaces have forced draft exhaust systems, meaning a blower propels exhaust gases out the flue to the outdoors. *Atmospheric* furnaces, which have no forced draft fan, are not as common due to federal efficiency requirements. However, some furnace manufacturers have been able to achieve the efficiency requirements in atmospheric units. Atmospheric furnaces should be isolated from the conditioned space.

Units located in closets or mechanical rooms inside the building, or in relatively tight crawl spaces and basements, may have problems. Furnace mechanical rooms must be well sealed from the other rooms of the building (Figure 8-3). The walls, both interior and exterior, should be insulated. Two outside-air ducts sized for the specific furnace should be installed from outside into the room, one opening near the floor and another near the ceiling, or as specified in the **FBC, Fuel Gas, Section 304 (IFGS) Combustion, Ventilation and Dilution Air** and the **FBC, Residential, Chapter 24 Fuel Gas**.

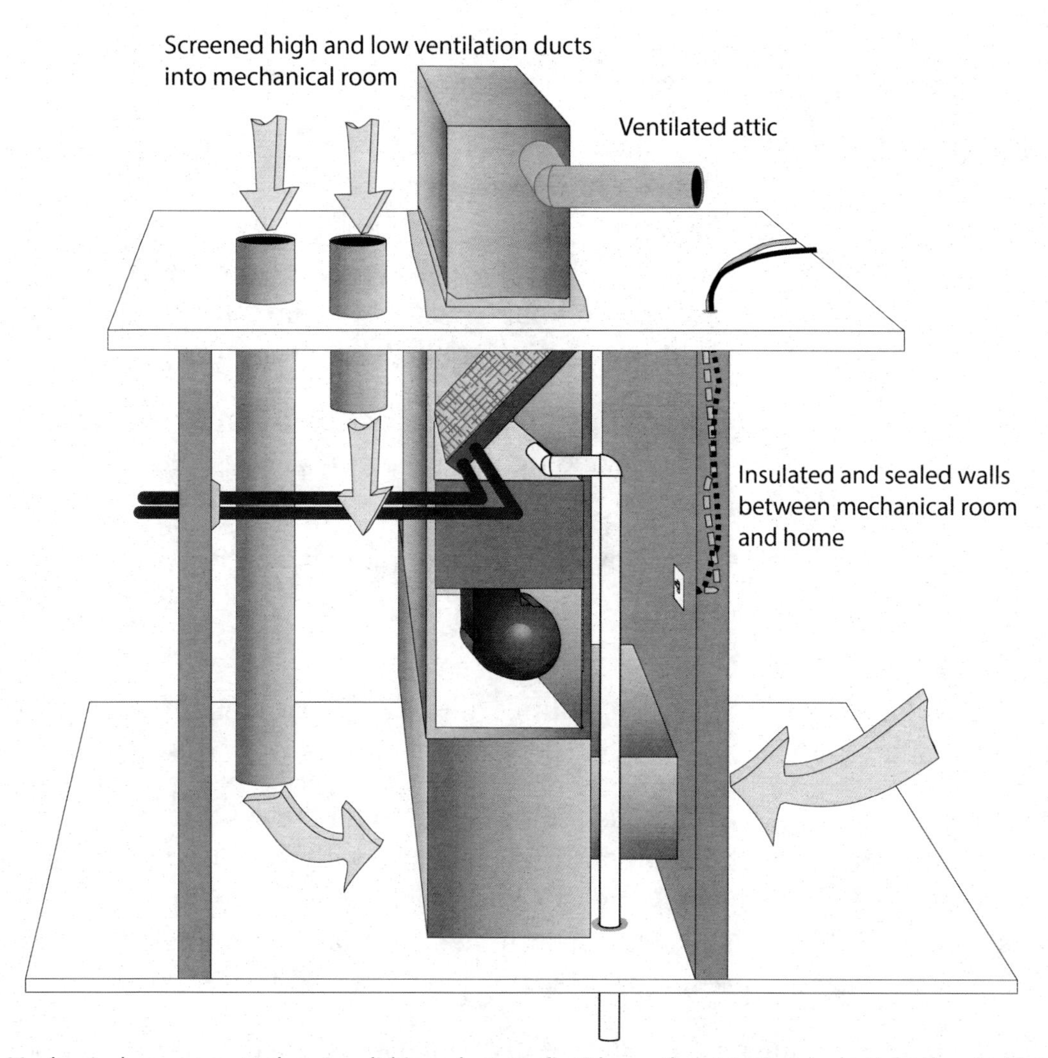

Figure 8-3 Sealed mechanical room design

Measures of efficiency for furnaces

The efficiency of a gas furnace is measured by the Annual Fuel Utilization Efficiency (AFUE), a rating which takes into consideration losses from pilot lights, start-up, and stopping. The minimum AFUE for most furnaces is approximately 80 percent, with efficiencies ranging up to 97 percent for furnaces with condensing heat exchangers.

Unlike Seasonal Energy Efficiency Ratio (SEER) and Heating Season Performance Factor (HSPF) ratings, the AFUE does not consider the unit's electricity use for fans and blowers.

New design and manufacturing techniques have enabled many furnace manufacturers to produce equipment that have AFUEs in the 93 to 97%, without resorting to the complexity of a condensing furnace. The **FBC, Energy Conservation, Table C403.2.3.(4), Warm Air Furnaces and Combination Warm Air Furnaces/Air-Conditioning Units, Warm Air Duct Furnaces and Unit Heaters, Minimum Efficiency Requirements** lists minimum requirement efficiencies for furnaces.

HEATING AND COOLING LOAD CALCULATIONS

Since January 1, 1993, it has been *required* by law to perform HVAC load calculations for newly installed HVAC units in Florida. The Florida Energy Conservation Code states: "Heating and cooling equipment shall be sized in accordance with ACCA Manual S based on the equipment loads calculated in accordance with Manual J or other approved heating and cooling calculation methodologies, based on building loads for the directional orientation of the building." **(FBC, Energy Conservation, Section R403.7.1, Equipment sizing**.)

Many contractors incorrectly select air conditioning systems based on a rule such as 600 square feet of cooled area per ton of air conditioning (a ton provides 12,000 Btu per hour of cooling). Instead, use a sizing procedure such as:

- Calculations in *Manual J* published by the Air Conditioning Contractors Association
- Similar procedures developed by the American Society of Heating, Refrigeration, and Air Conditioning Engineers (ASHRAE)
- Software developed by electric or gas utilities, the U.S. Department of Energy, or HVAC equipment manufacturers, using referenced sizing standards

The heating and cooling load calculations should be based on the exact area and type of construction for each component of the building envelope, as well as the heat given off by lights, people, and equipment inside the building. If a zoned heating and cooling system is used, the loads in each zone should be calculated. An accurate load analysis will help to prevent the problems that occur when a house is equipped with an oversized HVAC system.

An oversized air conditioning system will rapidly cool a house, which results in a short operational cycle. Since it takes a certain length of time for the air conditioning coil to become cool enough for proper dehumidification, short cycling will limit the amount of time that the system is effectively removing moisture from the dwelling.

For example, a building equipped with an oversized air conditioner on the hottest days of the year will see the compressor operate 20 minutes each hour. The first 3 minutes of this period are required to develop low coil temperatures for adequate dehumidification. During an eight-hour period proper dehumidification is occurring for only 2¼ hours.

(20 min. – 3 min. = 17 min.   17 min. × 8 = 136 min.   136 min. ÷ 60 = 2.26 hr.)

A building equipped with a properly sized air conditioner on the hottest days of the year could see the compressor operate up to 50 minutes or more, each hour. During an eight-hour period, proper dehumidification is occurring for 6¼ hours. This is nearly three times as much dehumidification as the over-sized case.

(50 min. – 3 min. = 47 min.   47 min × 8 = 376 min.   376 min. ÷ 60 = 6.26 hr.)

It is important to size heating and air conditioning systems properly. Not only does oversized equipment cost more, but it can waste energy and may decrease comfort. A properly sized air conditioning system should operate nearly continuously on the hottest days of the year.

A properly sized air conditioning system will:

- Minimize construction costs
- Provide maximum dehumidification
- Minimize energy costs

The example in Table 8-1 compares the size of heating and cooling systems for two homes with identical floor areas. The more efficient home reduces the heating load 35% and the cooling load 26%. Thus, the additional cost of the energy features in the more efficient home is offset by the savings from reducing the size of the HVAC equipment.

Table 8-1 Equipment Sizing and Cost Comparison Example

Type of House	More Efficient House	Less Efficient House
INSULATION R-VALUES AND AREAS:		
R-30 Attic	2,000 sq ft	1,000 sq ft
R-25 Cathedral Ceiling	0 sq ft	1,000 sq ft
Wall Area / R-value	1,750 sq ft / R-21 / 6" walls	1,600 sq ft / R-16 / 4" walls
Window Area / R-value / U-factor	250 / R-3 / U = 0.33	400 / R-2 / U = 0.5
Floors	2,000 / Slab-on-grade	2,000 / Slab-on-grade
Air Leakage (ACH)*	0.35	0.60
Duct Leakage (CFM25)**	50	250
HVAC SYSTEM SIZING:		
Heating (Btu/hour)	22,000	34,000
Cooling (Btu/hour)	24,000	27,000
Estimated tons of cooling***	2	2.5
Square feet/ton	700	800
TYPICAL EQUIPMENT COST:		
Lower Efficiency	$1,800 / ton	$1,680 / ton
Higher Efficiency	$2,000 / ton	$1,880 / ton

* *ACH means the number of natural air changes per hour the home has due to air leakage.*

** *CFM25 is the duct leakage rate at a pressure of 25 Pascals—a standard number used during a duct leakage test.*

*** *There are 12,000 Btu/hour in a ton of cooling.*

Oversimplified rules would have provided an oversized heating and cooling system for the more efficient building. The oversized unit would have cost more to install. In addition, the operating costs would be higher, it would suffer greater wear, and it may not provide adequate dehumidification.

Proper sizing includes designing the cooling system to provide adequate dehumidification. In Florida's humid climate it is critical to calculate the *latent load*—the amount of dehumidification needed for the building. If the latent load is ignored, the building may become uncomfortable due to excess humidity.

The Sensible Heat Ratio (SHR) designates the portion of the cooling load for reducing indoor temperatures (*sensible cooling*). For example, in a HVAC unit with a .75 SHR, 75% of the energy expended by the unit goes to cool down the temperature of indoor air. The remaining 25% goes for latent heat removal—taking moisture out of the air in the building. If the designer of a HVAC system accurately estimates the cooling load, they will also calculate the desired SHR and thus, the latent load.

Many buildings in Florida have design SHR's of approximately 0.7—70% of the cooling will be sensible and 30% latent. Systems that deliver less than 30% latent cooling may fail to provide adequate dehumidification in summer.

Rather than relying on "Rules of Thumb" to estimate latent loads, these loads should be calculated based on the actual structure and its intended use. Most of the load analysis tools currently in use are based on Manual J (see example, next page), which is produced by the Air Conditioning Contractors Association (ACCA). Manual J provides latent load calculation techniques in these categories:

- **Ventilation**: Mechanical ventilation is required if a 50 Pascal blower door test produces a measured flow equal to or less than 3 air changes per hour.
- **People Loads**: People produce a latent load that must be removed by the HVAC system. This load should be associated with the normal load for the house and not related to the "entertainment" loads that occur once or twice a year. For example: a family of four that has an annual family reunion for 25 people should include a latent load associated with 4 people not 25.
- **Infiltration**: The use of blower door test equipment has advanced the ability to quantify whole-house infiltration loads. Estimation techniques designed by organizations such as ASHRAE and ACCA can be used for new home construction. Current construction techniques can result in new homes infiltrating less than 0.25 air changes per hour.

As homes are built more tightly and are more energy efficient, sensible cooling loads will decrease and the need for mechanical ventilation will increase. Sensible heat ratios will be driven lower. Professionals in the HVAC industry will have to be able to install properly matched equipment with SHR's 0.75 or less.

Manual J Example

	Type of Exposure		Construction Number	Panel Faces	HTM Heating	HTM Cooling	Area or Length	Btuh Heating	Btuh S-Cooling	Btuh L-Cooling
1	Name of Room: Smith Residence						Entire House			
2	Running Feet or Exposed Wall						2 × (56 + 32) = 176			
3	Ceiling Height at Walls (Ft) and Gross Wall Area (Sq Ft)						8 & 10	1,408 + 749 = 2, 157		
4	Room Dimensions (Ft) and Floor Plan Area (Sq Ft)						56 × 32	1,792		
5	Ceiling Slope (Degrees) and Gross Ceiling Area						0	1,792		
6a	Windows and Glass Doors	a	Unit A = 1G	N	37.2	11.1	43.75	1,629	487	
		b	Unit A = 1G	E/W	37.2	36.9	43.75	1,629	1,614	
		c	Unit B = 1G	N	33.4	11.2	14.00	468	157	
		d	Unit B = 1G	S	33.4	15.9	28.00	936	444	
		e	Unit C = 1G	W	41.0	39.8	58.00	2,380	2,310	
		f	Unit D = 1G	S	41.0	17.4	47.13	1,934	819	
		g	Unit E = 1G	N	31.9	12.6	10.31	329	130	
		h	Unit E = 1G	S	31.9	23.0	10.31	329	237	
		i	Kit. Door Glass	W	Line 7	70.0	3.00	0	210	
6b	Skylights	a	Unit 1 = 8E	N	98.4	100.7	8.00	787	806	
		b	Unit 2 = 8E	S	69.0	92.9	32.00	2,207	2,974	
		c								
7	Wood and Metal Doors	a	11N		26.6	9.1	21.0	559	191	
		b	11N		26.6	9.1	21.0	559	191	
		c								
8	Above Grade Walls and Partitions	a	14A-8		6.92	1.16	1,207	8,347	1,395	
		b	15A-4sffc wall		10.41	2.10	653	6,798	1,369	
		c	15A-4sffc part		0.90	0.18	96	87	17	
		d								
		e								
9	Below Grade Walls	a	15A-4sffc-4		6.00		231	1,387		
		b	15A-4ffc-10		4.33		224	970		
		c								
10	Ceilings	a	16B-30ad		2.43	1.60	1,752	4,261	2,803	
		b								
11	Floors	a	19B-osp		2.43	0.48	736	1,788	352	
		b	22B-5ph		44.73		64	2,865		
		c	21A-32		1.52		544	827		
		d								
12	Infiltration	Heat Loss			11,237 Btuh		WAR	11,237		
		Sensible Gain			1,054 Btuh		1.00		1,054	
		Latent Gain			1,651 Btuh					1,651
13	Internal	a	Occupants at 230 and 200 Btuh				#4		920	800
		b	Scenario Number		1				2,400	
		c	Default Adjustments		None					
		d	Individual Appliances		NA					
		e	Plants		None					
14	Subtotals	Sum lines 5 through 12						52,313	20,877	2,451
15	Duct Loads	ELF-Loss and ELF-Gain			0.060	0.028		3,165	575	
		Latent Gain								735
16	Ventilation Loads Vent CFM			70	Exh	70		1,987	459	1,755
17	Winter Humidification Load Gal/Day					7.1		2,614		
18	Piping Load									
19	Blower Heat								1,707	
20	Total Load	Sum Lines 13 through 19					60,078	23,619	4,941	

(Entire form not shown)

8: HVAC

Equipment Selection

Once an accurate load calculation is performed, the HVAC equipment can be selected. The selection process should include the following topics:

- What source of energy will be used to power the equipment? Electricity, natural gas, both, other?
- Equipment efficiency
- Cooling equipment selection
- Equipment location

Source of Energy

Electricity is probably the most convenient energy source used today to power the many appliances that are included in a modern home. Natural gas is also a significant source of energy for home consumption. When constructing a new home or renovating an existing one, a final choice for the energy source will depend on availability and comparative cost. Natural gas is available throughout many, but not all, parts of the state. If multiple fuels are available, a cost comparison can be performed to assist in making the right economic choice. A common basis for energy source comparison is the dollar cost per million BTUs. A example of this comparison can be found in Appendix III.

Equipment Efficiency

Natural gas and propane furnace efficiencies are rated by their Annual Fuel Utilization Efficiency (AFUE) number. This was explained in an earlier section. The heating efficiency of a heat pump is measured by its *Heating Season Performance Factor* (HSPF), which is the ratio of heat provided in Btu per hour to watts of electricity used. This factor considers the losses when the equipment starts up and stops, as well as the energy lost during the defrost cycle.

Typical values for the HSPF are 6.8 for standard efficiency, 7.2 for medium efficiency, and 8.0 for high efficiency. Variable speed heat pumps have HSPF ratings as high as 9.0, and geothermal heat pumps have HSPFs over 10.0. The HSPF averages the performance of heating equipment for a typical winter in the United States, so the actual efficiency will vary in different climates.

The cooling efficiency of a heat pump or an air conditioner is rated by the *Seasonal Energy Efficiency Ratio (SEER)*, a ratio of the average amount of cooling provided during the cooling season to the amount of electricity used. (National legislation mandated a minimum SEER 14 for all units manufactured after January 2015 and the Florida Building Code mandates the installation of a minimum 14 SEER for single-phase air-cooled air conditioners less than 65,000 Btu/hour after January 30, 2015. Efficiencies of some units can be as high as SEER 23.0.)

Builders should be aware that the SEER rating is a national average based on equipment performance in Virginia. Some equipment may not produce the listed SEER in actual operation in Florida's buildings. One of the main problems has been the inability of some higher efficiency equipment to dehumidify buildings adequately.

If units are not providing sufficient dehumidification, the typical owner response is to lower the thermostat setting. Since every degree the thermostat is lowered increases cooling bills 3 to 7%, systems that have nominally high efficiencies, but inadequate dehumidification may suffer from higher than expected cooling bills.

As illustrated in Table 8-2 and Figure 8-4, poorly functioning "high" efficiency systems may actually cost more to operate than a well designed, moderate efficiency unit. Make certain that the contractor has used proper and approved load calculation techniques so that the air conditioning system meets both sensible and latent (humidity) loads at the manufacturer's claimed efficiency.

Ductless, Mini Split-System Air-Conditioners and Heat Pumps

Ductless, mini split-system air-conditioners and heat pumps (mini splits) have numerous potential applications in residential, commercial, and institutional buildings. The most common applications are in multifamily housing or as retrofit add-ons to houses with "non-ducted" heating systems, such as hydronic (hot water heat), radiant panels, and space heaters (wood, kerosene, propane). They can also be a good choice for room additions and small apartments, where extending or installing distribution ductwork (for a central air-conditioner or heating systems) is not feasible. Applications in other types of buildings include: school classrooms; perimeter cooling for office buildings; additional cooling for restaurant kitchens; and cooling for small offices within larger spaces, such as arenas, warehouses, and auditoriums.

Like central systems, mini splits have two main components: an outdoor compressor/condenser, and an indoor air-handling unit. A conduit, which houses the power cable, refrigerant tubing, suction tubing, and a condensate drain, links the outdoor and indoor units.

COOLING EQUIPMENT SHR EXAMPLE

Table 8-2 shows an equipment chart for an example air conditioning system. The system provides a wide range of outputs, depending on the blower speed and the temperature conditions. The SHR in the chart is the Sensible Heat Ratio—the fraction of the total output that cools down the air temperature. The remainder of the output dehumidifies the air—provides latent cooling (see Figure 8-4 for a pie chart of a typical SHR).

Example system with 80° return air, effective SEER = 12.15

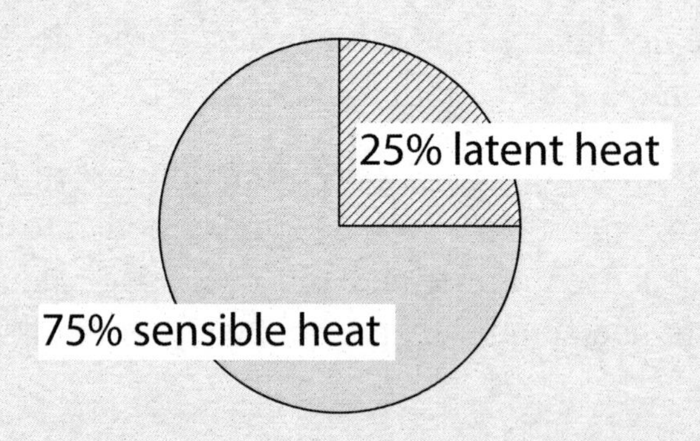

Figure 8-4 Typical Sensible Heat Ratio (SHR)

- Note that the system provides about 36,000 Btu/ hour of cooling and the SEER value represents the "effective" efficiency of the unit after installation.
- At low fan speed, provides 35,800 Btu/hour, 0.71 SHR, and thus 29% latent cooling (10,382 Btu/hour of latent heat removal).
- At high fan speed, provides 38,800 Btu/hour, but a 0.81 SHR, and only 19% latent cooling—not enough dehumidification in many Florida buildings (7,372 Btu/hour of latent heat removal).

Table 8-2 Example Cooling System Data Effective SEER = 12.15

Total Air Volume (cfm)	Total Cooling Capacity (Btu/hr)	Sensible Heat Ratio (SHR) Dry Bulb Temperature		
		75°F	80°F	85°F
950	35,800	0.58	0.71	0.84
1,200	37,500	0.61	0.76	0.91
1,450	38,800	0.64	0.81	0.96

Equipment Location

Central heating and air conditioning systems have a component called an air handling unit or AHU. These are typically located in the attic, garage, or within conditioned space. The choice of location depends on many, sometimes conflicting, requirements. A brief listing of some of the advantages and disadvantages might be helpful:

Disadvantages of air handler placement in attic:

- Putting a cold item in a hot attic is not energy efficient
- It is more difficult to access for maintenance
- Any leaks in the cabinet are surrounded by unconditioned space (which can contain dust, animal, and insect debris)
- Any overflow of the condensate system may cause building damage

The FBC includes a series of requirements for installing the AHU in the attic. See Appendix III for further details.

Disadvantages of air handler placement in garage:

- The air handler is still in unconditioned space
- The garage is not a central location designed to minimize duct lengths
- Leaks in the return side of the cabinet will draw garage air into the house. *This air could contain fumes from gasoline, solvents, pesticides, insecticides, and carbon monoxide from car exhaust.*

Advantages of air handler placement in conditioned space:

- The air handler is in a more benign environment
- The central location can minimize duct lengths and optimize air flow
- There is easier access for maintenance
- Any leaks are to conditioned space

The FBC has air handler location multipliers that are used in calculation of the code compliance rating. The multipliers provide a significant benefit for location in interior spaces.

Safety Notes:

Any home that uses fuel (natural gas, propane, fuel oil, etc) should be equipped with a carbon monoxide alarm. Remember to check the batteries per the manufacturer's recommended schedule.

The FBC has numerous requirements for the proper installation of fuel-fired appliances. Abiding by combustion air source and exhaust requirements will result in a safer and more efficient system if followed correctly.

PROPER INSTALLATION

Unlike a refrigerator or an air conditioner in an automobile, the air conditioner in a house is assembled on-site. Major components, such as the air handler and the condenser are joined together for the first time at the construction site. The efficiency and reliability of the entire system is directly related to the care and quality of the work that goes into the installation of the complete system.

Proper installation techniques include detailed work in the following areas:

- All gauges and measuring equipment must be properly maintained and calibrated. A quick glance at a calibration sticker should not reveal a date many years in the past. Proper refrigerant charging depends on accurate measurements.
- Manufacturer's procedures should be followed regarding the proper joining of refrigerant piping.
- Evacuation is often required before refrigerant charging. The manufacturer's procedures for this process should be followed precisely. Any air that invades the system at the job site will reduce system efficiency and also contain moisture that will accelerate internal corrosion.
- Charging the system with refrigerant should also be completed by following the manufacturer's recommendations.

Note: Refrigerants containing chlorofluorocarbons (CFCs) began phasing out in 1996. As of January 1, 2010, hydrochlorofluorocarbon (HCFC) refrigerants were not permitted for charging new equipment; production and import of HCFCs will be banned by January 1, 2020. As a result of these phase-outs, equipment is now designed for hydro fluorocarbon (HFC) refrigerants.

Note: CFCs and HCFCs are believed to contribute to the depletion of the earth's ozone layer.

Improper refrigerant charging seriously degrades system performance. The textbook *Refrigeration & Air Conditioning*, published by the Air Conditioning, Heating & Refrigeration Institute, provides the following information:

- A refrigerant under-charge of 5% causes a 2.4% decrease in system efficiency.
- A refrigerant under-charge of 10% causes an 8.3% decrease in system efficiency.
- A refrigerant under-charge of 15% causes a 19.6% decrease in system efficiency.

Refrigerant under-charging of these percentages often represents refrigerant amounts of just a few ounces. This should illustrate the necessity of proper charging and measuring techniques.

Another often-neglected area of installation concerns is the placement of the outside unit. Homeowners often want to hide the condenser from view behind a screen, small fence or shrubs. Care must be taken to prevent any blockage of air flow from the unit. Manufacturer's recommendations for proper clearance distances should be followed. Limiting the air flow to the condenser will cause the unit to operate less efficiently.

Ventilation and Indoor Air Quality

All buildings need ventilation to remove stale interior air and excessive moisture. There has been considerable concern recently about how much ventilation is required to maintain the quality of air in buildings. While there is substantial disagreement on the severity of indoor air quality problems, most experts agree that the solution is not to build an inefficient, "leaky" building.

Research studies show that standard buildings are as likely to have indoor air quality problems as energy efficient ones. Some building researchers believe that no building is so leaky that the occupants can be relieved of concern about indoor air quality. They recommend mechanical ventilation systems for all buildings.

The amount of ventilation required depends on the number of occupants and their lifestyle, as well as the design of the building. The ASHRAE standard, "Ventilation for Acceptable Indoor Air Quality" (ASHRAE 62) recommends that houses have 0.35 air changes per hour (ACH).

Older, drafty houses can have infiltration rates of 1.0 to 2.5 ACH. Standard buildings built today are tighter and usually have rates of from 0.5 to 1.0 ACH. New, energy efficient buildings may have 0.35 ACH or less.

The problem is that infiltration is neither continuous nor constant; it is unpredictable, and rates for all buildings vary. For example, infiltration is greater during cold, windy periods and can be quite low during hot weather. Thus, pollutants may accumulate during periods of calm weather even in drafty houses. These buildings will also have many days when excessive infiltration provides too much ventilation, causing discomfort, high energy bills, and possible deterioration of the building envelope.

Concerns about indoor air quality are leading more and more building owners to install controlled ventilation systems for providing a reliable source of fresh air. The simplest approach is to provide spot ventilation of bathrooms and kitchens to control moisture. Nearly all exhaust fans in standard construction are ineffective—a prime contributor to interior moisture problems in Florida buildings (Figure 8-5). Bath and kitchen exhaust fans should vent to the outside—not just into an attic or crawl space. General guidelines call for providing a minimum of 50 cubic feet per minute (cfm) of air flow for baths and 100 cfm for kitchens. Manufacturers should supply a cfm rating for any exhaust fan.

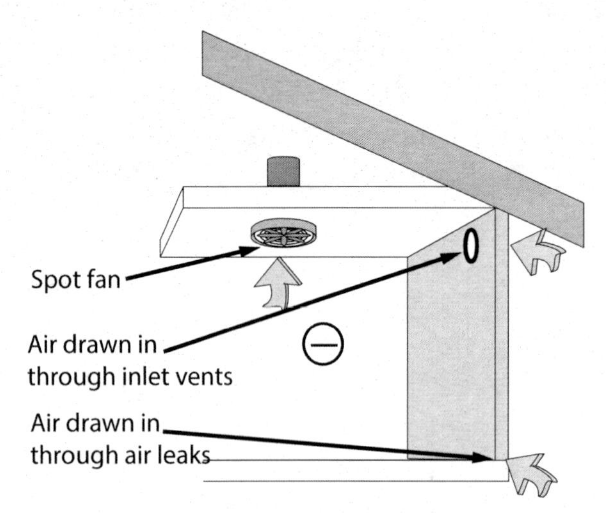

Figure 8-5 Ventilation with spot fans

The cfm rating typically assumes the fan is working against an air pressure resistance of 0.1 inch of water—the resistance provided by about 15 feet of straight, smooth metal duct. In practice, most fans are vented with flexible duct that provides much more resistance. Some fans are also rated at pressures of 0.25 to 0.30 inches of water—the resistance found in most installations. Most ventilation experts suggest choosing a fan based on this rating.

While larger fans cost more, they are usually better constructed and therefore last longer and run more quietly. The level of noise for a fan is rated by *sones*. Choose a fan with a sone rating of 2.0 or less. Top quality models are often below 0.5 sones.

Many ceiling- or wall-mounted exhaust fans can be adapted as "in-line" blowers located outside of the living area, such as in an attic. Manufacturers also offer in-line fans to vent a single bath or kitchen, or multiple rooms (Figure 8-6). Distancing the in-line fan from the living area lessens noise problems.

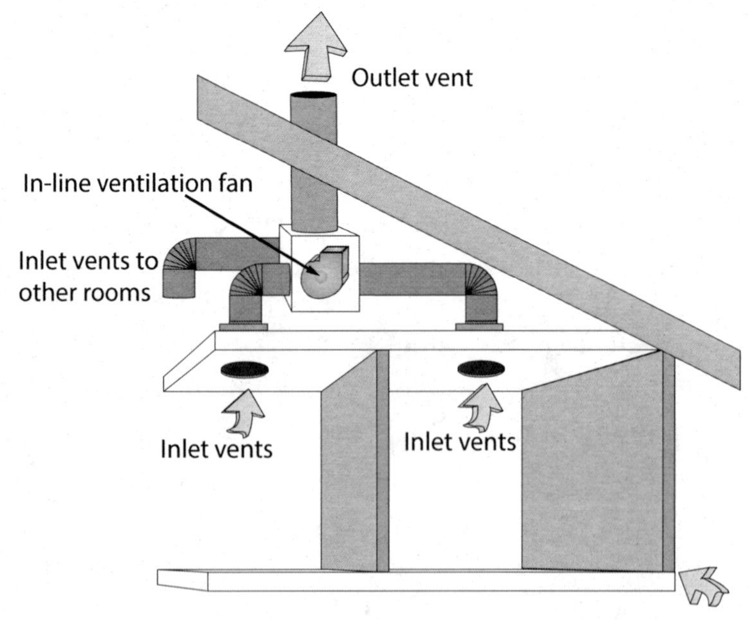

Figure 8-6 In-line ventilation fans

While improving spot ventilation will certainly help control moisture problems, it may not provide adequate ventilation for the entire building. A *whole house ventilation system* can exhaust air from the kitchen, all baths, and perhaps the living area or bedrooms.

Whole house ventilation systems usually have large single fans located in the attic. Ductwork extends to rooms requiring ventilation. These units typically have two-speed motors. The low speed setting gives continuous ventilation—usually 10 cfm per person or 0.35 ACH. The high speed setting can quickly vent moisture or odors.

Supplying Outside Air

From air leaks

The air vented from the building by exhaust fans must be replaced by outside air—either through air leakage or a controlled inlet. Relying on air leaks requires no extra equipment; however, the occupant has little control over the air entry points. Plus, many of the air leaks come from undesirable locations, such as crawl spaces or attics. If the building is too airtight, the ventilation fans will not be able to pull in enough outside air to balance the air being exhausted. It will generate a negative pressure in the building, which may cause increased wear on fan motors. Plus, the exhaust fans may threaten air quality by pulling exhaust gases from flues and chimneys back into the building.

From inlet vents

Providing fresh outside air through inlet vents (Figures 8-7 and 8-8) is another option. These vents can often be purchased from energy specialty outlets by mail order. They are usually located in exterior walls. The amount of air they allow into the building can be controlled manually or by humidity sensors.

Via ducted make-up air

Outside air can also be drawn into and distributed through the building via the ducts for a forced-air heating and cooling system. This type of system usually has an automatically controlled outside air damper in the return duct system.

The blower for the ventilation system is either the air handler for the heating and cooling system or a smaller unit that is strictly designed to provide ventilation air. A disadvantage of using the HVAC blower is that incoming ventilation air may have sufficient velocity to affect comfort conditions during cool weather.

The ductwork for the heating and cooling system may be connected to a small outside air duct that has a damper which opens when the ventilation fan operates. The resulting reduced air flow should not adversely affect comfort. Special controls are available to ensure that the air handler runs a certain percentage of every hour, thus drawing fresh air in on a regular basis.

Dehumidification Ventilation Systems

Florida buildings are often more humid than desired. A combined ventilation-dehumidifier system can bring in fresh, but humid outdoor air, remove moisture, and supply it to the building. These systems can also filter incoming air. Because these systems require an additional mechanical device—a dehumidifier installed on the air supply duct—they should be designed for the specific needs of the building.

Heat Recovery Ventilators

Air-to-air heat exchangers, or heat recovery ventilators (HRV), typically have separate duct systems that draw in outside air for ventilation and distribute fresh air throughout the house (Figure 8-9). Winter heat from stale room air is "exchanged" to the cooler incoming air. Some models, called *enthalpy* heat exchangers, can also recapture cooling energy in summer by exchanging moisture between exhaust and supply air. The term *enthalpy* is referring to the energy content due to the sensible temperature difference plus the latent heat due to the moisture content as well.

While energy experts have questioned the value of the heat saved in Florida buildings for the $700 to $2,000 cost for an HRV, recent studies on enthalpy units indicate their dehumidification benefit in summer offers an advantage over ventilation-only systems. The value of any heat recovery ventilation system should not be determined solely on the cost of recovered energy—the controlled ventilation and improved quality of the indoor environment must be considered as well.

SAMPLE VENTILATION PLANS

Upgraded Spot Ventilation

This relatively simple and inexpensive whole house ventilation system (Figure 8-7) integrates spot ventilation using bathroom and kitchen exhaust fans with an upgraded exhaust fan (usually 100 to 150 cfm) in a centrally located bathroom. When the fan operates, outside air is drawn through inlets in closets with louvered doors. The fan is controlled by a timer set to provide ventilation at regular intervals. Interior doors are undercut to allow air flow to the central exhaust fan. The fan must be a long-life, high-quality unit that operates quietly. In addition to the automatic ventilation provided by this system, occupants can turn on all exhaust fans manually as needed.

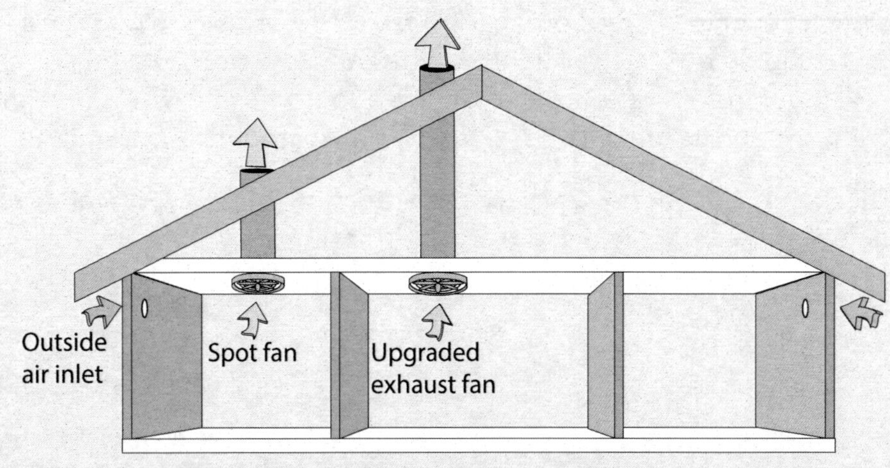

Figure 8-7 Upgraded spot ventilation

Whole House Ventilation System

This whole house ventilation system uses a centralized two-speed exhaust fan to draw air from the kitchen, bath, laundry, and living area (Figure 8-8). The blower is controlled by a timer. The system should provide approximately 0.35 ACH on low speed and 1.0 ACH on high speed. Outside air is supplied by a separate dampered duct connected to the return air system. When the exhaust fan operates, the outside air damper opens and allows air to be drawn into the house through the forced-air ductwork.

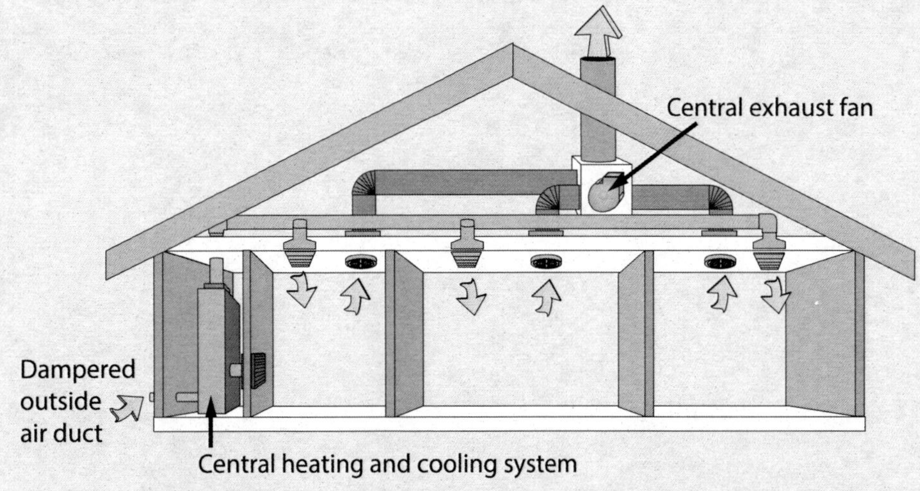

Figure 8-8 Whole house ventilation system

continued...

8: HVAC

continued...

SAMPLE VENTILATION PLANS

Heat Recovery Ventilation (HRV) System

An enthalpy air-to-air heat exchanger draws fresh outside air through a duct into the heat exchange equipment and recaptures heating or cooling energy from stale room air as it is being exhausted (Figure 8-9). The system also dries incoming humid air in summer — a particular benefit in the Southeast. Fresh air flows into the house via a separate duct system, which should be sealed as tightly as the HVAC ductwork. Room air can either be ducted to the exchanger from several rooms or a single source. Some HRV units can be wall-mounted in the living area, while others are designed for utility rooms or basements.

Check the **FBC, Energy Conservation, Section R405.7.7, Installation criteria for homes claiming the heat recovery unity (HUR) option**, for specific details on installation if using the Simulated Performance Alternative option.

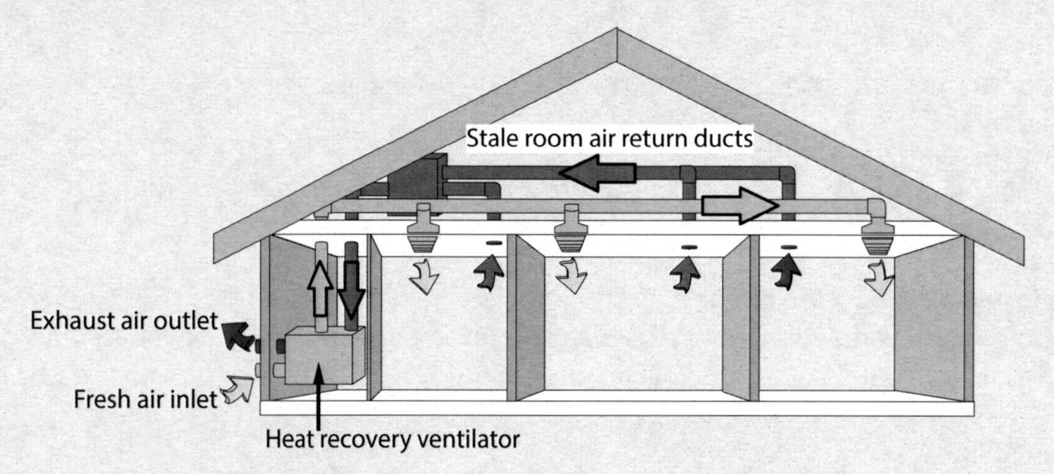

Figure 8-9 Heat recovery ventilation system

Resources

Note: Web links were current at the time of publication, but can change over time.

Air Conditioning Contractors of America Association, Inc. (ACCA):

(2016). *Manual J - Residential Load Calculation (Eighth Edition)*. American Society of Heating, Refrigerating, and Air-Conditioning Engineers, Inc. (ASHRAE). Retrieved November 21, 2017, from http://www.acca.org/standards/technical-manuals

(n.d.-a). *Manual D® - Residential Duct Systems*. Retrieved from http://www.acca.org/standards/technical-manuals

(n.d.-b). *Manual N® - Commercial Load Calculation (Fifth Edition)*. Retrieved from http://www.acca.org/standards/technical-manuals

Air-Conditioning, Heating, & Referigeration Institute (AHRI). (n.d.). Retrieved August 20, 2015, from http://www.ahrinet.org/site/1/Home

American Association of Heating, Refrigerating and Air-Conditioning Engineers (ASHRAE). (n.d.). *ANSI/ASHREA Standards 62.1 & 62.2 - Ventilation for Acceptable Indoor Air Quality*. Retrieved August 11, 2015, from https://www.ashrae.org/resources--publications/bookstore/standards-62-1--62-2

Hurm, Mark. (Personal communication, April 2002). Thompson Sheet Metal, Inc.

Lstiburek, J. (2005). Builder's Guide to Hot-Humid Climates. Building Science Corporation. Retrieved November 21, 2017, from https://buildingscience.com/bookstore/ebook/ebook-builders-guide-hot-humid-climates

http://buildingsciencepress.com/bookstore/books/builders-guide-hot-humid-climates

My Florida Home Energy. (n.d.). Retrieved from http://www.myfloridahomeenergy.com/

A useful resource with a wide array of information on energy and water efficiency, including The Energy Efficient Home series of fact sheets, available at http://www.myfloridahomeenergy.com/help/library

Tiller, J. S., & Creech, D. B. (1999). *A Builder's Guide to Energy Efficient Homes in Georgia (Third Edition)*. Atlanta, GA: Georgia Environmental Facilities Authority. Retrieved from http://www.southface.org/ez/media/georgiabuildersguide.pdf

8: HVAC

9
Duct Design and Sealing

The Problem of Duct Leakage

Studies conducted throughout the country have found that poorly sealed ductwork is often the most prevalent and yet easily solved problem in new construction. Duct leakage contributes 10 to 30% of heating and cooling loads in many homes. In addition, duct leakage can lessen comfort, and endanger health and safety.

Locating ducts in conditioned space eliminates many problems with leakage. They are often installed in *chases*—framed air passageways situated behind the ceiling or wall finish. However, these chases are often connected more directly to unconditioned space than interior space. It is important to completely seal these areas from unconditioned spaces, and insulate them properly, which is required by the **FBC, Energy Conservation, Section R403.3 Ducts**.

A common practice sometimes locates the air-handling unit in the garage. Supply and return air ducts are connected to the unit through the main supply duct and a return air plenum. While this practice saves interior space by locating the equipment in the garage, it can cause other problems. The garage is a storage area for gasoline, paints, solvents, and insecticides. Starting a car also fills the garage with by-products of combustion. Any leaks in the return side of the air handling unit can draw these fumes into the air conditioning system. This can create indoor air quality problems that could be serious.

The heating and cooling contractor should use proper materials when sealing ductwork, which is mandatory, in accordance with **FBC, Energy Conservation, Section C403.2.9 Duct and plenum insulation, construction, and sealing**, of the Florida Energy Conservation Code. **FBC, Energy Conservation, Section C403.2.9.3.7 Approved closure systems** lists approved closure systems, and **Table C403.2.9.2 Duct System Construction and Sealing** lays out duct construction and sealing. Duct insulation does not provide an airtight seal. To ensure ducts are tight, have your HVAC contractor conduct a duct leakage test per **FBC, Energy Conservation, Section R403.3.2. Sealing (Mandatory)**.

Duct Leaks and Air Leakage

Forced-air heating and cooling systems should be *balanced*—the amount of air delivered through the supply ducts should be equal to that drawn through the return ducts. If the two volumes of air are unequal, then the pressure of the building can be affected. Pressure imbalances can increase air leakage into or out of rooms in the building.

Pressure imbalances can create dangerous air quality conditions in homes, including:

- Potential backdrafting of combustion appliances such as fireplaces, wood stoves and gas burners. An improperly balanced system can create negative pressures within a structure. This imbalance or negative pressure can pull hot, humid attic air into a home. In homes with gas or propane fired appliances and furnaces this imbalance can prevent combustion gases from exiting the flue. This can cause a build up of carbon monoxide within the home, which can cause illness and even death.
- Increasing air leakage from the crawl space or slab to the home, which may draw in dust, radon gas, molds, and humidity.
- Pulling pollutants into the air handling system via return leaks.

Typical causes and concerns of pressure imbalances, addressed more fully in Chapter 4, include:

- HVAC systems with excessive supply leaks can cause homes to become depressurized, which may cause backdrafting of combustion appliances in the home.
- HVAC systems with excessive return leaks can cause homes to become pressurized and create negative pressures around the air handling unit. The negative pressures may cause combustion appliances near the air handling unit to backdraft.
- Homes with central returns can have pressure imbalances when the interior doors to individual rooms are closed. The rooms having supply registers and no returns become pressurized, while the areas with central returns become depressurized. Often the returns are open to living rooms with fireplaces or combustion appliances. When these spaces become sufficiently depressurized, the flues will backdraft. To correct this problem, the Florida Mechanical Code has a requirement that limits the allowable pressure imbalance for rooms with closable doors. The **FBC, Residential, Section M1602.3 Balanced return air**, and the **FBC, Mechanical, Section 601.6 Balanced return air**, limits the pressure differential to 2.5 Pascals or 0.01 inch WC.
- Tighter homes with effective exhaust fans, such as kitchen vent hoods, clothes dryers, and attic ventilation fans, may experience negative pressures whenever these ventilation devices are operating.

Sealing Air Distribution Systems

Duct leakage must be eliminated. In standard construction, many duct seams are not sealed or are poorly sealed with ineffective materials such as cloth "duct tape," unrated aluminum tape, or similar products with lower quality adhesives not designed to provide an airtight seal over the life of the home. The following products are examples of proper materials to be used to seal duct systems:

- Duct sealing mastic with fiber glass mesh tape—highly preferred—may add to the cost of the system, but can, if applied correctly, provide a lifetime, airtight seal (Figure 9-1).
- Pressure-sensitive tape with a UL 181A, Part I rating or heat-activated tapes labeled in accordance with UL 181A, Part II may be used for assembling fibrous ductboard—but must be installed properly to be effective. The duct surface must be clean of oil and dirt, and the tape must fully adhere to the duct with no wrinkles. A squeegee must be used to remove air bubbles from beneath the taped surface. There are some tapes that meet the UL standard that are not aluminum.
- Aerosol sealant installed by manufacturer-certified installers.
- Some duct manufacturers are listing closure products that they allow to be used with their products.

Proper sealing and insulation of the ductwork requires careful attention to detail and extra time on the part of the heating and air conditioning contractor. The cost of this extra time is well worth the substantial savings on energy costs, improved comfort, and better air quality that an airtight duct system offers.

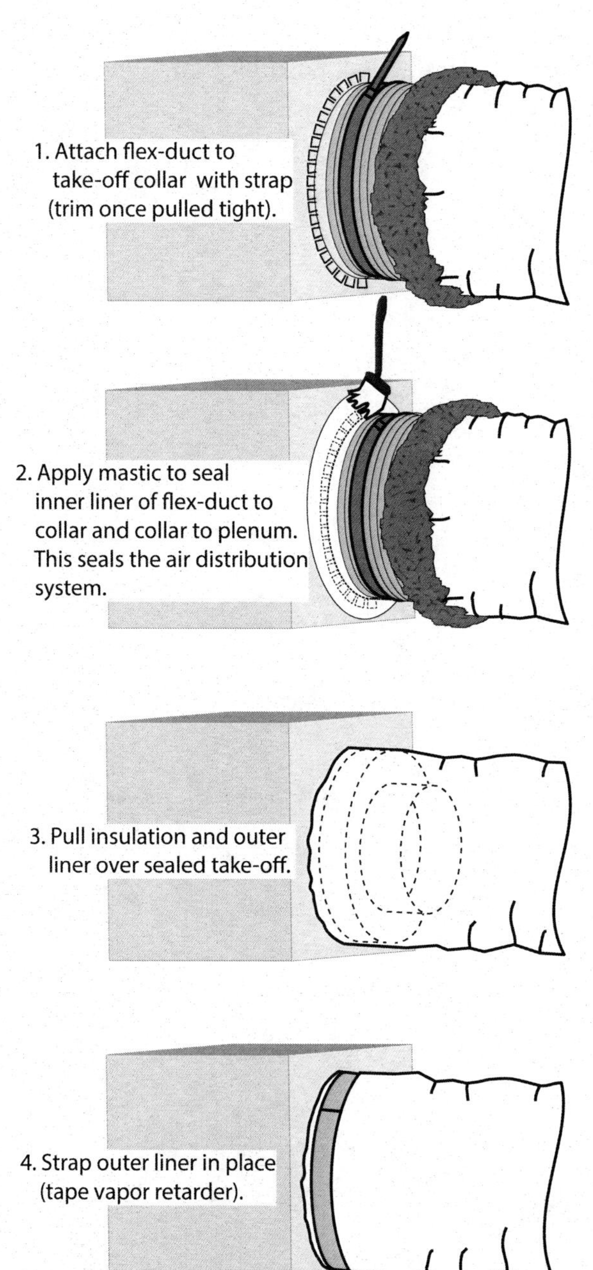

Figure 9-1 Sealing flex-duct collar with mastic

The Florida Building Code, Energy Conservation has compliance methods that include a tested duct option by an approved party. The tested duct leakage shall be determined by an energy rater certified in accordance with Section 533.993(5) or (7), *Florida Statutes*, or as authorized by Florida Statutes. (Also see **FBC, Energy Conservation, Section R403.3.2 Sealing (Mandatory).**)

The easiest answer to the question of where to seal air distribution systems is "everywhere." **FBC, Energy Conservation, Section R403.3.2 Sealing (Mandatory)**, states "All ducts, air handlers, filter boxes and building cavities which form the primary air containment passageways for air distribution systems shall be considered ducts or plenum chambers, shall be constructed and sealed in accordance with Section C403.2.9 of Commercial Provisions of this code and shall be shown to meet duct tightness criteria" in the same section. Note under Exceptions that *if in compliance* with **Section R405 Simulated Performance Alternative (Performance)**, duct testing is not mandatory. A list of the key locations is as follows:

High priority leaks

- Disconnected components, including takeoffs that are not fully inserted, plenums or ducts that have been dislodged, tears in flex-duct, and strained connections between ducts (visible when the duct bends where there is no elbow) (Figure 9-2).

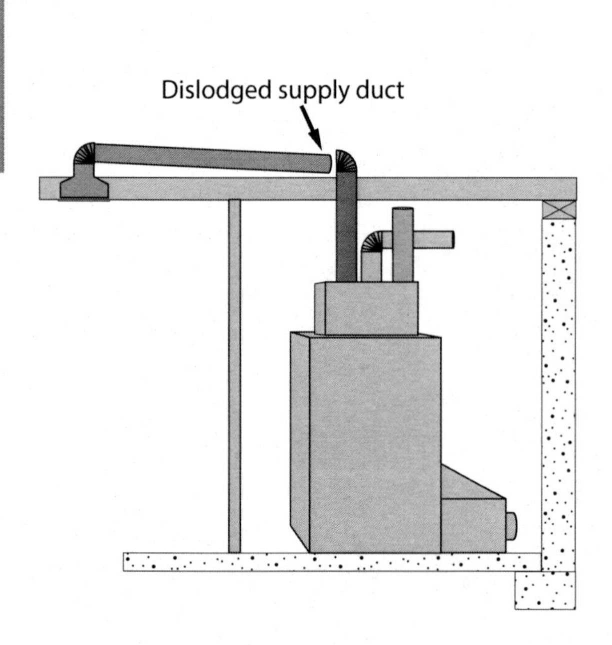

Ducts can become disconnected during initial installation, maintenance, or even normal operation. They should be checked periodically for problems.

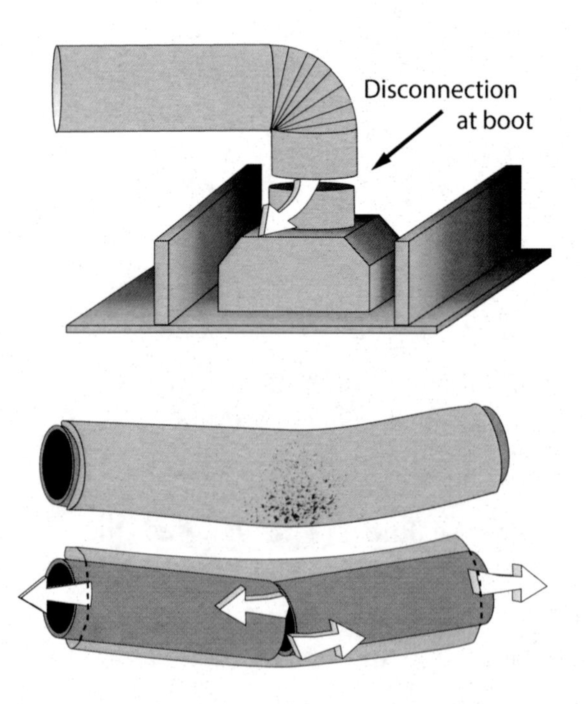

Sometimes, disconnected ducts can be hidden behind the insulation. Look for patterns of dust or dirt on the insulation that indicate air leaks, or kinks or curves where there is no elbow.

Figure 9-2 Disconnected ducts are high priorities

- The connections between the air handling unit and the supply and return plenums.
- Leaks in supply ducts and leaks into unconditioned spaces. This can include leaks into wall cavities, which, though in conditioned space, can force air into unconditioned spaces like attics (Figure 9-3).

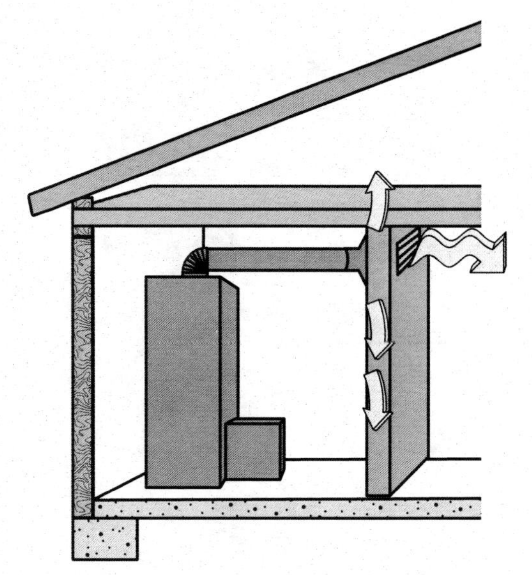

Although this supply duct is theoretically in conditioned space, the supply leaks pressurize the interior wall cavity and air leaks to unconditioned space.
The best solution: seal all duct leaks and leaks into unconditioned spaces.

Figure 9-3 Duct leaks in inside spaces

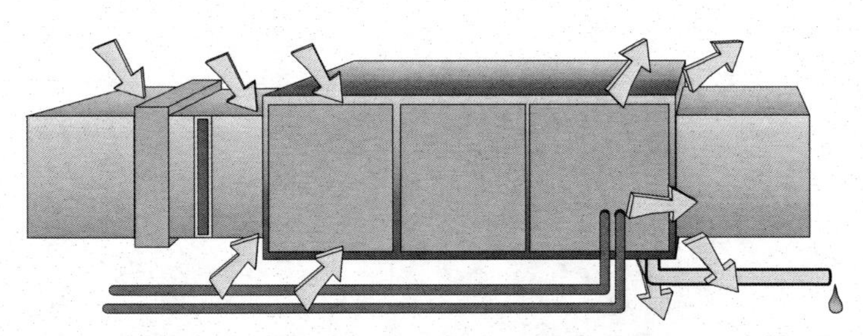

Many air handling cabinets come from the factory with leaks, which should be sealed with duct-sealing mastic.
Removable panels should be sealed with rated aluminum tape.

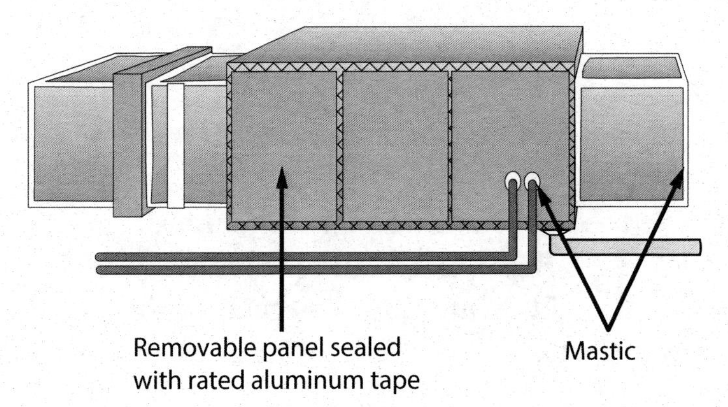

Figure 9-4 Seal all leaks in air handling unit

- All of the seams in the air handling unit, plenums, and rectangular ductwork—look particularly underneath components and in any other tight areas (Figures 9-4 and 9-5). Also seal the holes for the refrigerant, thermostat, and condensate lines. Use approved tape rather than mastic to seal the seams in the panels of the air handling unit so they can be removed during servicing. After completion of service and maintenance work, such as filter changing, make sure the seams are retaped.

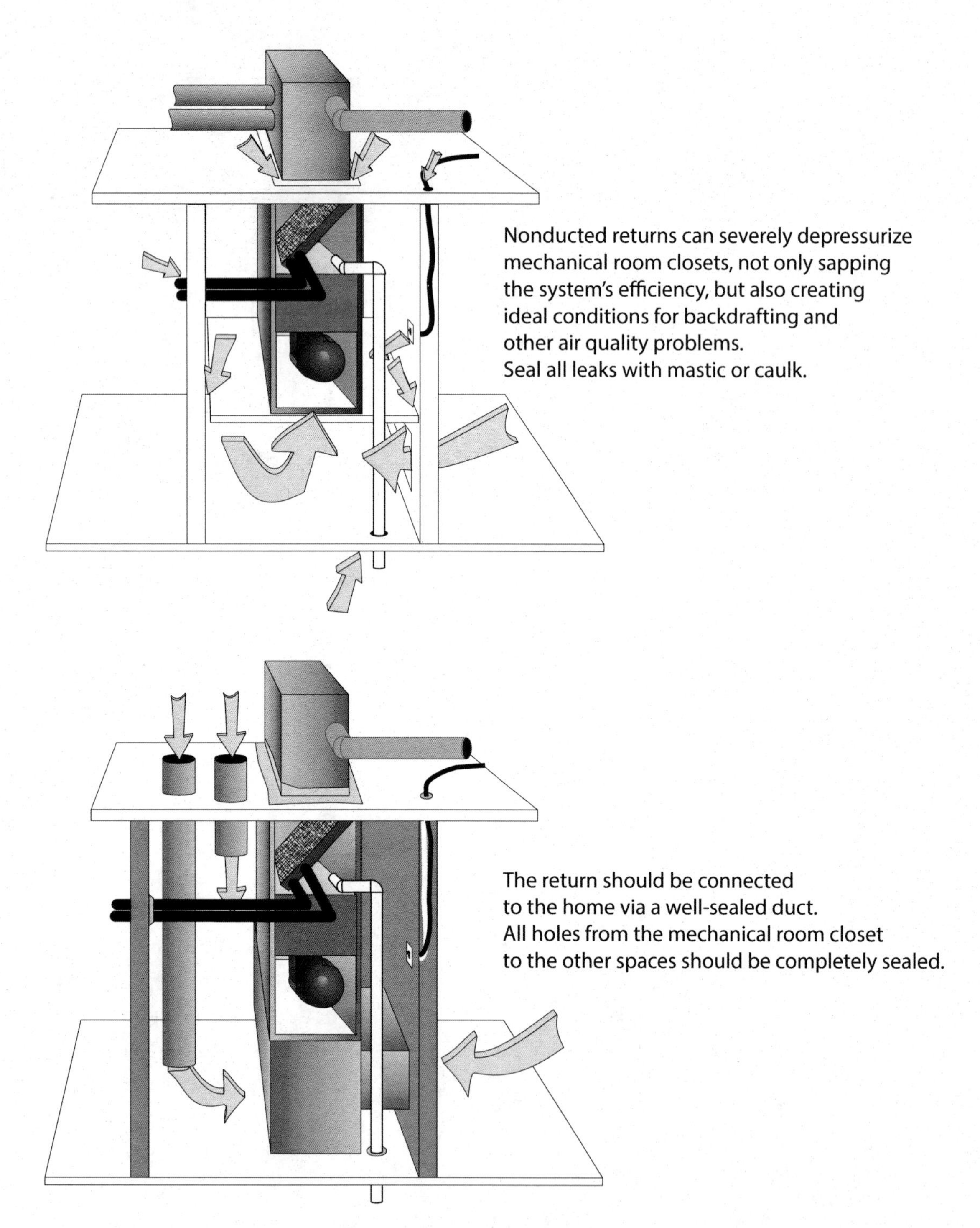

Figure 9-5 Shelf-mounted systems with and without ducted returns

Mechanical closets and building cavities can't be used to transport supply or return air according to **FBC, Energy Conservation, Section R403.3.5 Building cavities (Mandatory)**. **FBC, Energy Conservation, Section C403.2.9.4 Cavities of the building structure** does permit it if lined with a continuous sealed air barrier or insulated air duct insert. This insert must also be sealed.

- The return takeoffs, elbows, boots, and other connections (Figure 9-6). If the return is built into an interior wall, all connections and seams must be sealed carefully. Look especially for unsealed areas around site-built materials.
- The takeoffs from the main supply plenum or trunk line.

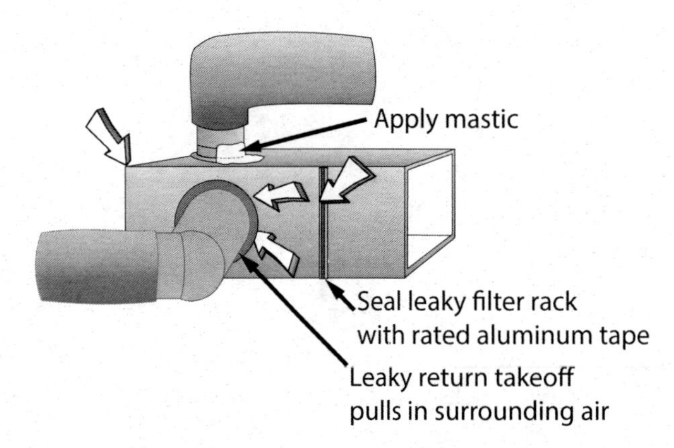

Figure 9-6 Seal all leaky takeoffs

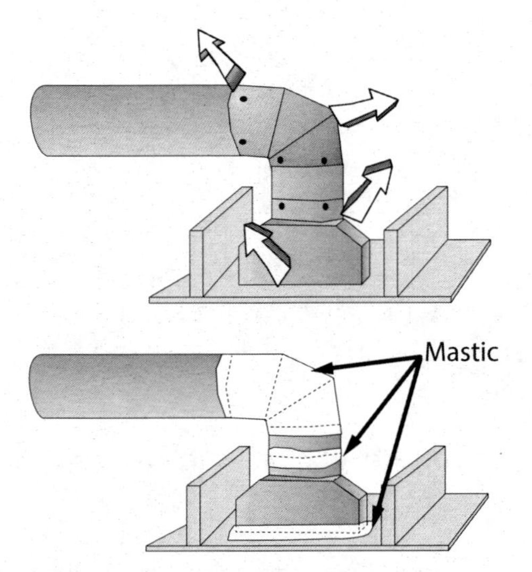

Use mastic to completely seal all leaky seams and holes. Use mesh tape with mastic to cover cracks over 1/8-inch wide.

Figure 9-7 Sealking leaky boots

- The connections near the supply registers—between the branch ductwork and the boot, the boot and the register, the seams of the elbows, and all other potential leaks in this area (Figure 9-7).

Moderate priority leaks

- The joints between sections of the branch ductwork.

Low priority leaks

- Longitudinal seams in round metal ductwork.

The Florida Building Code provides extensive direction for sealing duct systems and other components of a typical home HVAC system. These code directions can be found in particular sections of **FBC, Energy Conservation, Table C403.2.9.2 Duct System Construction and Sealing** and in **FBC, Mechanical, Chapter 6: Duct Systems**.

TESTING FOR DUCT LEAKAGE

The best method to ensure airtight ducts is to pressure test the entire duct system, including all boot connections, duct runs, plenums, and air handler cabinet. Much like a pressure test required for plumbing, ductwork can be tested during construction so that problems can be easily corrected.

In most test procedures, a technician temporarily seals the ducts by taping over the supply registers and return grilles. Then, the ducts are pressurized to a given pressure—usually 25 Pascals. This pressure is comparable to the pressure the ducts experience when the air handler operates.

The ducts are usually tested for tightness using a duct testing fan. Measuring the airflow through the fan gives an estimate of the air leakage through unsealed seams in the ductwork.

Some energy efficiency programs require that the cubic feet per minute of duct leakage measured at a 25 Pascal pressure (CFM25) be less than 3% of the floor area of the house. For example, a 2,000 square-foot house should have less than 60 CFM25 of duct leakage.

Another test is to use a blower door (described in Chapter 4) and a duct testing fan together to measure duct leakage after construction is complete. This procedure gives the most accurate measurement of duct leakage to the outside of the home. A duct leakage test can usually be done in about one hour for an average sized home.

See **FBC, Energy Conservation, Section R403.3.2. Sealing (Mandatory)** if prescriptive compliance, for Florida testing options: Postconstruction or Rough-in.

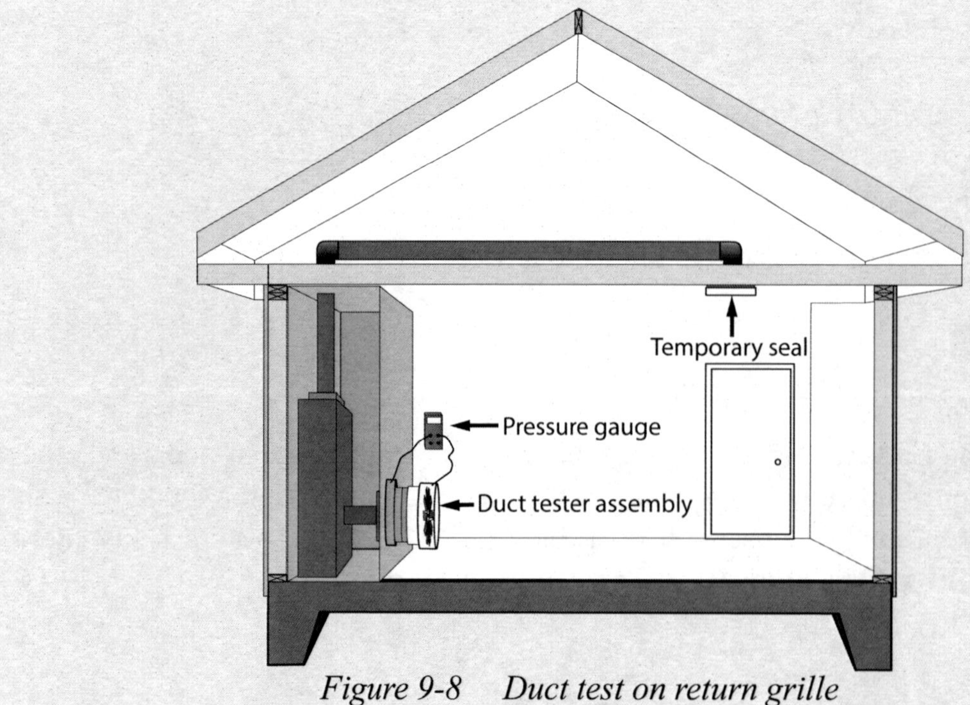

Figure 9-8 Duct test on return grille

Duct Design

Duct Materials

The three most common types of duct material used in building construction are *metal, fiber glass ductboard*, and *flex-duct* (Figure 9-9). Both metal and fiber glass ductboard are rigid and installed in pieces, while flex-duct comes in long sections.

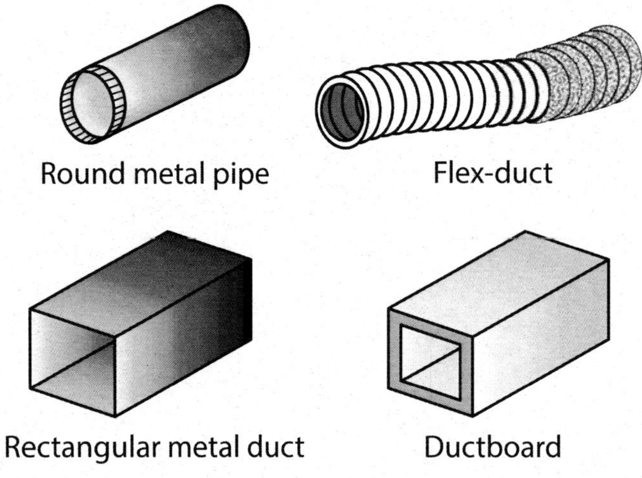

Figure 9-9 Types of ductwork

Flex-duct is usually installed in a single, continuous piece between the register and plenum box, or plenum box and air handler. While it has fewer seams to seal, it is important that the soft lining material not be torn. The flex-duct must also not be pinched or constricted (Figure 9-10). Long flex-duct runs can severely restrict air flow, so they must be designed and installed carefully (Figure 9-11). Flex-duct takeoffs, while often airtight in appearance, can have substantial leakage and should be sealed with mastic.

Round and rectangular metal duct must be sealed with mastic and insulated during installation. It is important to seal the seams first, because the insulation does not stop air leaks. Rectangular metal duct used for plenums and larger trunk duct runs is often insulated with duct liner, a high density material that should be at least 2 inches thick.

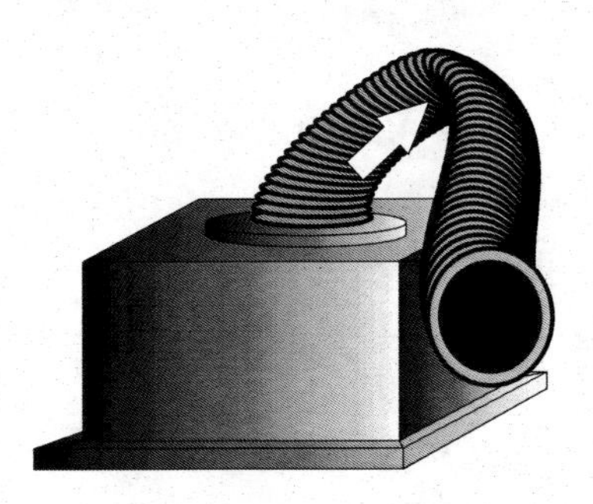

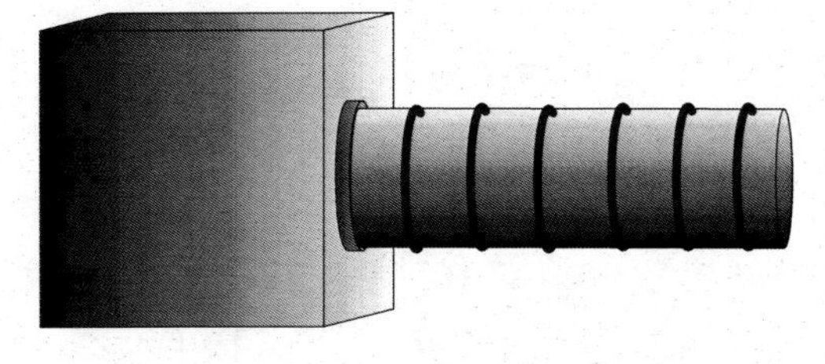

Figure 9-10 Duct details

Metal ducts often use fiber glass insulation having an attached metal foil vapor retarder. According to **FBC, Energy Conservation, Section R403.3.1, Insulation (Prescriptive)**, "Supply and return ducts in attics shall be insulated to a minimum of R-8 where 3 inches in diameter and greater and R-6 where less than 3 inches in diameter. **Exception:** Ducts or portions thereof located completely inside the *building thermal envelope.*" The vapor retarder should be installed to the outside of the insulation—facing away from the duct. The seams in the insulation are usually stapled together around the duct and then taped. Duct insulation in homes at least two years old provides great clues about duct leakage—when the insulation is removed, the lines of dirt in the fiber glass often show where air leakage has occurred.

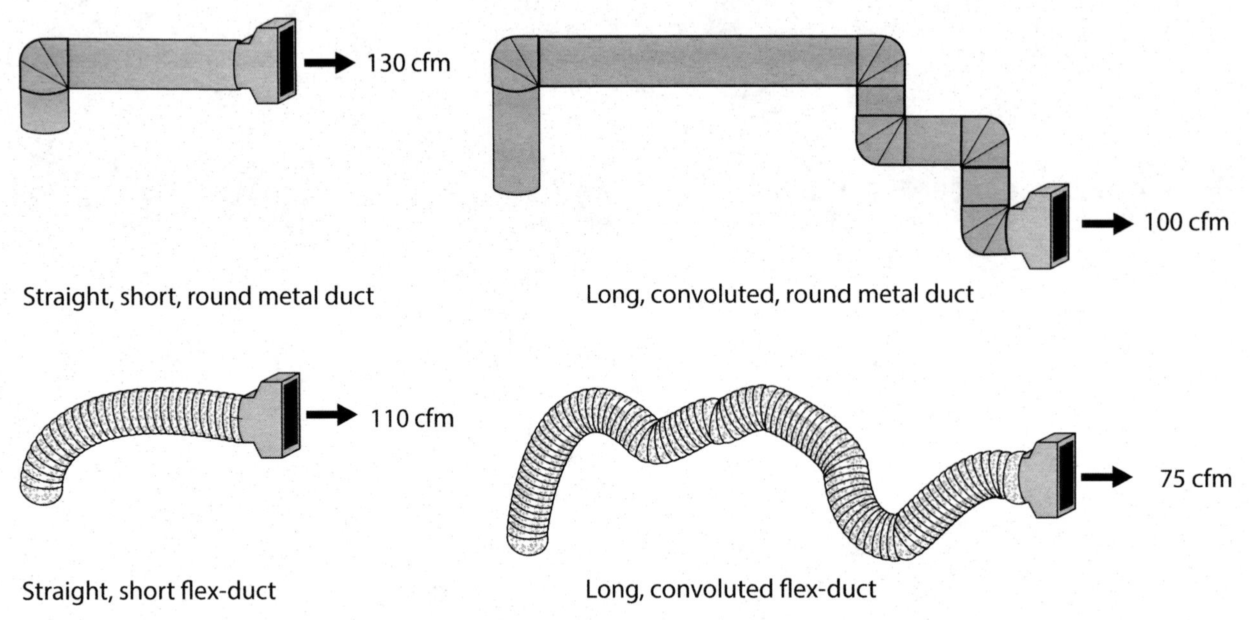

Figure 9-11 Comparison of air flow in different 6-inch ducts

Sizing and layout

The size and layout of the ductwork affects the efficiency of the heating and cooling system and comfort levels in the home. The proper duct size depends on:

- The estimated heating and cooling load for each room in the house.
- The length, type, and shape of the duct.
- The operating characteristics of the HVAC system (such as the pressure, temperature, and fan speed).

The lower temperature of the heated air delivered by a heat pump affects the placement of the registers. A heat pump usually supplies heated air between 90°F and 110°F. At these temperatures, air leaving registers may feel cool. It is important that they are placed so as to avoid blowing air directly onto people. Fuel-fired furnaces typically deliver heated air at temperatures between 110°F and 140°F, 40°F to 70°F greater than room temperature, so placement of the supply registers is less important to maintain comfort.

In standard duct placement and design, supply registers are almost always located on outside walls under or above windows, and return registers are placed towards the interior, typically in a central hallway.

Some builders of energy efficient homes have found little difference in temperature between interior areas and exterior walls because of the extra energy features. Locating the supply registers on exterior walls is not as necessary to maintain comfort. These builders are able to trim both labor and material costs for ductwork by locating both supply and return ducts near the core of the house.

Many duct systems are incorrectly designed by using "rules of thumb," such as using 6-inch round flex or metal duct for all supply ducts and having only one return.

These rules work for some homes, but can create operating problems for others, including:

- Too much heating and cooling supplied to small rooms, such as bathrooms and bedrooms with only one exterior wall.
- Inadequate airflow, and thus, insufficient heating and cooling in rooms located at a distance from the airhandler.
- Overpressurization of rooms when interior doors are closed.

The heating and cooling industry has comprehensive methods to size supply and return ductwork properly. These procedures are described fully in *Manual D, Duct Design* published by the Air Conditioning Contractors' Association.

Unfortunately, few residences have ductwork designed via *Manual D*. Most HVAC contractors use improper duct design methods in the home. The primary "design" is determining, usually via intuition, how many registers should be installed in each room.

CHECKING SYSTEM AIR FLOW

Use this simple form to check the ductwork for proper sizing:

Step 1: Find the system's cooling capacity in tons.

Step 2: Multiply the tonnage by 400 to get the desired total air flow in cubic feet per minute (cfm) = 400 × ________ tons = ____________ cfm total

Step 3: Check the supply air flow:

a. Determine the number of supply registers connected to 4", 6", 8", and 10" branch ducts.

b. Fill in column 2 in the chart below. Then multiply the number of ducts by the air flow and put the result in Column 4. Add the flows in Column 4. If the total is within 10% of the actual air flow (from Step 2), the supply ductwork is probably adequate.

Step 4: Check the return air velocity:

a. Measure the total area of all return grilles = ________ square inches

b. Multiply the total area in 4a by 70% = _______ square inches

c. Divide the answer to 4b by 144 to get square feet of area = ________ square feet

d. Write the total air flow here = ________ cfm (total cfm in chart above)

e. Divide the air flow in 4d by the area in 4c to get the estimated return air velocity: airflow _______ / area ________ = ________ ft/minute

If the velocity is over 650 ft/ minute, add a return or increase the size of a return.

Example: Home with 2.5 ton system, one 4-inch, seven 6-inch, and one 8-inch branch supply ducts.

1. Branch Duct Size	2. Number of Supply Registers	3. Air Flow per Register (cfm)	4. Duct Air Flow (cfm) Step 2 × Step 3
4"		50	
6"		100	
8"		200	
10"		400	
Total Air Flow			**cfm**

continued...

continued...

Step 1: 2.5 tons

Step 2: $400 \times 2.5 = 1{,}000$ cfm

Step 3:

1. Branch Duct Size	2. Number of Supply Registers	3. Air Flow per Register (cfm)	4. Duct Air Flow (cfm) Step 2 × Step 3
4″	*1*	50	*50*
6″	*7*	100	*700*
8″	*1*	200	*200*
10″	*0*	400	*0*
Total Air Flow			***950*** **cfm**

Since 950 cfm is within 10% of system air flow, there should be enough supply ducts.

Figure 9-12 shows the size ductwork *Manual D* would specify for a small home. The design is vastly different than the typical, all 6-inch system. The advantage of proper design is that each room receives air flow proportionate to its heating and cooling load, thus increasing overall comfort and efficiency.

The following recommendations, while no substitute for a *Manual D* calculation, should improve system performance:

- If two rooms have similar orientation, window area, and insulation characteristics, but one room is considerably farther from the air handling unit than the other, consider increasing the size of the ductwork going to the farthest room.
- Bonus rooms over garages often need additional or larger supplies.
- Rooms with large window areas may warrant an extra supply duct, regardless of the room size.
- Likewise, large rooms with few windows, only one exterior wall, well insulated floor, and conditioned space above may need only one small duct.
- Provide return air capability for bedrooms when bedroom doors are shut. (As stated earlier, this is a requirement of the **FBC, Residential, Section M1602.3 Balanced return air** and the **FBC, Mechanical, Section 601.6 Balanced return air**.)

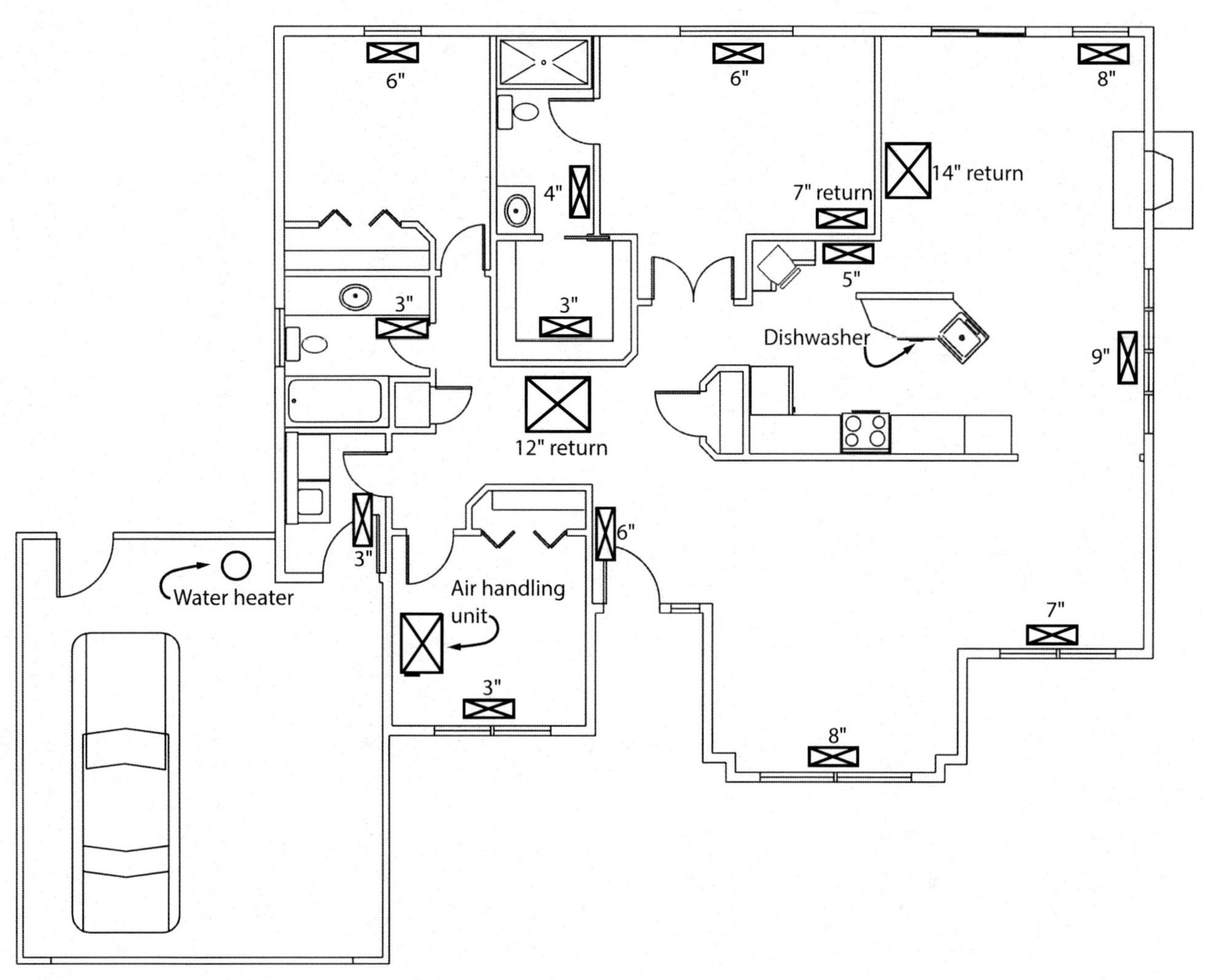

Figure 9-12 Duct design using Manual D

(In standard duct installation, all supply registers would be 6 inches in diameter, and there would be a single 14-inch to 16-inch return.)

Table 9-1 Ductwork Summary for Figure 9-12

Supply		Return	
Size	**Number**	**Size**	**Number**
3″	4	7″	1
4″	1	12″	1
5″	1	14″	1
6″	3		
7″	1		
8″	2		
9″	1		

Resources

Note: Web links were current at the time of publication, but can change over time.

Air Conditioning Contractors of America Association, Inc. (ACCA)

(2016). *Manual J - Residential Load Calculation* (Eighth Edition). Air Conditioning Contractors of America, Inc. (ACCA). Retrieved from http://www.acca.org/standards/technical-manuals

(n.d.). *Manual D® - Residential Duct Systems*. Air Conditioning Contractors of America, Inc. (ACCA). Retrieved from http://www.acca.org/standards/technical-manuals

Air-Conditioning, Heating, & Referigeration Institute (AHRI). (n.d.). Retrieved August 20, 2015, from http://www.ahrinet.org/site/1/Home

Hurm, M. (2002, April). Personal communication. Thompson Sheet Metal, Inc.

Lstiburek, J. (2005). *Builder's Guide to Hot-Humid Climates*. Building Science Corporation. Retrieved from http://buildingsciencepress.com/bookstore/books/builders-guide-hot-humid-climates

My Florida Home Energy. (n.d.). Retrieved July 28, 2015, from http://www.myfloridahomeenergy.com/

A useful resource with a wide array of information on energy and water efficiency, including The Energy Efficient Home series of fact sheets, available at http://www.myfloridahomeenergy.com/help/library

10 DOMESTIC WATER HEATING

Water heating is an important end-use that accounts for roughly 14% to 25% of the total energy consumption in the home. Efficiency standards on water heaters and hot water-using appliances and equipment (e.g., showerheads, faucets, dishwashers and clothes washers) can and will substantially affect the energy use of water heaters in the future. Water heater efficiency minimums are listed in **FBC, Energy Conservation, Section R403.5.6 Water heater efficiencies (Mandatory)**.

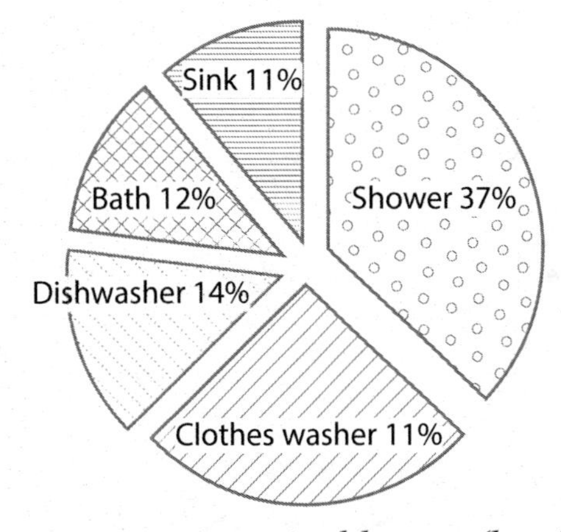

Figure 10-1 Hot water use in typical homes (by place of use)
Source: United States Department of Energy

The typical U.S. homeowners' hot water consumption, by place of use, is shown in Figure 10-1. The average amount of hot water used in various household tasks is shown in Table 10-1.

Table 10-1 Average Hot Water Use

Activity	Gallons per Use
Clothes washing	25–40
Showering	10–25
Bathing	20–36
Automatic dishwashing	6–16
Preparing food	5
Hand dishwashing	8–27

Source: American Council for an Energy-Efficient Economy

10: WATER HEATING

Energy Conservation For Water Heating

No matter what type of energy source is used to heat water, be certain to take advantage of the savings from conservation measures:

- Lower the temperature setting on the water heater to 120ºF.
    - ◊ Saves energy; for each 10ºF reduction in water temperature, water-heating energy consumption can be reduced 3% to 5 %.
    - ◊ Reduces the risk of injury from scalding.
    - ◊ Provides plenty of hot water.
    - ◊ If hotter temperatures are needed for dish washing, select dishwashers with booster heaters. (Most new dishwashers have this feature.)

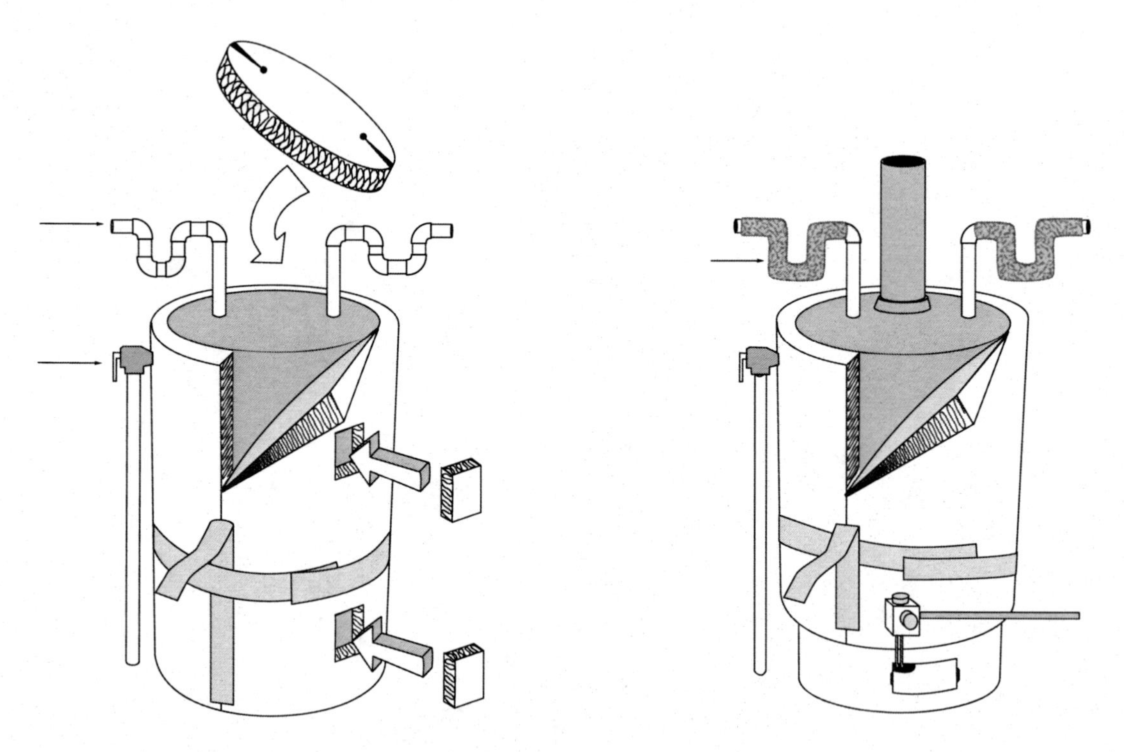

Figure 10-2 Insulating jackets for electric and gas water heaters

- If needed, wrap the outside of the water heater tank with an insulation jacket (Figure 10-2). This is especially useful for older water heaters, as new water heaters are already well insulated.
    - ◊ Simple to install.
    - ◊ On all types, do not cover the emergency pressure relief or drain valve; pressure buildup due to continuously heated water will force open this valve, allowing hot water to flow out of the tank through the pressure relief pipe to an outlet outside.

◊ On electric water heaters, do not cover the thermostat or access panels.

◊ On gas water heaters, do not cover the top, the controls, or the lower third of the tank—the air inlet or the flue vent on the top. Check with the installer for specific instructions, because if insulation is placed over any of these areas, the controls may overheat or the unit may become starved for air—resulting in hazardous explosions. Keep in mind that some models have the temperature/pressure relief valve and the overflow pipe on the side of the tank instead of on the top.

- Heat can be lost from hot water tanks through convection (as warm water rises and cold water sinks). To prevent this, install heat traps to control the direction of flow. They can be one-way valves as shown in Figure 10-3 or u-shaped traps at least 3.5 inches deep and as close as practical to the storage tank, as shown in Figure 10-2. Most new water heaters already have them integrated into the tank. If not, **FBC, Energy Conservation, Section R403.5.5 Heat traps (Mandatory)** requires that they be installed as described on inlet and outlet pipes.

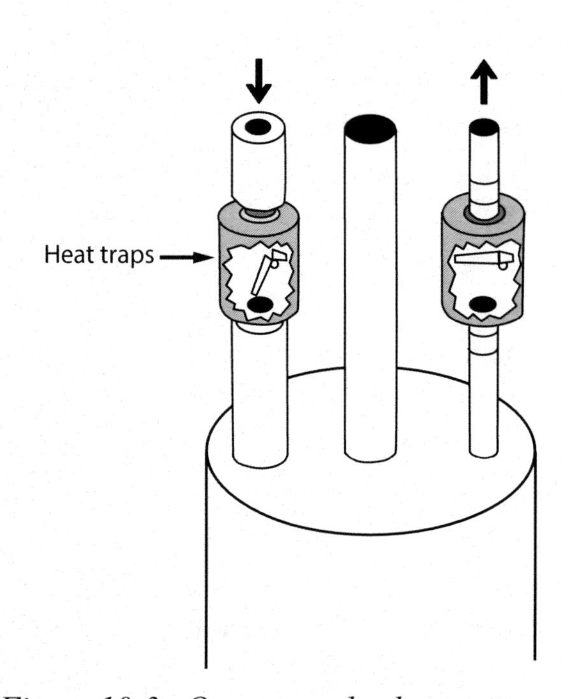

Figure 10-3 One-way valve heat traps
(U.S. Department of Energy, Energy Efficiency and Renewable Energy)

- Insulate the first 6 feet of the cold and hot water pipes connected to the unit.
- Install low-flow showerheads with well-designed fixtures that deliver water at about 2.0 gallons per minute or less and still provide plenty of force (2.5 gallons per minute maximum allowed by Code (404.8.1)).
- Put in shutoff valves on low-flow showerheads and kitchen faucets, which are designed to dribble when closed, so water in the pipe stays at the selected temperature while soaping, shaving or shampooing. These valves are built into many low-flow heads.
- Include low-flow aerators or laminar flow controls on sink and lavatory faucets.

◊ Saves on energy bills as aerators add air to the water stream and make a light flow feel heavier.

◊ Laminar flow controls work by producing dozens of streams of water very close to one another. Many believe they make the water feel that it is flowing at a higher rate than in actuality.

◊ Kitchen or utility sinks may need a higher volume flow faucet (2.2 gallon-per-minute) for washing and for filling larger volume containers.

- Minimize the piping runs to the bathroom and kitchen.
- Consider installing a drain water waste heat recovery system; the Department of Energy showed energy savings of 25–30%.
- One way to save money and energy is to wash clothes in cold water. Laundry detergents are formulated to work well in cold water under normal circumstances. Washing in warm or hot water is a habit, but is really not necessary. Hot water wash and rinse cycles can consume about 32 gallons of hot water. Front-loading washing machines use up to 25% less water and less laundry detergent than top loaders. The majority of the energy used by a washing machine goes to heat the water; motors use much less. Cold water washing can save about $60 per year for an average family.
- Put in or recommend ENERGY STAR® washing machines; most of the energy consumed in washing clothes is required to heat the water, with some of the newer energy-efficient models using about one-third the hot water.
- Install ENERGY STAR dishwashers; as with clothes washers, most of the energy consumed by dishwashers is used to heat the water. Virtually all energy efficient dishwashers are equipped with a temperature boost feature that raises the incoming water temperature by 20ºF.

Temperature Settings

Some water heater manufacturers supply units pre-set to 140°F; however, most residential heaters should be reduced to 120°F for several reasons. Lowering the setting to 120°F will save approximately 6%–10% of energy costs, will reduce mineral deposition inside the tank and piping, and eliminate the risk of scalding which can occur at the higher temperature.

When changing the setting of an electric water heater, for safety, first turn off its power supply if a panel must be removed to access the thermostat. Some electric water heaters may have two thermostats controlling upper and lower heating elements.

Other Possible Ways to Conserve Water

Hot Water Demand Systems (HWDS)

Several companies manufacture a pumping system that saves both water and energy by utilizing the cold water line as a return line to the water heater. The cold water connection is made at the furthest fixture from the water heater. In one system, the unit is placed between the cold and hot water line located under the cabinet at the most distant fixture. When the system is activated, an electronic valve opens and the pump moves the cold water in the hot water side rapidly across the valve to the cold water side. The cold water in the line is then replaced by the hot water that is now moving from the water heater through the hot water pipe towards the most distant fixture. Upon the arrival of the hot water, a thermal sensor at the location senses a temperature rise, thus quickly closing the valve and automatically shutting down the pump. This system can be activated in a number of different ways including push button, remote control, light beam, motion sensor, etc. The savings are a result of getting hot water to your farthest fixture faster without the loss of any water down the drain. The energy requirement to run the pump is minimal since it only operates when activated and turns off automatically.

Additional Water Heater

Some builders add a small (2 to 6 gallon) water heater at the end of a long plumbing run. By being connected in a series with the main water heater, the owner has instant hot water and before the small heater runs out of water, the main unit is activated. Energy is spent to save water. A tankless instantaneous heater may be a better option to serve rooms that are distant from the water heater. The best solution of all is to minimize the problem by centrally locating the water heater.

Selecting an Efficient Water Heater

Sizing

Above all, the "capacity" of a water heater should be judged by its first hour rating (FHR), not its tank size. The peak-hour demand capacity or the FHR, required on the EnergyGuide label, is actually more important that the size of the storage tank. The FHR is a measure of how much hot water the heater will deliver during a busy hour. The size of tank needed will depend on the number of people living in the home and the patterns of usage. Selecting an oversized water heater, besides raising the purchase cost, will result in increased energy costs due to excessive cycling and standby losses.

Gas water heaters have higher FHRs than electric water heaters of the same storage capacity. Therefore, it may be possible to meet the water-heating needs with a gas unit that has a smaller storage tank than an electric unit with the same FHR. More efficient gas water heaters use various nonconventional arrangements for combustion air intake and exhaust. These features, however, can increase installation costs.

All other things being equal, the smaller the water heater tank, the higher the efficiency. Compared to small tanks, large tanks have a greater surface area, which increases heat loss from the tank and decreases the energy efficiency somewhat, as mentioned above.

Selecting an oversized water heater, besides raising purchase cost, will result in increased energy costs. The American Council for an Energy-Efficient Economy (ACEEE) website has information on choosing a water heater (see Resources at the end of this chapter). It also has a regularly-updated publication for sale, *Consumer Guide to Home Energy Savings*. In addition, the Air Conditioning, Heating and Refrigeration Institute's (AHRI) *Consumer's Directory of Certified Efficiency Ratings* provide good, simple guidance on proper sizing of water heaters. Finally, assistance in comparing life cycle costs of several types of water heaters and selecting appropriate size is available on the Energy Saver website operated by U.S. Department of Energy.

Fuel Type

One of the first steps in choosing a water heater is to determine the appropriate fuel type. Natural gas water heaters are generally less expensive to operate, but not always—check local rates. Also check rebates from local utility companies.

Energy Factor

Once you have decided what type of water heater best suits your needs, determine which water heater in that category is the most fuel efficient. The best indicator of a heater's efficiency is its Energy Factor (EF), which is based on recovery efficiency (i.e., how efficiently the heat from the energy source is transferred to the water), standby losses (i.e., the percentage of heat lost per hour from the stored water compared to the heat content of the water), cycling losses, and the average household use of 64 gallons of hot water per day. The higher the EF, the more efficient the water heater. Table 10-2 contains the minimum energy factors of water heaters set by the Office of Energy Efficiency and Renewable Energy, Department of Energy. Electric resistance water heaters have EFs ranging from 0.7 to 0.97 (the most efficient electric storage water heaters all have energy factors between 0.94 and 0.97). Heat pump water heaters use less than half as much electricity as conventional electric resistance water heaters. The most efficient gas-fired storage water heaters have energy factors ranging from 0.60 to 0.67 with some high-efficiency models ranging around 0.8; and heat pump water heaters from 1.8 to 2.5. The efficiency of a dual integrated system is given by its combined annual efficiency, which is based on the AFUE of the space heating component and the energy factor of the water heating components.

Table 10-2 Minimum Energy Factors of Water Heaters* - Current as of January 1, 2017

Product class	Rated storage volume and input rating (if applicable)	Draw pattern	Uniform energy factor
Gas-fired Storage Water Heater	≥ 20 gal and ≤ 55 gal	Very Small	0.3456 – (0.0020 × Vr*)
		Low	0.5982 – (0.0019 × Vr)
		Medium	0.6483 – (0.0017 × Vr)
		High	0.6920 – (0.0013 × Vr)
	> 55 gal and ≤ 100 gal	Very Small	0.6470 – (0.0006 × Vr)
		Low	0.7689 – (0.0005 × Vr)
		Medium	0.7897 – (0.0004 × Vr)
		High	0.8072 – (0.0003 × Vr)
Oil-fired Storage Water Heater	≤ 50 gal	Very Small	0.2509 – (0.0012 × Vr)
		Low	0.5330 – (0.0016 × Vr)
		Medium	0.6078 – (0.0016 × Vr)
		High	0.6815 – (0.0014 × Vr)
Electric Storage Water Heaters	≥ 20 gal and ≤ 55 gal	Very Small	0.8808 – (0.0008 × Vr)
		Low	0.9254 – (0.0003 × Vr)
		Medium	0.9307 – (0.0002 × Vr)
		High	0.9349 – (0.0001 × Vr)
	> 55 gal and ≤ 120 gal	Very Small	1.9236 – (0.0011 × Vr)
		Low	2.0440 – (0.0011 × Vr)
		Medium	2.1171 – (0.0011 × Vr)
		High	2.2418 – (0.0011 × Vr)
Tabletop Water Heater	≥ 20 gal and ≤ 120 gal	Very Small	0.6323 – (0.0058 × Vr)
		Low	0.9188 – (0.0031 × Vr)
		Medium	0.9577 – (0.0023 × Vr)
		High	0.9884 – (0.0016 × Vr)
Instantaneous Gas-fired Water Heater	< 2 gal and > 50,000 Btu/h	Very Small	0.80
		Low	0.81
		Medium	0.81
		High	0.81
Instantaneous Electric Water Heater	< 2 gal	Very Small	0.91
		Low	0.91
		Medium	0.91
		High	0.92
Grid-Enabled Water Heater	> 75 gal	Very Small	1.0136 – (0.0028 × Vr)
		Low	0.9984 – (0.0014 × Vr)
		Medium	0.9853 – (0.0010 × Vr)
		High	0.9720 – (0.0007 × Vr)

** Note:* *Vr is the Rated Storage Volume, which equals the water storage capacity of a water heater, in gallons, as determined pursuant to 10 CFR 429.17.*

Source: *Federal Register. Title 10, Energy. Chapter II—Department of Energy. Subchapter D—Energy Conservation. Part 430—Energy Conservation Program for Consumer Products. Subpart C—Energy and Water Conservation Standards. Section 430.32: Energy and water conservation standards and their compliance dates.*

Note: In April 2010, new standards for residential water heaters were set by the US Department of Energy, but take effect April 15, 2016 (Table 10-3).

Table 10-3 Minimum Energy Factors of Water Heaters* - 2016

Product Class	Energy Factor as of April 15, 2016
Gas-fired Water Heater	For tanks with a Rated Storage Volume at or below 55 gallons: EF = 0.675 - (0.0015 × Rated Storage Volume in gallons) For tanks with a Rated Storage Volume above 55 gallons: EF = 0.8012 - (0.00078 × Rated Storage Volume in gallons)
Oil-fired Water Heater	EF = 0.68 - (0.0019 × Rated Storage Volume in gallons).
Electric Water Heater	For tanks with a Rated Storage Volume at or below 55 gallons: EF = 0.960 - (0.0003 × Rated Storage Volume in gallons) For tanks wwith a Rated Storage Volume above 55 gallons: EF = 2.057 - (0.00113 × Rated Storage Volume in gallons)
Instantaneous Gas-fired Water Heater	EF = 0.82 - (0.0019 × Rated Storage Volume in gallons).

* *EF is the "energy factor," and the "Rated Storage Volume" equals the water storage capacity of a water heater (in gallons), as specified by the manufacturer.*

Source: *Federal Register. Final Rule—Energy Conservation Standards for Residential Water Heaters, Direct Heating Equipment, and Pool Heaters. April 16, 2010.*

EnergyGuide label

In the U.S., all water heaters are sold with a bright yellow and black EnergyGuide label to indicate the estimated annual energy consumption and operating cost of the appliance at a given rate. These labels provide an estimated annual energy consumption on a scale showing a range for similar models. By comparing a model's annual operating cost with the operating cost of the most efficient model, you can compare efficiencies. Be sure to check the rates in your area for comparison purposes.

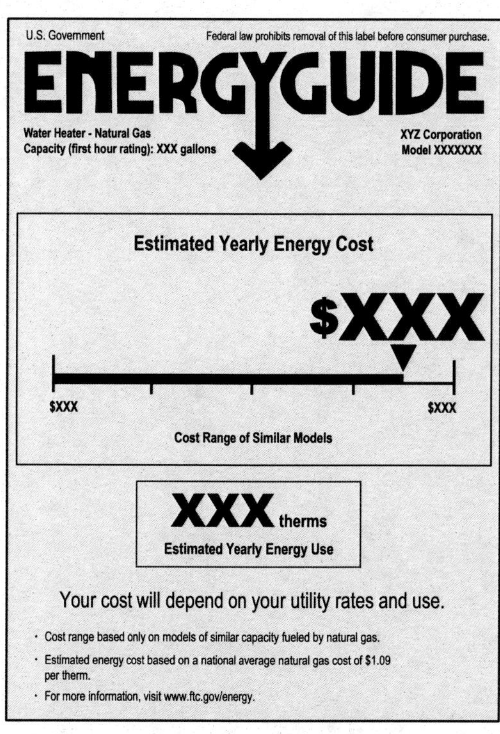

Figure 10-4 An example of an EnergyGuide label

10: Water Heating

ENERGY STAR certification

In 2008 the U.S. Department of Energy established ENERGY STAR criteria for water heaters. Efficiency and performance for ENERGY STAR certified water heaters may be evaluated using two different methods. A water heater can be certified—and thus carry the current ENERGY STAR label—by meeting ***either***

1. the Energy Factor (EF), Standby Loss, and/or Thermal Efficiency criteria (listed in Table 10-4), ***or***
2. the Uniform Energy Factor (UEF) criteria (listed in Table 10-5).

Although these performance metrics are different, certification as ENERGY STAR by either method requires equivalent efficiency. (EF values should not be directly compared to UEF values.)

Table 10-4 ENERGY STAR Criteria: Energy Factor, Standby Loss, Thermal Efficiency

Type	Energy Factor (EF) or Solar Energy Factor (SEF) or Thermal Efficiency (TE)	First-hour Rating or Gallons per Minute or Standby Loss	Warranty Requirements	Safety
Electric storage	≤ 55 gallons — EF ≥ 2.00 > 55 gallons — EF ≥ 2.20	≥ 50 gallons/hour at 135°F outlet temperature	≥ 6 years on sealed system	UL 174 and UL 1995
Gas storage	≤ 55 gallons — EF ≥ 0.67 > 55 gallons — EF ≥ 0.77	≥ 67 gallons/hour at 135°F outlet temperature	≥ 6 years on system (including parts)	ANSI Z21.10.1 / CSA 4.1
Gas instantaneous	EF ≥ 0.9	2.5 gallons/minute with 77°F increaser	≥ 6 years on heat exchanger ≥ 5 yeares on parts	ANSI Z21.10.3 / CSA 4.3
Gas high-capacity storage (light duty EPACT-covered)	TE ≥ 0.90	Standby loss ≤ 1889 Btu/Hr × (TE - 0.73)	≥ 6 years on system / CSA 4.3	ANSI Z21.10.3 / CSA 4.3
Solar	SEF ≥ 1.8 for electric backup SEF ≥ 1.2 for gas backup	NA	≥ 10 years on collector ≥ 6 years sealed system ≥ 2 years on controls ≥ 1 year on parts	NA

Water heaters that meet these criteria report Uniform Energy Factor (UEF) but are not required to meet the UEF criteria.

Table 10-5 ENERGY STAR Criteria: Uniform Energy Factor

Type	Uniform Energy Factor (UEF)		First-hour Rating (FHR) or Maximum Gallons/ Minute (GMP)	Warranty Requirements	Safety
Electric storage*	≤ 55 gallons — UEF ≥ 2.00 > 55 gallons — UEF ≥ 2.20		FHR ≥ 45 gallons/hour at 125°F outlet temperature	≥ 6 years on sealed system	UL 174 and UL 195
Gas storage	≤ 55 gallons	Medium draw pattern UEF ≥ 0.64	FHR ≥ 67 gallons/hour at 135°F outlet temperature	≥ 6 years on system (including parts)	ANSI Z21.10.1 / CSA 4.1
		High draw pattern UEF ≥ 0.68			
	> 55 gallons	Medium draw pattern UEF ≥ 0.78			
		High draw pattern UEF ≥ 0.80			
Gas instantaneous	UEF ≥ 0.87		Max GPM ≥ 2.9 over a 67° rise	≥ 6 years on heat exchanger ≥ 5 years on parts	ANSI Z21.10.3 / CSA 4.3
Gas high-capacity storage (storage residential-duty commercial)	UEF ≥ 0.80		NA	≥ 6 years on system	ANSI Z21.10.3 / CSA 4.3

** Electric storage water heaters must also report the Lower Compressor Cut-Off Temperature - the ambient temperature (°F) below which the compressor cuts off and electric resistance-only operation begins*

Water heaters that meet these criteria are not required to meet Energy Factor, Thermal Efficiency, or Standby Loss criteria.

Results from testing of different types of water heaters sponsored by the U.S. DOE at the Florida Solar Energy Center indicated that advertised EF ratings for all but one type of water heater exceeded the actual performance. The ratings are determined with colder inlet water temperatures than we generally have in Florida. We use less energy heating our water, but have similar heat losses, so the losses represent a greater proportion of the total energy consumed.

The research showed that tankless gas systems used 27% less natural gas compared to standard storage gas units, but tankless electric units saved only 5% over standard electric units and required very high initial electric demand.

Active solar systems with flat plate collectors tested saved about 60% of water heating energy, with no significant difference whether they employed PV controlled pumping (direct loop circulation) or differential control (direct loop circulation).

The passive solar system (ICS) test unit raised water temperature about 15-25 degrees in winter and 30-35 degrees in summer. It would be expected to save about 26% compared to a standard electric system in Florida.

Tax Credits

Federal tax credits for energy efficient hot water heaters ended in 2011, but many utilities offer rebate programs for solar thermal water heating and some include energy efficient water heaters in general energy efficiency rebate programs. You can check the Database of State Incentives for Renewables & Efficiency® created by the US Department of Energy for eligible rebates or the appropriate local utility for the most reliable information.

Types of Available Water Heaters

The following types of water heaters are now on the market: conventional storage, gas condensing, tankless gas/demand, heat pump, heat recovery units, demand, solar, and indirect (tankless integrated water and space heating); however, these are not suitable for warm climates.

Conventional Storage Water Heaters

Ranging in size from 20 to 80 gallons (75.7 to 302.8 liters), storage water heaters remain a popular type for residential heating needs in the US, dominating sales with roughly 80% of the market. A storage heater operates by releasing hot water from the top of the tank when the hot water tap is turned on. To replace that hot water, cold water enters the bottom of the tank, ensuring that the tank is always full. Because the water is constantly heated in the tank, energy can be wasted even when no faucet is on.

Conventional Storage tanks: gas vs. electric.

If choosing a conventional storage tank, compare Energy Factors(EF). EF represents the efficiency of the unit in converting incoming energy to heat, and higher is better. A standard natural gas storage heater may have an EF in the range of .55 to .67, but a new high efficiency model can be around 0.8. An electric unit will have a higher EF, around 0.9 or more. The difference is primarily due to heat lost in the vented gasses in natural gas models. However, electricity is an indirect source of power, and at the point of use has already incurred as much as 3 times the efficiency losses of a gas system in the generation and transmission of electricity. Even though the energy factors would make you think the electric heater is more efficient, it is not, and the annual cost of electricity to operate a water heater is more than that of an equivalent capacity natural gas water heater. If purchasing a gas storage water heater, select one with automatic ignition on demand, rather than a unit with a pilot light. The energy required to burn the pilot light can negate the energy saved over an electric water heater.

Special Note on Fuel-Fired Water Heaters

Where natural gas is available on site, a gas water heater will usually be life-cycle cost effective relative to an electric model—but check local rates. For safety as well as energy-efficiency reasons, when buying gas-fired water heaters, look for units with sealed combustion closets or power venting to avoid back-drafting of combustion gases into the building. If located in conditioned space, a carbon monoxide alarm should be installed.

If fuel-fired water heaters are located in interior spaces, such as interior mechanical rooms connected to conditioned spaces or laundry rooms, they should include provisions for outside combustion air (Figure 10-5). The Florida Building Code includes requirements for fuel-fired appliances, such as water heaters and furnaces. In particular, see **FBC, Fuel Gas, Section 304 (International Fuel Gas - Standard) Combustion, Ventilation and**

Dilution Air. **FBC, Mechanical, Chapter 6: Duct Systems** and **Chapter 5: Exhaust Systems** also provide details for safe, energy-efficient installation of fuel-fired devices.

More sophisticated energy features found on high efficiency furnaces, such as electronic ignition, flue dampers, and condensing heat exchangers, are being introduced into domestic water heaters.

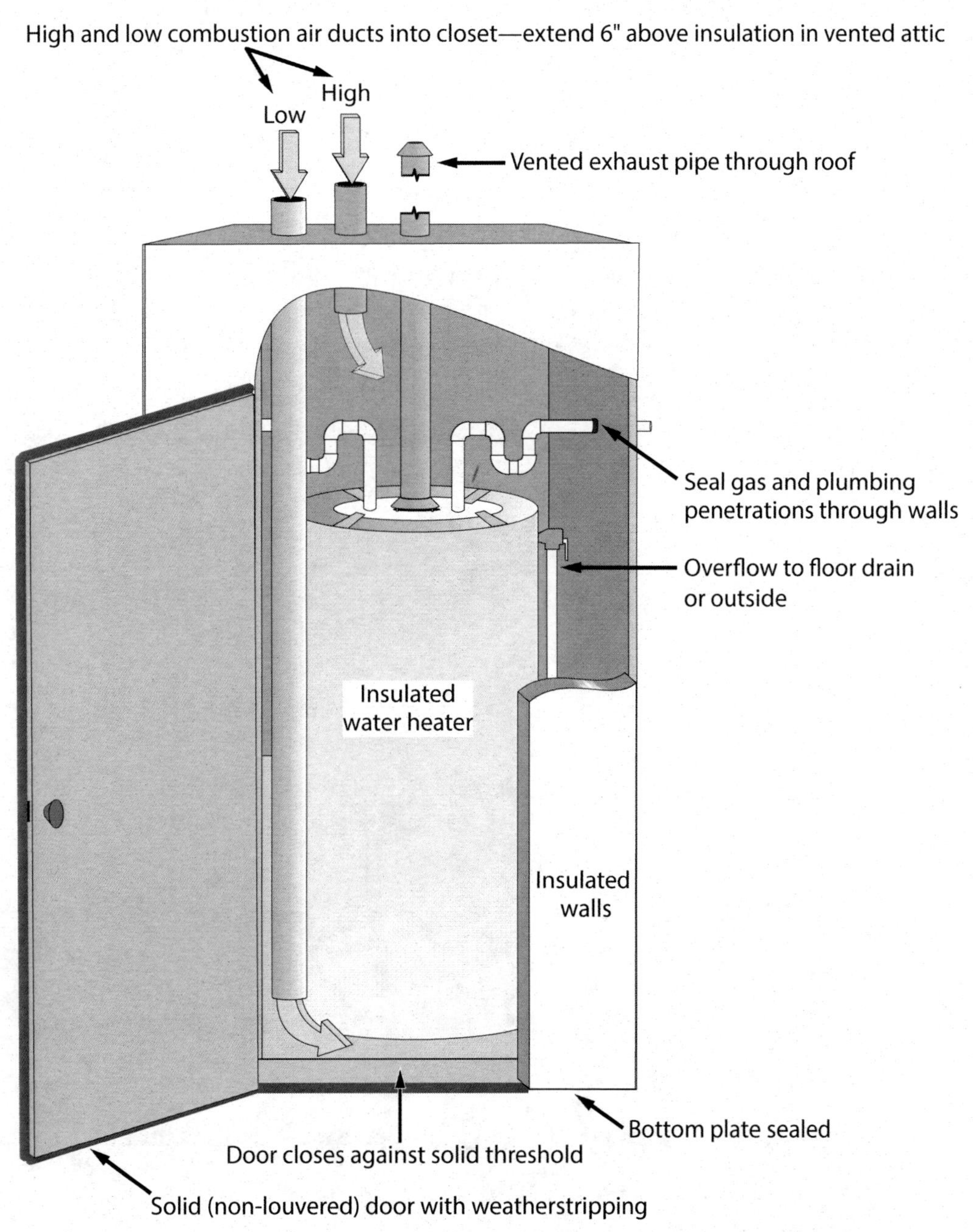

Figure 10-5 Combustion closet for fuel-fired water heater

Gas Condensing Water Heaters

Gas condensing water heaters are the most efficient type of hot water heaters with greater than 90% efficiency. They are common in Europe, where they are called condensing boilers or combi boilers and cylinders. They are also used in commercial buildings in the US.

They boil water by burning natural gas, propane or electric power, the same as their conventional counterparts. The difference is they extract heat that would have been wasted in the exhaust gasses by retaining and condensing the steam produced through boiling. Overall efficiency is increased up to 98%. They are somewhat larger than standard tank heaters and produce an acidic condensate that is neutralized and drained through a PVC line. Venting pipe can also be plastic (PVC or ABS). Otherwise installation is very similar.

Tankless Water Heaters/ Demand Water Heaters

Tankless water heaters are compact—entire household, or smaller units located directly at a usage point at some distance from the main water heater. The smaller units are typically mounted under a bathroom sink, above a shower, or adjacent to an appliance that uses a lot of hot water, such as a clothes washer. Units can be gas fired (natural gas or propane) or electric. (Caution: For electric units, the initial electric demand is very high, and utility companies do not recommend them; they are reported to have caused power outages.) No separate storage tank is needed because water is heated only as needed. Once the hot water tap is closed, the sensor shuts off the gas.

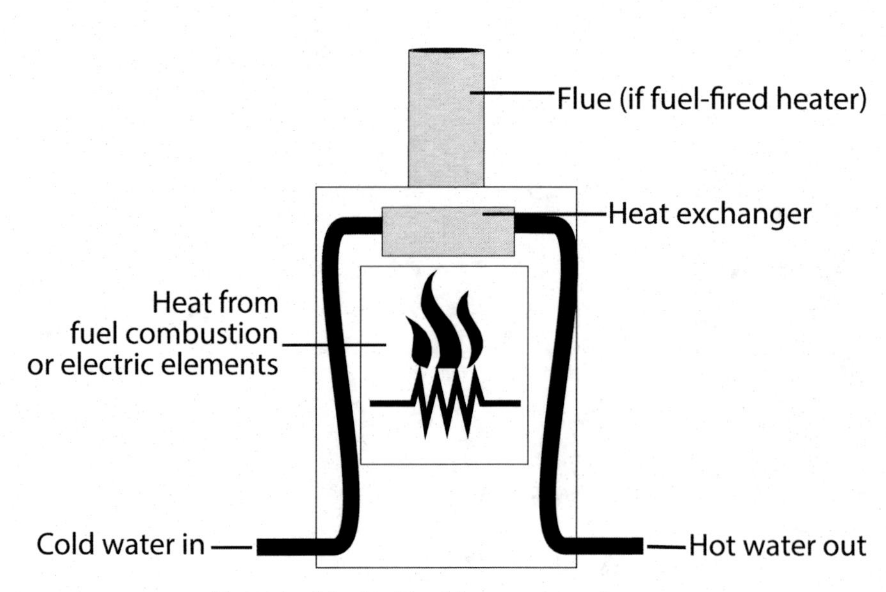

Figure 10-6 Tankless water heater

Sizing tankless heaters

Tankless water heaters are sized by their flow rate, in gallons per minute (gpm). Small units provide 2 to 5 gpm and whole house units commonly provide 6-10 gpm. Two small units can be installed in parallel to meet a higher peak demand rate. The peak rate should consider likely co-incident loads, i.e. more than 1 shower or dishwasher and washing machine drawing hot water simultaneously.

Consider the desired flow rate, the incoming water temperature, and the desired exit temperature. A small unit located in a bathroom may only need to provide 1.5 to 2 gpm, and increase the water temperature from about 65°F to 110°F, a difference of 45°F. The temperature is affected by both the flow rate and the influent temperature. Gas fired demand heaters generally can increase water temperature by about 70 F at a flow rate of 5 gpm, and electric units can provide the same increase for a flow of about 2 gpm. Some models control the outlet temperature via a thermostat.

There are several other issues to consider with tankless models, including the minimum flow rate for activation. Activation rates are frequently in the range of 0.4—0.75 gpm, but some are higher. Lower rates will not produce hot water. A low flow sink fixture may limit flow to 0.5 gpm, so match the unit to the anticipated need.

Maintenance of tankless water heaters may require removing scale deposits caused by hard water from the heat exchanger to prevent clogging of their small diameter pipes. Also, units and fixtures may need periodic flushing to remove accumulated debris. However, high initial expense, not maintenance problems, is the major drawback of tankless heaters.

Whole house tankless units will have a longer lag time than those located adjacent to the point of use. Lag time is the travel time for water to flow from the heater to a fixture. For long pipe runs to a distant usage point, consider installing a separate unit.

Condensing Tankless Water Heaters

Condensing tankless water heaters are also available. They are rated by the flow rate at which they produce hot water, generally about 4 -10 gallons per minute. The maximum flow rate decreases if water is being heated over a larger temperature range, but in Florida, influent water is relatively warm and reduced flow rates should not be a problem. Check the rated flow capacity at the expected temperature rise.

Expect the price to be higher than storage units, however their longer life and increased efficiency could make them cost effective over the long term. The life expectancy of tankless water heaters is about 20 years, and for storage type water heaters is generally 10 to 15 years. Periodic maintenance can extend tankless heaters life without significant loss of efficiency; however, outdoor installations in Florida coastal areas may see shorter than predicted lifetimes due to high humidity and salt deposition.

Heat-pump Water Heaters

Unlike the conventional water heater, which uses only resistance heating coils to heat water, the air-source heat pump water heater works by taking heat from the surrounding air and transferring it to the water in the tank. It uses the same principle as refrigerators but in reverse. Heat-pump water heaters have back-up heating coils to ensure ample hot water supply should the heat pump not produce the necessary amount of hot water.

Heat-pump water heaters are available in two styles: self-contained and add-on. A self-contained unit looks like a taller version of a standard electric water heater. This type completely replaces the conventional electric water heater. An add-on heat pump unit can be purchased separately and installed on top of, alongside, or even several feet away from your traditional electric water heater (Figure 10-7). The heat-pump unit (about the size of a window air conditioner) provides the energy, but uses your standard water heater for storage.

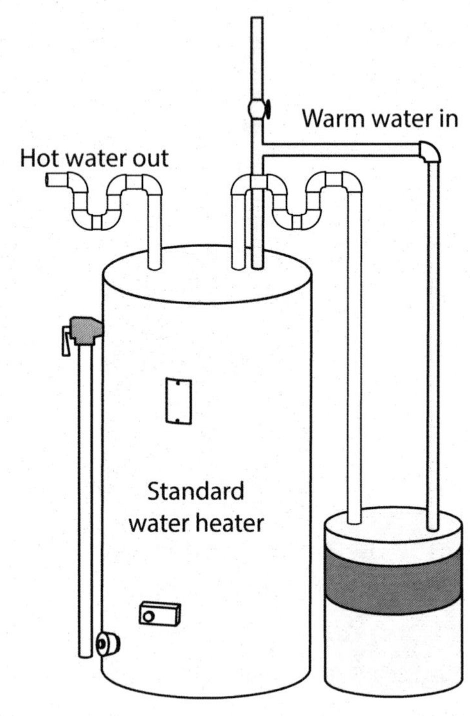

Figure 10-7 Heat pump water heater

They have essentially the same performance as electric resistance storage water heaters, except that efficiencies are typically 2 to 2.5 times higher. Energy Factor (EF) ranges from 1.8 to 2.5, compared to 0.7 to 0.97 for electric resistance systems.

Heat pump water heaters cool and dehumidify the air surrounding the evaporator coil. This can be an advantage where cooling is desirable, a disadvantage when cooling is undesirable. Some heat pump water heaters are designed to recover waste heat from whole house ventilation systems.

Although a heat-pump water heater may have a high initial cost, it can save up to 50% of your water heating bill in moderate climates. They require installation in locations that remain in the 40° to 90°F range year-round. To avoid damaging the equipment, never install a heat pump water heater in areas where the temperature drops below freezing. Since some operating heat-pump water heaters are about as loud as an air conditioner, either visit a site that has a unit in operation to determine if the noise level is tolerable or locate the unit where noise will not be a problem.

Heat Recovery Units

Heat recovery units, also called desuperheaters, most often extract waste heat from AC or heat pump systems. They contain a heat exchanger which captures some of the heat from hot refrigerant gases, raising water temperature in the exchanger by about 15°F. They also increase the efficiency of the AC compressor by reducing its load and potentially extending its functional life.

An average AC system is capable of producing about 5 gallons of hot water per hour for each rated ton. So a 3 ton unit could supply about 15 gallons for each hour it is operating. This works best when AC units run frequently during the time the hot water is needed.

AC systems with high SEER ratings operate more efficiently and have less waste heat to be recovered, reducing the benefit of a heat recovery system, although some systems are being designed to work with higher SEER units. Check the suitability of any system being considered.

Another type of heat recovery system recaptures heat from hot water going down drain pipes (Figure 10-8). These can be placed on sink, shower, dishwasher or clothes washer drains. They can operate simultaneously with use (for example, to pre-heat incoming water during showers). Or, a storage tank of fresh water can be preheated by drain water passing through coils at the bottom of the tank. The warmer water rises and is withdrawn for hot water supply.

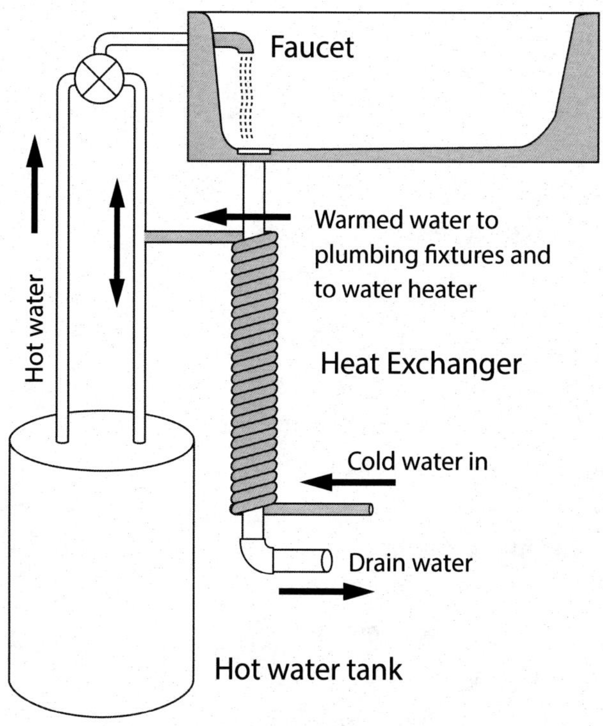

Figure 10-8 Drain heat recovery system
(U.S. Department of Energy, Energy Efficiency and Renewable Energy)

Preheating cold water lines close to usage location can reduce the volume of hot water withdrawn and can allow the water heater temperature setting to be lowered.

10: Water Heating

Solar Water Heaters

For homes that use a large amount of hot water and receive full sun year-round, solar water heaters may be economical. Most solar water heaters operate by preheating water for a standard water heater. Normally, gas or electric water heaters bring incoming cold water to a desired temperature of about 120°F. A solar water heater uses sunlight to preheat cold water and stores it, often at temperatures well above 120°F.

If the solar-heated water is hot enough, the standard water heater does not need to add more heat. If the water is cooler than needed, the standard water heater will operate as a backup to increase the temperature. Thus, the temperature or availability of hot water is never affected.

Of course, even when the solar-heated water is at temperatures below 120°F, the backup unit will use less energy than it would to heat incoming cold water.

A variety of solar water heaters are available commercially, most of which should last 15 years or longer. They are divided into *active* and *passive* types. Active water heaters include both direct and indirect circulation; passive include thermosyphon and integral collector/storage (ICS)—sometimes referred to as batch.

Active systems (Figure 10-9) imply that there is active pumping of fluid through the system. The fluid can be household water (direct circulation) or some other fluid that will not freeze (indirect circulation). For indirect circulation a heat exchanger transfers the heat to influent water.

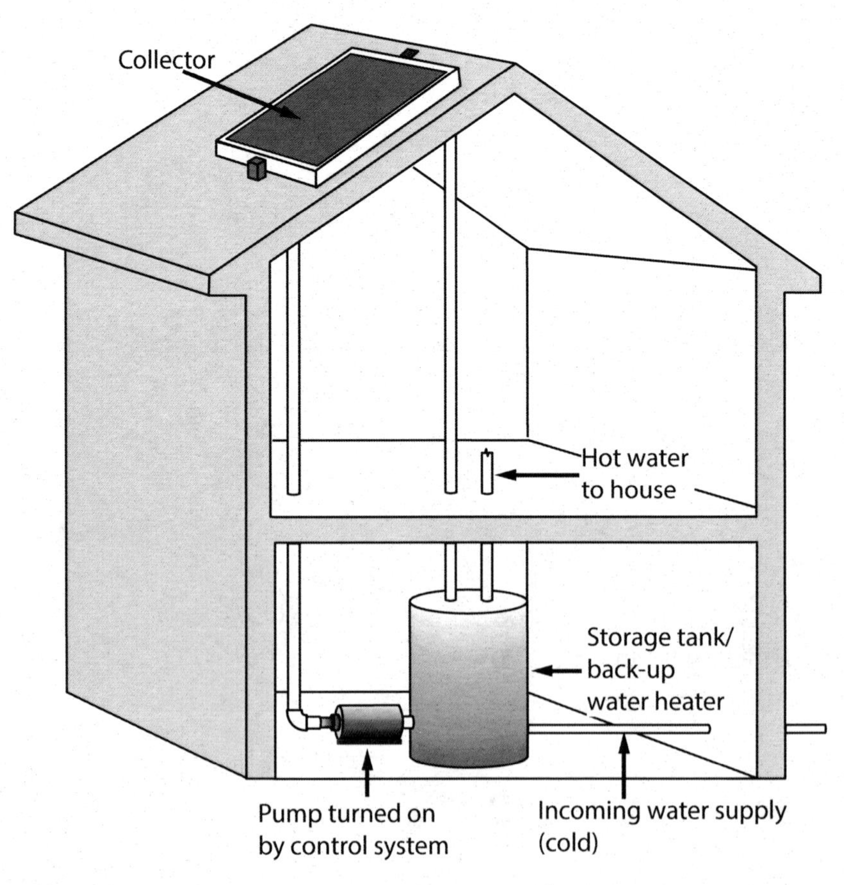

Figure 10-9 Active solar water heating system

Passive systems have several advantages. They have no pump, require no electricity, and tend to be less expensive than active systems and very durable (since there are no moving parts to wear out). They use the natural tendency of hot water to rise to circulate water between the collectors to the storage tank, rather than pumping it. However, they sacrifice some efficiency when compared to active solar systems.

10: WATER HEATING

In thermosyphon systems, incoming cold water enters directly into a storage tank mounted on the roof, above the collectors. into which it flows by gravity. Water in the collectors increases in temperature and becomes lighter than the cold water above it. As gravity pulls the cold water down, the heated water is pushed out through piping into a storage tank in the home. A thermally operated valve opens to drain water and prevent freezing in the collector. Since these systems require the storage tank to be located on the roof, make certain that the roof is strong enough to support the extra weight. Thermosyphon systems can cost more than passive integral collector-storage designs.

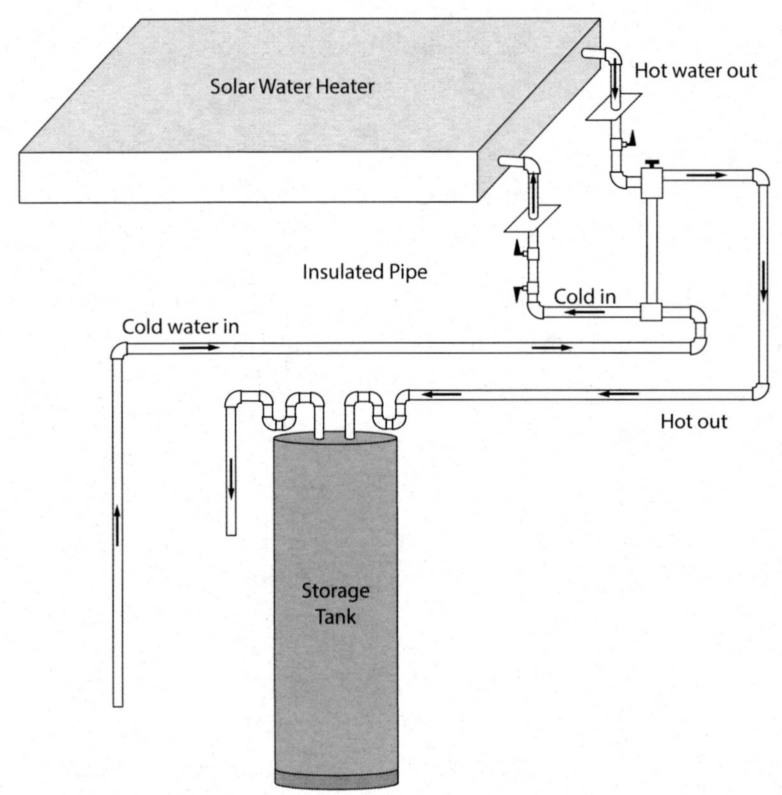

Figure 10-10 Batch solar water heating system

Integrated Collector systems (ICS) incorporate the collector and the storage tank into one unit (Figure 10-10). Cold water flows into the bottom of the collector box, where it is heated and rises to the top of the collector. Demand from a fixture or appliance inside the house draws heated water from the top of the collector down into a standard storage tank, where it is further heated if required.

The rooftop collectors have insulated sides, one or more clear covers and large tubes that absorb the solar energy and also act as the storage tank. Sunlight travels through the glazing of the batch unit and is absorbed by special coatings that absorb the light, but reduce heat loss from the tank. ICS units preheat inlet water before it is stored in a standard hot water tank. One unit tested in central Florida raised water temperature about 15-25 degrees in winter and 30-35 degrees in summer. It saved about 26% in energy usage compared to a standard electric system at the same location.

ICS solar heaters are manufactured and sold commercially.Water inside the tanks of an ICS water heater will only freeze on bitterly cold nights. However, the water in the pipes that connect the batch heater to the interior of the house can freeze at 32 F. A special *freeze prevention drip valve* should be used on an ICS solar water heater.

Some solar water heaters use a single, large storage tank. The water is normally heated by the sun, but gas or electric power can be used as a backup. Other systems have two separate storage tanks, one for solar and the second for their backup system.

For all shingle and tile roofs in Florida that generally have pitches greater than 3 in 12 (i.e. 14 degrees), collectors should be mounted parallel to the roof. Collectors mounted in this manner are more aesthetically pleasing. However, for flat or very low-sloping roofs, collectors should be tilted at an angle (to the horizontal) that is approximately equal in degrees to the local latitude. Florida latitudes range between 25 degrees (in the Florida Keys) to 31 degrees (northern border). Since the sun is lower on the horizon during the winter months, tilting the collector at an angle up to 15 degrees greater than latitude will increase winter performance, which is desirable in most cases.

Solar water heaters must be protected from freezing. Active and thermosiphon systems use nonfreezing fluids or automatic drain systems to prevent freezing.

The collectors for any type of solar water heater should be located as close as possible to the water heater tank to minimize the connecting piping. The glazing should face within 45 degrees of due south.

Collectors are usually located on the roof, but they can be attached to supports on the side of a house or on the ground. Because ICS solar heaters combine collectors, storage tanks, and water they are heavy. Adequate structural support must be provided when they are located on the roof.

Water inside the tanks of an ICS water heater will only freeze on bitterly cold nights. However, the water in the pipes that connect the batch heater to the inside can freeze at temperatures around 32°F. A special *freeze prevention drip valve* should be used on an ICS solar water heater.

Solar water heating can provide year round savings. Households that use a large amount of hot water and can adapt the time when hot water is used to match when it is available will benefit the most. Savings will be greatest if laundry, dishes, and bathing are done between noon and early evening—after the sun has heated the water stored in the tank.

Indirect Water Heaters and Tankless Coil Water Heaters

These devices integrate water heating with indoor space heating. These are a very inefficient choice in Florida because heating is required for such a short portion of the year.

Circulating Hot Water Systems

Circulating hot water systems send hot water to all fixtures, so it is instantly available when a demand is made. When the water in the piping cools below a set temperature, a pump activates and replaces the cooled water with fresh hot water. Some systems send the cooled water down the drain, wasting it. Other types recirculate the water back to be re-heated. **Beware, because both types waste energy when they pump hot water that is not being used.** The use of a programmable timer can reduce the energy losses if the system operates only during a set period of time when hot water is needed. None of these are recommended for energy efficiency. There is a version of the system that only pumps hot water out on demand and returns cooled water to the tank, which prevents wasted energy and wasted water while the user is waiting for it to get hot. If a household wastes a significant amount of water this way, eliminating it may also reduce their wastewater charges.

The Florida Building Code requires that that all piping for circulating hot water piping be insulated to at least R-3. The systems must also include an automatic or readily accessible manual switch to turn off the hot-water circulating pump when the system is not in use. Pipe insulation buried underground must be specified by the manufacturer for underground use.

Resources

Note: Web links were current at the time of publication, but can change over time.

Amann, J. T., Wilson, A., & Ackerly, K. (2012). *Consumer Guide to Home Energy Savings* (Tenth edition). Gabriola Island, BC: New Society Publishers.

American Council for an Energy-Efficient Economy, Smarter House. (n.d.). Water Heating. Retrieved August 7, 2015, from http://smarterhouse.org/home-systems-energy/water-heating#fuel%20size

Florida Solar Energy Center. (n.d.). Solar Thermal. Retrieved August 12, 2015, from http://www.fsec.ucf.edu/en/research/solarthermal/

Koomey, J. G., Dunham, C., & Lutz, J. D. (1994). *The Effect of Efficiency Standards on Water Use and Water Heating Energy Use in the U.S.: A Detailed End-use Treatment* (No. LBL-35475, UC-000) (p. 21). Berkeley, CA: Lawrence Berkeley National Laboratory. Retrieved from http://enduse.lbl.gov/Info/LBNL-35475.pdf

My Florida Home Energy. (n.d.). Retrieved December 20, 2017, from http://www.myfloridahomeenergy.com/

A useful resource with a wide array of information on energy and water efficiency, including The Energy Efficient Home series of fact sheets, available at http://www.myfloridahomeenergy.com/help/library

North Carolina Clean Energy Technology Center at N.C. State University. (n.d.). Database of State Incentives for Renewables & Efficiency®. Retrieved August 3, 2015, from http://www.dsireusa.org/

U.S. Department of Energy (DOE) Energy Efficiency & Renewable Energy, Building Technologies Office. (n.d.). Appliances and Commercial Equipment Standards: Residential Water Heaters. Retrieved August 12, 2015, from http://www1.eere.energy.gov/buildings/appliance_standards/product.aspx/productid/27

U.S. Department of Energy (DOE) Energy Saver:

(n.d.-a). Drain-Water Heat Recovery. Retrieved August 12, 2015, from http://energy.gov/energysaver/articles/drain-water-heat-recovery

(n.d.-b). Heat Pump Water Heaters. Retrieved August 12, 2015, from http://energy.gov/energysaver/articles/heat-pump-water-heaters

(n.d.-c). Solar Water Heaters. Retrieved August 12, 2015, from http://energy.gov/energysaver/articles/solar-water-heaters

(n.d.-d). Tankless Coil and Indirect Water Heaters. Retrieved August 12, 2015, from http://energy.gov/energysaver/articles/tankless-coil-and-indirect-water-heaters

(n.d.-e). Tankless or Demand-Type Water Heaters. Retrieved August 12, 2015, from http://energy.gov/energysaver/articles/tankless-or-demand-type-water-heaters

(n.d.-f). Water Heating. Retrieved August 7, 2015, from http://energy.gov/public-services/homes/water-heating

U.S. Department of Energy (DOE), Office of Energy Efficiency and Renewable Energy (EERE). (n.d.). Appliance and Equipment Standards Program. Retrieved August 12, 2015, from http://energy.gov/eere/buildings/appliance-and-equipment-standards-program

U.S. Department of Energy, Office of Energy Efficiency and Renewable Energy. *Energy Conservation Program for Consumer Products: Energy Conservation Standards for Water Heaters*, Pub. L. No. 10 CFR Part 430 (2004). Retrieved from http://www.gpo.gov/fdsys/pkg/FR-2001-01-17/pdf/01-1081.pdf

U.S. Environmental Protection Agency (EPA) ENERGY STAR:

(n.d.-a). Demand Hot Water Recirculating System. Retrieved August 12, 2015, from https://www.energystar.gov/index.cfm?c=water_heat.pr_demand_hot_water

(n.d.-b). How it Works — Heat Pump Water Heaters (HPWHs). Retrieved August 12, 2015, from https://www.energystar.gov/index.cfm?c=heat_pump.pr_how_it_works

(n.d.-c). Water Heaters. Retrieved August 12, 2015, from http://www.energystar.gov/index.cfm?c=water_heat.pr_water_heaters_landing

(n.d.-d). Water Heater, Solar for Consumers. Retrieved August 12, 2015, from http://www.energystar.gov/products/certified-products/detail/water-heater-solar

(n.d.-e). Water Heater, Whole Home Gas Tankless. Retrieved August 12, 2015, from https://www.energystar.gov/products/certified-products/detail/water-heater-whole-home-gas-tankless

(n.d.-f). Whole-Home Gas Tankless Water Heaters — How it Works. Retrieved August 12, 2015, from https://www.energystar.gov/index.cfm?c=gas_tankless.pr_how_it_works

Wilson, A. (2001, October). "Water Heating: A Look at the Options." *Environmental Building News*, 11(10). Retrieved from http://www2.buildinggreen.com/article/water-heating-look-options

11
Appliances and Lighting

Overview

In the past, the combination of cooling, hot water, and heating were usually the biggest portion of energy needs in Florida homes. However, with increasing stringency of building codes and appliance standards over time, the efficiency of Florida's homes, installed HVAC systems and water heaters has greatly improved while the share of residential demands for other energy services has grown. According to the Energy Information Administration, in 2009, appliances (including refrigerators, clothes washers, clothes dryers, ovens, dishwashers, etc.), electronics (computers, televisions, and other "plug loads"), and lighting accounted for about 50 percent of all energy consumed in residences within Florida.

Minimum standards of energy efficiency for many major appliances were established by the U.S. Congress in Part B of Title III of the Energy Policy and Conservation Act (EPCA), Public Law 94-163, as amended by the National Energy Conservation Policy Act, Public Law 95-619, by the National Appliance Energy Conservation Act (NAECA), Public Law 100-12, by the National Appliance Energy Conservation Amendments of 1988, Public Law 100-357, by the Energy Policy Act of 1992, Public Law 102-486, by the Energy Policy Act of 2005, Public Law 109-58 (EPACT 2005), and the Energy Independence and Security Act of 2007, Public Law 110-140 (EISA 2007).

All of these laws are codified in the United States Code, Title 42, Chapter 77, Subchapter III, Part A – Energy Conservation Program for Consumer Products Other Than Automobiles and Part A-1 – Certain industrial Equipment. The U.S. Department of Energy's (DOE) Appliances and Commercial Equipment Standards Program is responsible for developing the energy efficiency standards for residential appliances and commercial equipment. These standards, and the procedures with which appliances and equipment are tested to meet the standards, are published in the Code of Federal Regulations (10 CFR Chapter II, Parts 430 and 431). The DOE periodically issues new standards or rulemakings for specific appliances. These are published in the Federal Register (FR). Information on the rulemaking process and FR notices is available at the Department of Energy's Appliances and Commercial Equipment Standards Program Web site. While the Code of Federal Regulations (CFR) and the Federal Register are available on the Web, print versions can be purchased from the U.S. Government Printing Office (GPO), and libraries may have print copies in their reference section.

Energy Efficient Appliances

Appliance manufacturers must produce products that either meet the minimum level of energy efficiency, or that consume no more than the amount of energy that the standard allows. These rules do not affect the marketing of products manufactured before the standards went into effect. Any products already made and in stock can be sold. The new standards stimulate energy savings that benefit the consumer, and reduce fossil fuel consumption, thus reducing air pollution emissions.

Appliances that operate efficiently may cost more to buy, but the energy savings they provide may make them a good investment. When choosing appliances, it is important to consider their operating costs—how much energy they require to run—as well as the purchase price and the various features and conveniences they offer.

For example, according to the EPA, a refrigerator over 10 years old could be costing about $100 per year to run whereas a new ENERGY STAR® certified refrigerator would cost about $60 per year to operate. In addition to saving money on operating costs, energy efficient appliances may also give off less waste heat than standard models. Therefore, they may help keep rooms inside the building cooler during warm weather.

ENERGY STAR

In 1992 the U.S. Environmental Protection Agency (EPA) introduced ENERGY STAR—a voluntary labeling program designed to identify and promote energy-efficient products to reduce greenhouse gas emissions. Computers and monitors were the first labeled products followed by additional office equipment products and residential heating and cooling equipment in 1995. In 1996, EPA partnered with the U S Department of Energy for particular product categories and now the ENERGY STAR label is on major appliances, office equipment, lighting, home electronics, homes, and more. Note that in 2016, national market share for new site-built single-family ENERGY STAR certified homes reached 10%, but this varies widely by state. In Arizona, for example, ENERGY STAR certified homes constitute 53% of site-built single family homes; in Florida, it drops to 6%. Standards for commercial and industrial buildings are also now included.

An item receives the ENERGY STAR certified rating if it is more energy efficient than the minimum government standards, as determined by standard testing procedures. The amount by which an appliance must exceed the minimum standards is different for each product rated and depends on available technology. ENERGY STAR certified products are always among the most efficient available today, typically exceeding federal efficiency standards (Table 11-1). Look for the ENERGY STAR certification mark (Figure 11-1), which appears as a label on products, homes and buildings that meet or exceed ENERGY STAR performance guidelines.

Do all types of appliances and electronics have ENERGY STAR guidelines or specifications?

As of July 2017, the following major appliance types have ENERGY STAR guidelines to meet certification requirements:

Figure 11-1 Sample ENERGY STAR logo for certified products

- Air purifiers (cleaners)
- Clothes dryers
- Clothes washers
- Dehumidifiers
- Dishwashers
- Freezers
- Refrigerators

Categories of electronics that offer ENERGY STAR certified products include:

- Audio/video
- Digital media players
- Professional displays
- Set-top boxes
- Slates and tablets
- Telephones
- Televisions

Categories of office equipment that offer ENERGY STAR certified products include:

- Computers
- Data center storage
- Displays
- Enterprise servers
- Imaging equipment
- Large and small network equipment
- Monitors
- Uninterruptible power supplies
- Voice over Internet Protocol (VoIP) phones

The ENERGY STAR website includes a Product Finder that offers a real-time list of ENERGY STAR Certified products, www.energystar.gov/productfinder. Using the vertical and horizontal scroll bars, you can look for specific manufacturer brand names, model numbers, and compare a variety of product specifications and energy-related performance metrics. Individual columns can be filtered using the column specific "Menu" icons adjacent to their "Information" icons. The entire dataset can be searched using the "Magnifying Glass" icon.

Table 11-1 National Appliance Energy Conservation Act (NAECA) and ENERGY STAR® Comparisons for Selected Appliances (Residential)

Appliance	Notes	NAECA Current Criteria	ENERGY STAR Current Criteria
Clothes washers (top- and front-loading)	The energy efficiency of qualified clothes washers includes both the Integrated Modified Energy Factor (IMEF) and the Integrated Water Factor (IWF). ENERGY STAR-certified clothes washers use 25% less energy and about 45% less water than standard models.	EEffective January 1, 2018: ≥ 1.6 ft^3 capacity (top-loading) IMEF ≥ 1.57 IWF ≤ 6.5 ≥ 1.6 ft^3 capacity (front-loading) IMEF ≥ 1.84 IWF ≤ 4.7	Effective March 7, 2015: ≤ 2.5 ft^3 capacity IMEF ≥ 2.07 IWF ≤ 4.2 > 2.5 ft^3 (front-loading) IMEF ≥ 2.38 IWF ≤ 3.7 > 2.5 ft^3 (top-loading) IMEF ≥ 2.06 IWF ≤ 4.3

Appliance	Type and Capacity	Minimum Combined Energy Factor (CEF) (Manufactured or distributed into commerce on or after Januay 1, 2015)	ENERGY STAR Current Minimum CEF Criteria (Established May 19, 2014)*
Clothes dryers (Efficiency measured in terms of the combined energy factor (CEF), which is lbs/kWh.)	Vented electric, standard ≥ 4.4 ft^3 capacity	3.73	3.93
	Vented electric, compact (120 v) < 4.4 ft^3 capacity	3.61	3.80
	Vented electric, compact (240 v) < 4.4 ft^3 capacity	3.27	3.45
	Vented gas	3.30	3.48
	Ventless electric, compact (240 v) < 4.4 ft^3 capacity	2.55	2.68
	Ventless, electric combination washer/dryer	2.08	N/A

** Maximum cycle time of 80 minutes*

Appliance	Notes	NAECA Current Criteria	ENERGY STAR Current Criteria
Dishwashers (standard sized) ≥ 8 place settings + 6 serving pieces	ENERGY STAR dishwashers are about 14% more energy efficient and 23% more water efficient than minimum federal government standards.	Effective May 30, 2013: ≤ 307 kWh/year ≤ 5.0 gallons/cycle	Effective January 29, 2016: ≤ 270 kWh/year ≤ 3.5 gallons/cycle
Refrigerators	Federal minimum efficiency standards (NAECA) for refrigerators went into effect on July 1, 2001.	Effective September 15, 2014: Dependent upon product class (i.e., top mount freezer vs. bottom mount, etc.)	10% or more energy efficient than the federal government standards (for refrigerators and freezers)

Room air conditioners

The following minimum standards (NAECA) and ENERGY STAR criteria for room air conditioners went into effect as of June 1, 2014 and October 26, 2015, respectively. The ENERGY STAR criteria for room air conditioners is 15 percent above the NAECA criteria. Currently, units with louvered sides and those without louvered sides (commonly referred to as "built in" or "through-the-wall"), the casement product classes, and some reverse cycle units are considered for ENERGY STAR status. The numbers in the following table represent the Energy Efficiency Ratios (EER) that should be used to determine qualification for the ENERGY STAR label.

<table>
<tr><th rowspan="2">Capacity (Btu/Hr)</th><th colspan="2">With louvered sides</th><th colspan="2">Without louvered sides</th></tr>
<tr><th>Federal Standard EER</th><th>ENERGY STAR EER</th><th>Federal Standard EER</th><th>ENERGY STAR EER</th></tr>
<tr><td>< 6,000 and 6,000 – 7,999</td><td>≥11.0</td><td>≥ 12.1</td><td>≥ 10.0</td><td>≥ 11</td></tr>
<tr><td>8,000 – 10,999</td><td rowspan="2">≥ 10.9</td><td rowspan="2">≥ 12.0</td><td>≥ 9.6</td><td>≥ 10.6</td></tr>
<tr><td>11,000 – 13,999</td><td>≥ 9.5</td><td>≥ 10.5</td></tr>
<tr><td>14,000 – 19,999</td><td>≥ 10.7</td><td>≥ 11.8</td><td>≥ 9.3</td><td>≥ 10.2</td></tr>
<tr><td>20,000 –24,999</td><td>≥ 9.4</td><td>≥ 10.3-</td><td rowspan="2">≥ 9.4</td><td rowspan="2">≥ 10.3</td></tr>
<tr><td>≥ 25,000</td><td>≥ 9.0</td><td>≥ 9.9</td></tr>
</table>

	Federal Standard EER	ENERGY STAR EER
Casement only	≥ 9.5	≥ 10.5
Casement slider	≥ 10.4	≥ 11.4

<table>
<tr><th rowspan="2">Capacity (Btu/Hr)</th><th colspan="4">Reverse Cycle</th></tr>
<tr><th>Federal Standard EER</th><th>ENERGY STAR EER</th><th>Federal Standard EER</th><th>ENERGY STAR EER</th></tr>
<tr><td>< 14,000</td><td rowspan="3">≥ 9.8</td><td rowspan="2">N/A</td><td>≥ 9.3</td><td>≥ 10.2</td></tr>
<tr><td>≥ 14,000</td><td rowspan="3">≥8.7</td><td>≥ 9.6</td></tr>
<tr><td>< 20,000</td><td>≥ 10.8</td><td rowspan="2">N/A</td></tr>
<tr><td>≥ 20,000</td><td>≥ 9.3</td><td>≥ 10.2</td></tr>
</table>

Note that the EPA has a designated category, ENERGY STAR Most Efficient 2017, for some of the appliances/devices (Figure11-2). This label recognizes products that deliver cutting-edge energy efficiency along with the latest in technological innovation. The year included on the label designates that the device/appliance meets the criteria for the year indicated.

Figure 11-2 Sample ENERGY STAR Most Efficient logo for use on qualified products

EnergyGuide Label

The U.S. government established a mandatory compliance program in the 1970s requiring that certain types of new appliances bear a label to help consumers compare the energy efficiency among similar products (water efficiency was added later). In 1980, the Federal Trade Commission's Appliance Labeling Rule became effective.

As of July 2015, the FTC's Appliance Labeling Rule, now referred to as the "Energy Labeling Rule," requires the placement of the EnergyGuide label on any new product in the following product lines:

- Boilers
- Central air conditioners
- Clothes washers
- Dishwashers
- Furnaces
- Heat pumps
- Pool heaters
- Refrigerators
- Refrigerator-freezers and freezers
- Room air conditioners
- Televisions
- Water heaters (some types)

Note that the Rule also includes labeling for plumbing products and ceiling fans as well as labeling requirements for certain types of light bulbs.

EnergyGuide labels for appliances, in addition to other things, contain three key pieces of information. First, the labels show an estimate of how much electricity the appliance uses in a year based on typical use, as determined from standard DOE tests. Second, some labels include a "range of comparability" indicating the highest and lowest energy consumption of efficiencies for all similar models. Third, labels for most appliances must provide the estimated annual operating cost based on its electricity use and the national average cost of energy. Manufacturers arrive at this estimate by basing their calculations on figures published by the DOE.

The EnergyGuide label (Figures 11-3 and 11-4) is a bright yellow tag that the Federal Trade Commission (FTC) developed to help consumers more easily compare energy efficiency among similar products. Note that the label on the left (Figure 11-3) is an older version that has black numbers on a yellow background; the newer label on the right (Figure 11-4) features yellow numbers on a black background. This change reflects updated energy testing procedures and was required on new refrigerator-freezers in September 15, 2014, and new clothes washers manufactured beginning March 7, 2015.

When comparison shopping, make sure you are comparing models tested to the same energy-efficiency measures so your comparison is more accurate—compare models with labels having yellow numbers *only* to other models that have yellow number tags. Similarly, compare models labeled with black numbers only with other models tagged with black-on-yellow labels.

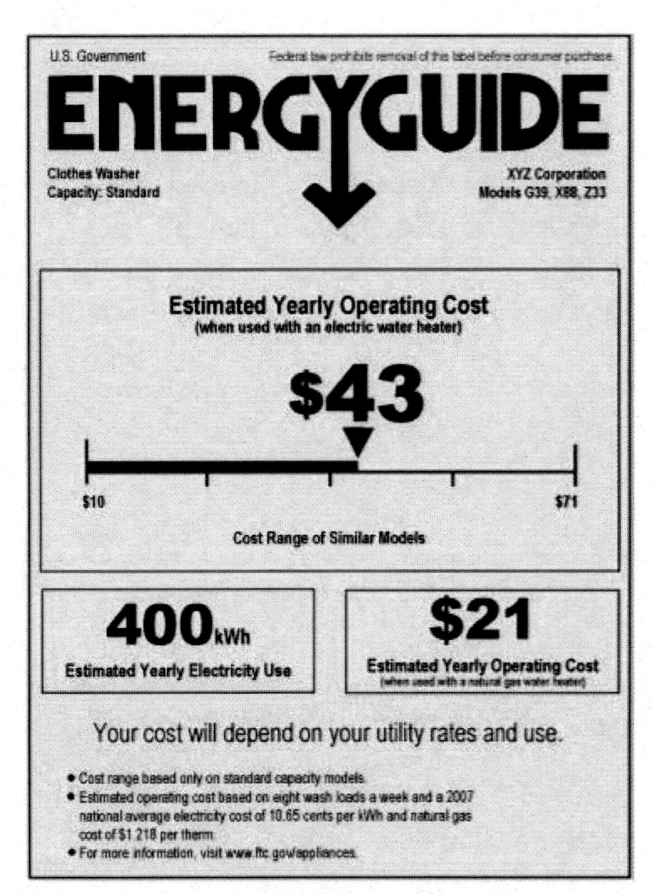

Figure 11-3 Sample of an older EnergyGuide label, for a cothes washer. This label has black numbers on a yellow background.

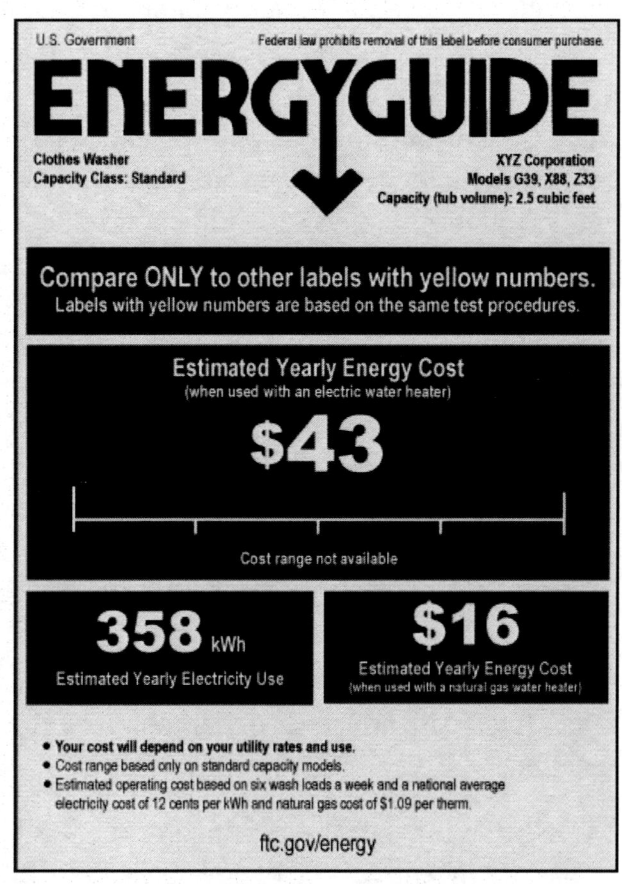

Figure 11-4 Sample of a more recent EnergyGuide label, also for a clothes washer. This label features yellow numbers on a black background.

Also, compare similar models with similar capacities. For example, comparing one top-loading clothes washer with another top-loader that handles the same sized batch of laundry will help in making a more informed decision than comparing models that lack such similarities. Remember, EnergyGuide labels won't indicate the best appliance to buy, but they do provide a lot of information to help in decision making. They also help consumers assess the trade-offs between the energy costs of their appliances and other expenditures.

The display of an EnergyGuide tag does not mean the appliance is ENERGY STAR certified. Some manufacturers are incorporating the voluntary ENERGY STAR logo on their certified appliance EnergyGuide labels, but if you don't see the ENERGY STAR logo on the bright yellow EnergyGuide tag, investigate further—the ENERGY STAR logo might be on the appliance itself, or perhaps the item hasn't earned the ENERGY STAR certification.

Keep in mind that similar to the federal miles-per-gallon ratings for automobiles, the actual amount of energy used and its cost will vary according to local prices and each family's size and life-style. Check consumer magazines for other information such as repair history.

What does the CEE Tier information refer to on some appliances?

The Consortium for Energy Efficiency (CEE) is a U.S. and Canadian consortium of gas and electric efficiency program administrators working together to accelerate the development and availability of energy efficient products and services for lasting public benefit. They have developed high efficiency specifications for some appliances and other equipment for residential, commercial and industrial use and report the efficiencies as being within certain tiers. Check out their website for specific details.

Appliance Shopping Checklist

All appliances

Use the EnergyGuide label and ENERGY STAR symbol, where applicable, to help select appliances. Find the savings in operating costs for the more efficient appliance. Divide the savings per year into the extra purchase price to get the payback period. Paybacks of less than five years are generally attractive. See Chapter 2, "Why Build Efficiently?" for information.

- Search for special offers and/or rebates by entering your zip code into the Database of State Incentives for Renewables and Efficiency (DSIRE) and at the Special Offers and Rebates from ENERGY STAR Partners website.
- Check consumer magazines, and appliance dealers, for performance and repair history.
- Many local utilities and energy-efficiency organizations support appliance recycling programs, refrigerators and freezers being common. Some programs offer cash to recycle old appliances while others offer utility-bill credits.

- When possible, purchase appliances from retailers that partner with EPA's Responsible Appliance Disposal (RAD) Program. By partnering with RAD, which is voluntary, retailers help protect the ozone layer and reduce greenhouse gas emissions.
- Look for Steel Recycling Institute's recyclers in your area. Note that the average 10 year or older refrigerator contains more than 120 pounds of recyclable steel.

Refrigerators and Freezers

The refrigerator is the only home appliance that operates continuously. In most households, the refrigerator is the single biggest energy consuming kitchen appliance. You can see the expected lifetime energy savings of an ENERGY STAR certified refrigerator or freezer by clicking on the *Refrigerator Retirement Savings Calculator* located on their website.

- An ENERGY STAR certified refrigerator, refrigerator-freezer or freezer must use at least 10 percent less energy than required by current federal standards.
- Currently, the most efficient models are in the most popular 16- to 20-cubic foot range. Generally, the larger the refrigerator, the greater the energy consumption.
- Leave space between the refrigerator and walls or cabinets so air can circulate around the condenser coils. Trapped heat increases energy consumption.
- Check door seals to be sure they are airtight.
- Refrigerator/freezers with top-mounted freezers use less energy than similarly-sized side-by-side models.
- Features such as automatic icemakers and through-the-door dispensers increase energy use and raise the purchase price.
- Units that are more square, rather than rectangular, also save energy, but may not be as convenient to use.
- Manual defrost units save considerably more than frost-free units, but create more work for the building occupant. Proper maintenance is necessary to realize the savings.
- Look for a model with automatic moisture control that has been designed to prevent moisture accumulation on the cabinet exterior. This is not the same as an "anti-sweat" heater that will consume more energy than models without this feature.
- Install refrigerators and freezers in a cool location—in particular, units should not receive direct sunlight or be positioned near heat-producing appliances. Do not install in un-conditioned space, such as the garage.
- The refrigerator should operate between 35°F and 38°F, and the freezer should be 0°F to 5°F. Readjust temperatures that are outside of this range.
- Recycle older refrigerators and freezers. Check out the ENERGY STAR Make a Cool Change: Recycle Your Old Fridge (or Freezer), campaign by the U.S. Department of Energy, for more information.

Dishwashers

- Water heating accounts for most of the energy use.
- Currently, an ENERGY STAR certified dishwasher is 12 percent more energy efficient and 15 percent more water efficient than standard models.
- Should have light, medium, and heavy wash cycle options.
- Be aware of capacity when looking at EnergyGuide labels. The standard model is 24-inches wide with a capacity greater than or equal to eight place settings plus six serving pieces. Compact models are about 18-inches wide and have a less than eight place setting plus six serving piece capacity. There are also drawer-style units that allow for washing a small load in one drawer or a full load in both.
- Should have an energy saving "air dry" or "no-heat dry" switch; these allow room air to be circulated through the dishwasher by fans rather than using an electric heating element to dry the dishes.
- For most uses, choose a unit that contains a supplemental or booster water heater that raises the water temperature to 140–145°F (most all units have this feature and all ENERGY STAR certified models do); then set the building's water heater to 120°F.

The Laundry Area

The laundry room can be a big consumer of energy. As reported by ENERGY STAR, clothes dryers use more energy per year (769 kWh) than any other standard household appliance, which includes the refrigerator (596 kWh), clothes washer (590 kWh), and dishwasher (206 kWh). Owning and using efficient washers and dryers save money not only by reducing energy and water costs, but also by extending the "life" of clothing.

The laundry room also generates a large amount of unwanted heat and humidity in summer, so it makes good sense to think about location options as well as the efficiency of the appliances. If possible, try to design buildings such that clothes washers and dryers are in un-conditioned space thereby decreasing the cooling and humidity load on the structure. If the laundry room is in conditioned space, install a closeable fresh-air intake vent to the dryer, along with a closeable dryer exhaust vent. These will effectively keep unconditioned outdoor air from entering the home and keep conditioned air from exiting the home when the dryer is not operating.

Clothes washers

- Water heating accounts for most of the energy use.
- ENERGY STAR qualified clothes washers use 25% less energy and about 45% less water than standard washing machines.

- To compare models, use the **Integrated Modified Energy Factor** (IMEF), which is a measurement of energy efficiency that considers the energy used during the washing process and while on standby, the energy used to heat the water, and the dryer energy required for the removal of the remaining moisture in the wash load. The higher the IMEF, the more efficient the clothes washer.
- To compare models for water efficiency, use the **Integrated Water Factor** (IWF), which is a measurement of water efficiency that considers the gallons of water used per cubic foot of capacity of the tub. The lower the IWF, the less water the washing machine uses and thus the more water efficient. Therefore, this is a measure of water efficiency that is independent of the capacity of the unit.
- According to the ENERGY STAR website, the typical U.S. household washes about 300 loads of laundry every year. Today, a typical clothes washer uses 23 gallons per load whereas full-sized washers earning the ENERGY STAR label use 13 gallons per load.
- Choose a machine that offers several wash and rinse cycles and several sizes of loads.
- ENERGY STAR certified clothes washers are either front-load or redesigned top-load designs. Front-loading models, similar to those found in laundromats, use a horizontal or tumble-axis basket to lift and drop clothing into the water. Redesigned top-loading models look like conventional models but many use sensors to monitor incoming water temperature and rinse clothes with repeated high-pressure spraying instead of soaking them in a full tub of water.
- Check tub size when comparing EnergyGuide labels.
- Faster spin speeds extract more water from clothing, thus reducing the energy required for drying.
- Special note: Be certain the homeowner knows the correct type of detergent to use and how to clean the inside of the machine. Front-loading clothes washers are often designed to use high-efficiency detergent as regular detergent may create too many suds affecting the washing ability and rinsing performance of the machine. The ENERGY STAR website warns that using the incorrect type of detergent could lead to odor and mechanical problems. They also recommend, especially on front-loading models, that the clothes-washer door be left open for an hour or two after use to allow moisture to evaporate, while making sure no children or pets have access to the machine. Lastly, some manufacturers recommend a periodic special rinsing or cleaning of the machine's tub to help reduce the risk of mold or mildew. In all situations, follow the product owner's manual.

Clothes dryers

- All clothes dryers manufactured on or after January 1, 2015, had to meet a higher federal minimum energy standard. See specific details as found in Table 11-1. For comparison, you can see the ENERGY STAR requirements in the Table as well.
- ENERGY STAR criteria for clothes dryers were first established on May 19, 2014. However, there is no EnergyGuide label for clothes dryers at this time.
- Clothes dryers are rated by a **Combined Energy Factor** (CEF), which is determined by using the test load size in pounds (8.45 for standard dryers and 3 for compact dryers) divided by the sum of the machine's electric energy use during operational and standby cycles measured in kilowatt hours (kWhs). The higher the CEF, the more energy efficient the clothes dryer. The rating for gas dryers is also provided in kilowatt hours, a measure of electricity usage, even though the primary source of fuel for these dryers is natural gas, which is more commonly measured in units called "therms."
- Dryer vents lacking dampers allow warm, humid outside air to enter the home in summer and cold air to enter in the winter even when the dryer is not in use. They could also allow unwanted "pests" to enter! (Note that heat-pump dryers do not require a vent.)
- Energy-saving switches and models that detect "dryness" and shut off automatically offer considerable energy savings.
- Be sure to vent the exhaust to the outside with as straight and short a duct as possible. Check to see that the duct has been correctly sized and installed according to the manufacturer's installation instructions (see **FBC, Mechanical, Section 504, Clothes Dryer Exhaust**; and **FBC, Residential, Section M1502, Clothes Dryer Exhaust**).
- Flexible plastic or vinyl duct should be avoided because they restrict air flow, can be crushed, may not withstand the high temperatures of the dryer or could cause a fire. Remember, dryer vents should also be equipped with a backdraft damper to prevent uncontrolled entry of outside air into the home.
- Consider common area laundry rooms in multi-housing properties (i.e., several smaller rooms vs. in-apartment laundry equipment).

Cooking

- Consumer behavior is the most important feature.
- Self-cleaning ovens are more efficient because they have more insulation. (The feature should be used sparingly.)

- Cooktops vary tremendously. Although energy efficiency is very important, issues such as type of cookware to use, ability to maintain low constant temperature, etc. are also important. Some electric choices include coil, solid disk, ceramic glass (radiant), halogen, and induction. With regard to energy, induction elements, which use electromagnetic energy to heat the pan, are the most efficient.
- All new gas ranges are required to have electronic ignition instead of the "always on" pilot light.
- Be careful with large kitchen exhaust fans. While they are important, oversized units can create considerable negative pressures in tight homes and may cause backdrafting of combustion appliances. Vent exhaust to the outside.
- Check to see that the exhaust/ventilation equipment has been correctly sized and installed (see **FBC, Mechanical, Section 505, Domestic Kitchen Exhaust Equipment, Section 506, Commercial Kitchen Hood Ventilation System Ducts and Exhaust Equipment**, and **Section 507, Commercial Kitchen Hoods**; and **FBC, Residential, Section M1503, Range Hoods**).
- There are no EnergyGuide labels or ENERGY STAR stickers on any cooktops, ovens or ranges specifically for home use but there is ENERGY STAR certified commercial food service equipment for businesses and operators. In fact, ENERGY STAR reports that outfitting a commercial kitchen with a suite of commercial food service equipment bearing the ENERGY STAR label could save operators about $4,800/year.

Lighting

Lamps

There are four primary families of lamps (a "lamp" is the term used in the lighting industry to describe what is commonly referred to as a light bulb):

- Incandescent (includes halogen)
- Fluorescent (includes CFL – compact fluorescent lighting)
- High intensity discharge (HID – high-pressure sodium, mercury vapor, and metal halide.)
- Solid-state (LED – light emitting diode and OLED - organic light emitting diode)

The U.S. Federal Trade Commission requires manufacturers of incandescent, compact fluorescent and LED lamps to use the "Lighting Facts Label" on consumer packaging to help customers choose the most efficient product for their needs. The FTC label required on the front of packaging will provide information on brightness (lumen output) and estimated annual energy cost. The FTC Lighting Facts label required on the side or back of packaging will provide information about: brightness, energy cost, the bulbs life expectancy, light appearance (warm or cool), wattage and whether the bulb contains mercury. The "Lighting Facts Label" will encourage purchasers to think about lumens—the measure of brightness—and the amount of light they want for a particular use, rather than the traditionally-used watts, a measure of energy consumed. An example of the label is shown in Figure 11-5.

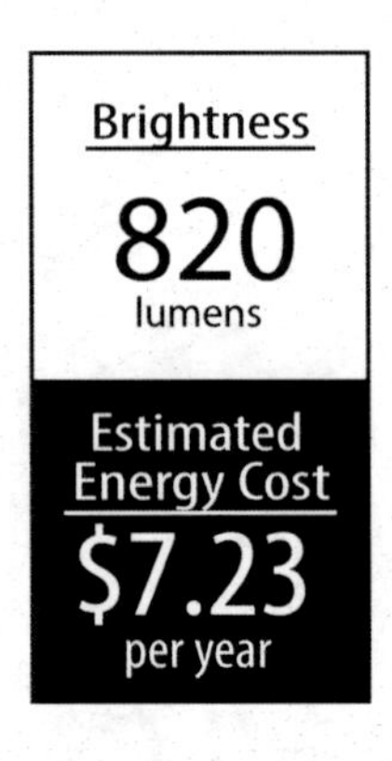

Lighting Facts Per Bulb

Brightness — **820 lumens**

Estimated Yearly Energy Cost — **$7.23**
Based on 3 hrs/day, 11¢/kWh
Cost depends on rates and use

Life
Based on 3 hrs/day — **1.4 years**

Light Appearance
Warm — Cool
2700 K

Energy Used — **60 watts**

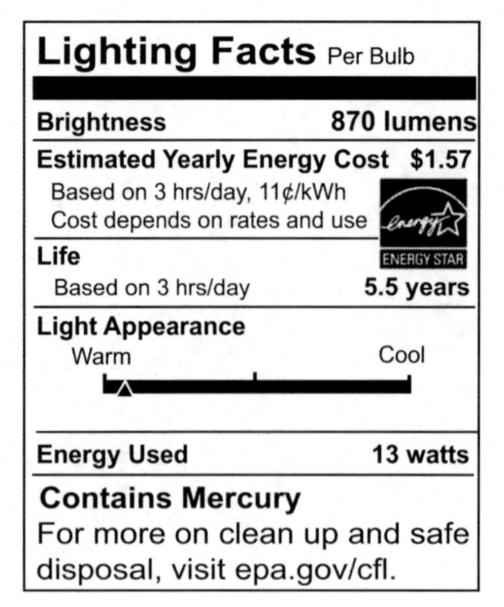

Figure 11-5 The Lighting Facts label: front of label, back of label, and special back label for mercury-containing lamps

There are other labels that may be found on a lamp's packaging including ENERGY STAR®, LED Lighting Facts Label®, The DesignLights™ Consortium label and Safety Certification labels (Table 11-2).

Energy efficient lighting design not only reduces the lighting portion of utility costs, but can also affect HVAC loads and costs by reducing the heating load created by inefficient lamps. Many advances have been made, and are continuing to be made in the lighting industry with regard to energy efficiency.

Remember that:

- Despite technological advances, efficiency and lamp life are dependent on the quality of the product
- Each lamp type has its place in the correct application.

Several light source options are compared in Table 11-3.

Table 11-2 Labeling and Certification for Lighting Products

Organization	Voluntary / Mandatory	Purpose	Information	Lighting Products Covered	Logo or Certification Mark	Contacts
ENERGY STAR® U.S. Environmental Protection Agency	Voluntary	To help consumers save money and protect the environment through energy efficient products and practices.	Labels products in more than 60 categories that meet strict energy efficiency guidelines set by the EPA and DOE. Requires third-party testing and certification.	Residential • CFL light bulbs • LED light bulbs • Residential luminaires • Decorative light strings • Ceiling fans with light kits	Logo use guidelines in place	www.energystar.gov For questions regarding ENERGY STAR lighting products, please contact: luminaires@energystar gov lamps@energystar.gov
LED Lighting Facts® Label U.S. Department of Energy (DOE)	Voluntary	Showcases LED products for general illumination from manufacturers who commit to testing products and reporting performance results according to industry standards.	For lighting buyers, designers, and energy efficiency programs. The label provides information essential to evaluating products and identifying the best options.	Commercial and residential Solid State Lighting products		www.lightingfacts.com For questions regarding the DOE LED Lighting Facts Label please contact: info@lightingfacts.com
Lighting Facts Label Federal Trade Commission (FTC)	Mandatory	The label on the package provides consumers with information about the bulb's brightness, estimated yearly energy cost, life, light appearance, and wattage.	Announced on June 18, 2010, the label became effective on product packagaing produced as of January 1, 2012.	Required for all general purpose medium-based (E26) lamps.		www.ftc.gov
The DesignLights™ (DLC) Consortium A collaboration of utility companies and regional energy efficiency organizations	Voluntary	DLC is committed to raising awareness of the benefits of efficient lighting in commercial buildings.	Products not generally covered under the ENERGY STAR program. Requires third party testing and fee to list products.	Commercial Products not generally covered under the ENERGY STAR program such as: • Roadway • Parking Garage	Logo use guidelines in place	www.designlights.org
Safety Certifications	Voluntary (but required by most customers)	Products with one or more of these safety designations have been tested to industry standards to ensure safe operation in normal environments.	**UL**: Underwriters Laboratories Inc. **CSA**: Canadian Standards Association **ETL**: Originally a mark of ETL Testing Laboratories, now a mark of Intertek Testing Services.	Available for all products		www.ul.com www.csagroup.org www.intertek.com

Table 11-3 Comparison of Various Light Sources

Lamp Type	Efficacy (lm/W)	CRI	Color Temp. (K)	Life (hours)	Typical Applications	Advantages	Disadvantages
Incandescent: Standard (A-19 screw base)	10-17	98-100 (excellent)	2700–2800 (warm)	750–2,500	• Residential buildings • Little–used socketed fixtures in commercial buildings • Indoors/outdoors	No mercury	• Very low efficacy; significant waste heat per unit of light • Short life contributes to solid waste • Efficacy lower with long–life bulbs
Energy-Saving Incandescent or Halogen	12–22	98–100 (excellent)	2900–3200 (warm to neutral)	1,000–4,000	• Residential downlight • Commercial applications where very high light quality and precise focus required • Indoors/outdoors	• Ability to focus light may allow energy savings in some applications; no mercury	• Low efficacy
Linear Fluorescent Lamp (LFLs) Also known as tube or straight tube fluorescents	65–110 (for simplicity, the efficacy range excludes the losses associated with the ballast)	50–95 (fair to good)	2700–6500 (warm to cool)	7,000–24,000	• Commercial buildings—mostly indoor • Fixtures with multiple T–5s for high–bay applications • Indoors/outdoors	• High efficacy (with better products) • Excellent controllability with dimming electronic ballasts offers additional savings	• Contains mercury • Potential for leaching from landfills or airborne emissions from incineration
Compact Fluorescent (CFLs)	33–70	77–88 (good)	2700–6500 (warm to cool)	10,000	• Replacement for incandescent lamps in homes • Use in hard–wired wall sconces, other fixtures in commercial buildings • Indoors/outdoors	• Much higher efficacy than incandescent lamps they replace	• Contains mercury • Potential for leaching from landfills or airborne emissions from incineration
Mercury Vapor (Being phased out)	25–60	15–50 (poor to fair)	3200–7000 (warm to cool)	16,000–24,000	• Outdoor where poor light quality is adequate • Outdoors	• Higher efficacy than incandescent • Slightly longer life than metal halide	• Lower efficacy than other HID light sources • Poor lumen maintenance • Contains mercury
Metal Halide	70–115	70 (fair)	3700 (cold)	5,000–20,000	• Outdoor applications • Indoor high–bay applications where dimming and rapid restrike are not required • Indoors/outdoors	• High efficacy with very good light quality	• Contains mercury

Table 11-3 Comparison of Various Light Sources *continued*

Lamp Type	Efficacy (lm/W)	CRI	Color Temp. (K)	Life (hours)	Typical Applications	Advantages	Disadvantages
High-Pressure Sodium	50–140	25 (poor)	2100 (warm)	16,000–24,000	• Outdoor applications where poor light quality is acceptable • Outdoors	• High efficacy	• Poor light quality reduces effective lumens, lowers "pupil efficacy" • Contains mercury
Low-Pressure Sodium	100–200	0 (poor)	1800 (warm)	18,000	• Outdoor applications where extremely poor color rendition is considered acceptable (e.g., near observatories) • Outdoors	• Very high efficacy	• Very poor light quality reduces effective lumens dramatically
Induction	65–90	80 (good)	3000 (warm)	60,000–100,000	• Locations where maintenance is difficult • Security lighting; • High-bay applications • Indoors/outdoors	• Exceptionally long life • Not affected by on/off cycling • Good lumen maintenance • Tolerates wide range of temperatures	• High cost • Contains mercury
Sulfur	90–100	80–85 (good)	6000 (cold)	Lamp: > 50,000 Magnetron: 15,000	• Outdoor or large indoor spaces (hangers, large warehouses) where very high light levels are needed • Often used with light pipes • Indoors/outdoors	• Good light quality • No mercury	• Uncommon • Fixtures complex • Operates at very high temperature
Cool White LEDs Warm White LEDs	60–94 27–88	70–90 (fair to good) 70–92 (fair to good)	5000 (cold) 3300 (neutral)	25,000–50,000 5,000–50,000	• 6" downlights, accent lighting, way-finding, decorative, case lighting • Recessed or directional lighting • Indoors/outdoors	• Potential for high efficacy • Ability to focus precisely • Long life • No mercury	• Technology continually developing • May have higher initial cost • Significant lumen depreciation in some products—choose quality

Compiled and adapted from: Energy.gov, Lighting Basics, http://energy.gov/eere/energybasics/articles/lighting-basics ; Environmental Building News, "Electric Lighting: Focus on Lamp Technologies," https://www2.buildinggreen.com/article/electric-lighting-focus-lamp-technologies/sidebar/2 ; and Lighting Retrofit and Relighting: A Guide to GreenLighting Solutions, by James R. Benya & Donna J Leban.

EFFICACY

A 100-watt lamp does not necessarily provide more illumination than a 75-watt lamp. Lumens measure *light output*. Watts measure *energy use*. Efficacy is measured as lumens per watt (LPW). LPW is similar to "miles per gallon" for an automobile and is a measure of how effective the light source is in converting the watts (or energy) into light. Therefore, efficacy is how much light (lumen) is put *out* compared to how much energy (watt) is put *in*.

- A traditional incandescent lamp with 1600 lumens output and 100 watts energy usage is 16 LPW.
- A halogen incandescent lamp with 1600 lumens output and 77 watts energy usage is 20 LPW.
- A compact fluorescent lamp (CFL) with 1600 lumens output and 23 watts energy usage is 70 LPW.
- A light emitting diode (LED) lamp with 1600 lumens output and 20 watts energy usage is 80 LPW.

A federal law passed in 1995 (EPAct) requires all lamp manufacturers to list the lumens and watts on the label.

LAMP LIFE

Lamp life is also important. All electric light sources experience a decrease in the amount of light they emit over time, a process known as lumen depreciation. "Research has shown that the majority of occupants in a space will accept light level reductions of up to 30% with little notice, particularly if the reduction is gradual. Therefore a level of 70% of initial light level could be considered an appropriate threshold of useful life for general lighting." (From Building Technologies Fact Sheet - *Lifetime of White LEDs.*) With old-style incandescent lamps, lumen depreciation is primarily due to the deposition of tungsten from the filament on the glass bulb, thereby darkening it and reducing the transmission of light. Fluorescent lamp life is dependent on many variables, such as lamp type, ballast type, operating environment (not too cold or too hot) and how often they are switched on and off. Light emitting diodes (LEDs) have the longest lamp life but lumen depreciation is effected by heat. LEDs do not emit heat as infrared radiation, so they must use a heat sink or ventilation to divert the heat. Lamp life is also dependent on the quality of the lamp product. See Table 11-3 for a comparison of lamp life and other characteristics among fluorescent, incandescent, and other types of light sources.

HOW LIGHT AFFECTS WHAT WE "SEE"

COLOR RENDERING

The color rendering index (CRI), measured on a scale from 0 to 100, describes how a light source makes the color of an object appear to human eyes and how well subtle variations in color shades are revealed. That is, CRI tells us how well a light source allows us to discriminate between colors when compared to a standard, or known, light source. The higher the number, the more natural and vibrant an object or color will appear. Standard incandescent bulbs have a CRI of 95+. A lower CRI means that some colors will look unnatural under the artificial light. The old standard cool white fluorescent lamp has a poor CRI of 62, which is why people complained in the past that fluorescents gave false colors. Today, many kinds of fluorescent lamps have a much higher CRI—80 and above—so check the packaging. It's not surprising that CRI is important for merchandising.

continued...

continued...

COLOR TEMPERATURE

The correlated color temperature (CCT) measures the appearance of the light itself, or how warm or cool a light seems. Color temperature is another common complaint about fluorescent lighting. Often, older fluorescent lighting was considered harsh compared to incandescent lighting. People perceive some light—with more reds/oranges/yellows—as "warm," and other light—with more blue—as "cool." A low CCT—below 3100 K—is a warm white light. For instance, old-style standard incandescent bulbs have a CCT of 2800. Many of today's fluorescents and LEDs have a CCT of 3000 and provide the same warm, white light that an incandescent bulb produces.

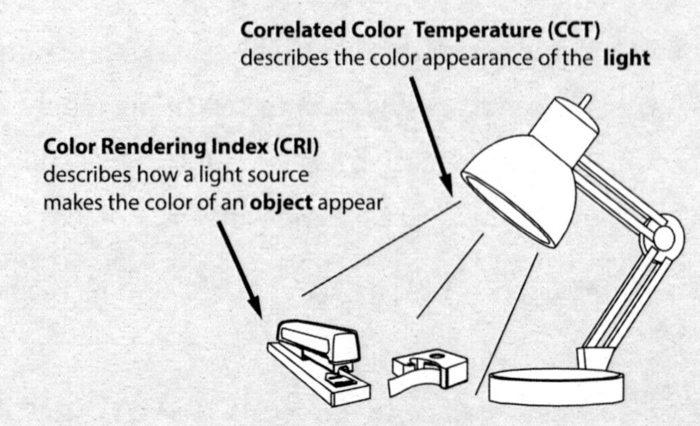

Figure 11-6 Correlated Color Temperature (CCT) and Color Rendering Index (CRI)

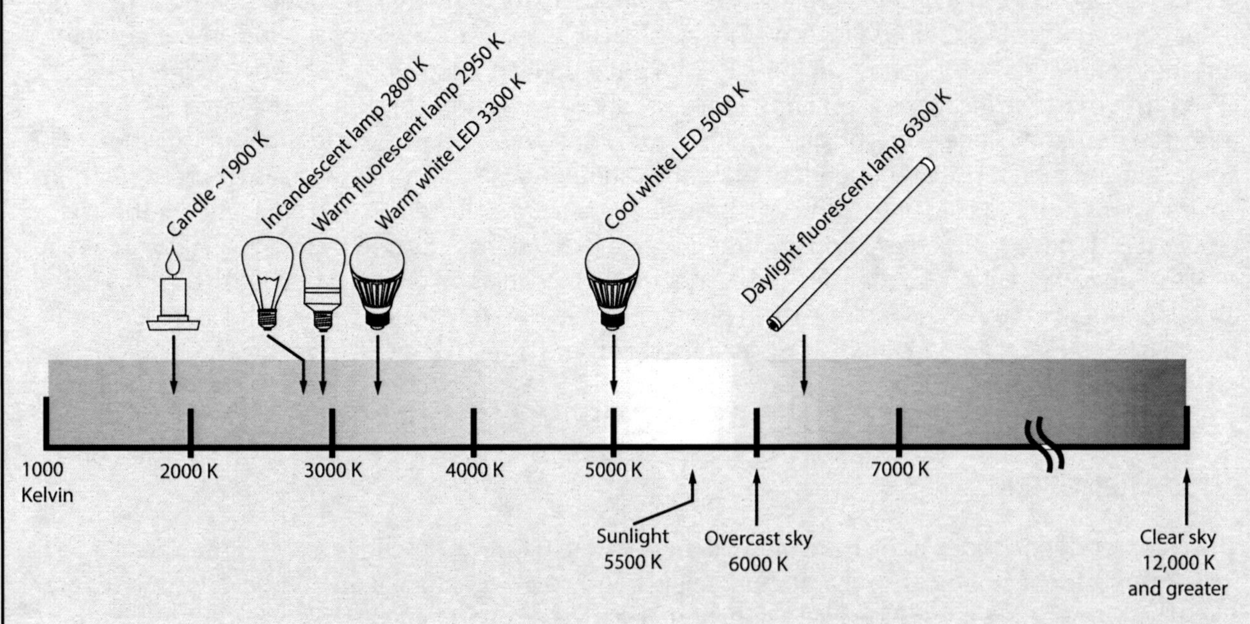

Figure 11-7 Some typical color temperatures

Table 11-3 compares the efficacy, CRI, and CCT, etc. of a few of the many kinds of lamps that are available. As this table shows, there are many alternatives. Choose the combination of efficacy, lamp life and cost, color rendering, and color correlated temperature that best fits each application.

Remember CRI and CCT. In places where color does not matter, such as the garage, the very inexpensive traditional fluorescent lamps will usually be fine. In a kitchen or bathroom, on the other hand, color is important.

THE ENERGY INDEPENDENCE AND SECURITY ACT (EISA)

Public Law 110-140 (2007), known as The Energy Independence and Security Act of 2007, was signed on December 19, 2007 by President Bush. It aims to move the United States toward greater energy independence and security. Section 321 of the Act establishes increased minimum energy efficiency standards for general service lamps. Phased in from 2012 through 2014, it requires roughly 25 percent greater efficiency for light bulbs (lamps). Lighting products affected include:

- General service incandescent and halogen lamps
- Incandescent and halogen reflector lamps
- General service linear fluorescent lamps
- Ballasts

The U.S. Department of Energy (DOE) is responsible for setting efficacy standards for a number of residential and commercial products. These minimum performance standards apply to new equipment manufactured for sale in the United States. See the Resources section at the end of the chapter for more information on the specific products and links to the EISA text.

1. Incandescent Lamps

As of 2010 standard incandescent lamps were the most common lighting source for most homes; however, they are a very inefficient lighting source. They convert only 10 percent of the electricity to light; the rest, i.e. 90 percent, ends up as heat. They also have the shortest life and are relatively inefficient when compared with other lighting types. The inefficient transformation of electrical energy to visible light can increase preventable expenses on a utility bill. A filament inside the lamp is heated by electricity until it glows and gives off light. Reduced wattage and long-life bulbs may or may not save energy or money and the savings are usually not great. The Energy Independence and Security Act (EISA) effectively bans the U.S. production and import of common incandescent lamps. Retailers are allowed to sell remaining inventory. Most big box stores now sell energy efficient halogen incandescents, compact fluorescents and Light Emitting Diodes (LEDs), and only sell specialty old-fashioned incandescent lamps (appliance, heavy duty, colored and three-way) which are EISA exempt.

For incandescent and halogen lamps, the format of a lamp's designation is to: first list its wattage; followed by the shape and size; followed by a slash; then list additional information. For example, 100A19/CL refers to a 19/8" diameter, 100-wattage arbitrary shape lamp that is clear; and 60G40/W refers to a 5" (40/8") diameter, 60-wattage globe shape lamp with a white envelope (Livingston, 2014).

Halogen Lamps

Tungsten halogen lamps are basically incandescent lamps, but with several significant design enhancements. The "tungsten" part of the name comes from the filament material—tungsten. The "halogen" part refers to the type of gas that fills the bulb, which gives the lamp its special properties. They operate at higher pressure and temperature than standard incandescent lamps, producing a whiter light; and because the halogen gas helps preserve the tungsten filament, they generally have a longer life. Halogen lamps are dimmable using the correct dimmer switch.

Not only do halogen lamps burn longer, they have a greater light output. As incandescent lamps burn, they have a tendency to slowly deposit tungsten onto the glass bulb, gradually darkening the bulb and eventually leading to failure. On the other hand, the halogen gas filling these lamps combines with the tungsten vapor and redeposits some of it on the filament, significantly reducing the deposition of evaporated tungsten onto the bulb—meaning a brighter light and a longer life for the lamp.

Halogen lamps are slightly more energy efficient than standard incandescent lamps meeting the federal minimum efficiency standard, but are not as energy efficient as some other types such as fluorescents and LEDs. Because halogens burn at such high temperatures—between 700° and 1,100° F—they can present a safety concern in some luminaires (fixtures). According to the Consumer Product Safety Commission (CPSC), halogen lamps could start a fire when they contact with flammable materials (https://www.cpsc.gov/content/cpsc-issues-warning-on-tubular-halogen-bulbs).

Reflector Lamps

Standard incandescent lamps emit light in all directions. The "R" lamp is a popular variation that has a built-in reflector to collect and focus the light from the filament, giving a directional characteristic to the beam. It uses a cone-shaped glass enclosure with a reflective coating (usually polished aluminum) to redirect light in a controlled manner. These lamps are mainly available in two beam spreads—spot (which concentrates the light) or flood (which spreads it).

Reflector lamps use a one-piece design of thin glass and are not for outdoor use.

A variation of the reflector lamp is the PAR (Parabolic Aluminized Reflector) lamp. They are often seen in retail store track lighting, in foyers or hallways in recessed downlights ("cans"). These lamps are formed from two pieces (lens and reflector) of pressed, thick, heat-resistant glass and are therefore able to be used outdoors in the rain or snow.

Note that there are reflector and PAR lamps on the market for compact fluorescents as well as LEDs. CFL PAR and reflector lamps may or may not be dimmable and do not focus light in a more concentrated way as halogens do. The CFLs do offer energy savings and longer lifetimes than halogen lamps. Most LEDs on the other hand are dimmable using the correct dimmer switch and because they are inherently directional, they are a good fit for reflector lamps. LEDs also offer energy savings and long life.

2. Fluorescent Lamps

Fluorescent lamps produce light by passing an electric arc through a mixture of an inert gas, such as argon, and a very small amount of mercury. The mercury radiates ultraviolet energy that is transformed to visible light by the phosphor coating on the bulb. All fluorescent lamps require assistance to operate in the form of a separate component called a ballast, discussed below. Fluorescent lamps outperform incandescents and halogens in energy efficiency. They also provide longer lamp life, ranging from 9000 to 20,000 hours. The combination of lower energy use, lower heating load, and longer lamp life means that fluorescents have a lower total operating cost than incandescent and halogen lamps. The old complaints about fluorescents—they give false colors, they are too harsh (cool), they buzz, flicker and hum, and they take too long to start—are no longer true when you choose quality products.

The label on a fluorescent lamp provides the information about its specifications but it may be in an unfamiliar format. Typically lamps are identified by a code, for example: FxxTy where F is for fluorescent, xx indicates either the power in watts or length in inches, T indicates that the shape of the bulb is tubular, and y is the diameter in eighths of an inch. Other letters or numbers may be included that indicate color (CW for cool white, WW for warm white) and mode of starting. Some manufacturers use numerical codes for color rendering and color temperature. Refer to the manufacturer's specification on label codes.

BALLASTS

Fluorescent and HID lamps produce light using an arc. Electric arcs will use large amounts of current if uncontrolled. The ballast limits the current flow to the proper amount. Ballasts also can enable the lamp to start quickly. Except for screwbase compact fluorescents, fluorescent luminaires come with either electromagnetic or electronic ballasts. Electronic ballasts are more energy efficient than electromagnetic ballasts. For many applications, you can select among three start-up modes—preheat, rapid start, and instant start. Preheat starts cause the lamp to flicker a few times before it starts, a feature that may be unacceptable to many clients for most indoor residential uses. In rapid start mode, there is a delay of 1 to 2 seconds before the lamp starts, but there is no flickering. In instant start mode, there is no delay or flickering. Most electronic ballasts are instant start.

Ballasts come in two forms, one a combined lamp and ballast, the other a separate ballast and lamp. Units with separate ballast and lamps are preferable because ballasts last much longer than lamps. In the combined type the consumer must replace—and pay for—the ballast every time the lamp burns out. Just remember when purchasing a separate compatible ballast for a luminaire that lamp type, number of lamps, and the line voltage should be known to achieve the rated light output and rated lamp life.

Fluorescent lamps come in a wide array of shapes, sizes, and lengths (Figure 11-8). They also come in several types, with varying characteristics:

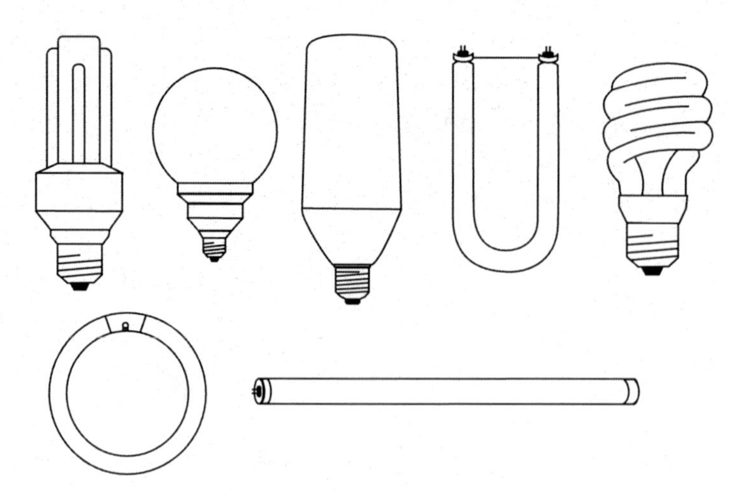

Figure 11-8 Examples of fluorescent lamps

- Dual Phosphor: not as long lived; lower CRI. Lumen depreciation occurs more quickly.
- Tri Phosphor: longer life, better CRI
- Rare Earth: longer life, better CRI, but may be perceived as distorting colors in individuals who are colorblind.

Linear fluorescents are the standard long, straight tube lamp. They are usually the least expensive to buy. However, they *are* long and straight, which does not lend itself to some situations—a pull-down lamp over a table, for example.

Compact fluorescent lamps (CFLs) are small-diameter fluorescent lamps folded for compactness. U-shaped fluorescents are simply a long tube bent into a "U." Circline fluorescents are a long tube bent into a circle. Some CFLs feature a round adaptor, allowing them to screw into common electrical sockets.

However, typical 60-100 watt incandescent lamps are no more than 5.3-inches long, while standard CFLs are longer than 6 inches. Therefore, sub-CFLs have been developed. No more than 4.5 inches long, sub-CFLs fit into most incandescent fixtures. Because of their energy efficiency, brightness, and low heat output, CFLs are often good replacements for halogen lamps in torchieres. Compact fluorescents come in many sizes and shapes and are designed to be used in applications that normally take an incandescent or halogen lamp. However, always read the label and follow manufacturer's directions.

Screwbase fluorescents with diffusers and reflectors are also available, and may be worth their extra cost in some applications. Lamps with diffusers, like tubular fluorescents, shed light in all directions. However, the diffuser does just what it says—diffuses the light. This effect can be important—in a pull-down light over a table, for example. Reflectors aim the light. Again, this can be important in some situations, such as accent lighting on a piece of artwork. In general, select these more expensive alternatives only when other fluorescents cannot produce the desired result.

Note: Again, do not use compact fluorescent bulbs in dimmable fixtures, recessed cans, enclosed, or outdoor fixtures unless the packaging specifically says the product can be used for that situation. Always follow the manufacturer's directions.

3. High Intensity Discharge (HID) Lamps

Consider high intensity discharge (HID) lamps if you need a large amount of light projected over a long distance. HID types include high-pressure sodium, mercury vapor, and metal halide. HID lamps use an electric arc to produce intense light and can save 75% to 90% compared to incandescent lamps. They have ballasts and usually take a few minutes to emit their rated light output because the ballast needs time to establish the arc. The most energy efficient types are metal halide and high pressure sodium. For the most part, HIDs provide ample illumination, but have very poor color rendition. However, this is rarely an issue for the applications where HIDs are used, such as parking lots, gyms, or playing fields.

4. Solid-State Lighting

Light emitting diodes (LEDs) and light-emitting polymers (organic LEDs or OLEDs) are included in this category. An LED is a very small electrical device that produces light through the semi-conducting properties of its metal alloys. An OLED is composed of small molecules or polymers shaped in the form of a film or transparent layer that emits light.

Breakthroughs in LED and OLED technology are catalyzing development of energy-efficient solid-state lighting (SSL). Once used for indicator lights on electronic equipment, SSL technology is now available for most lighting applications, including automobile brake lights, traffic signals, exit signs, flashlights, televisions, cell phones, computer screens, desk and task lighting, architectural directed-area lighting, under-cabinet lighting, outdoor lighting, and most residential lighting scenarios.

LEDs

LED lamps are particularly suitable for directional general and accent lighting, due to their ability to focus output exactly where the light is desired. Whereas, up to 50% of the light from incandescent and CFL lamps can be lost in directional fixtures, such as recessed ceiling lighting, or under-cabinet lighting, LEDs are easily dimmable and programmable. Occupancy sensors can be integrated into individual fixtures, as well as daylight sensors to dim lights on bright sunny days and increase light output on cloudy days.

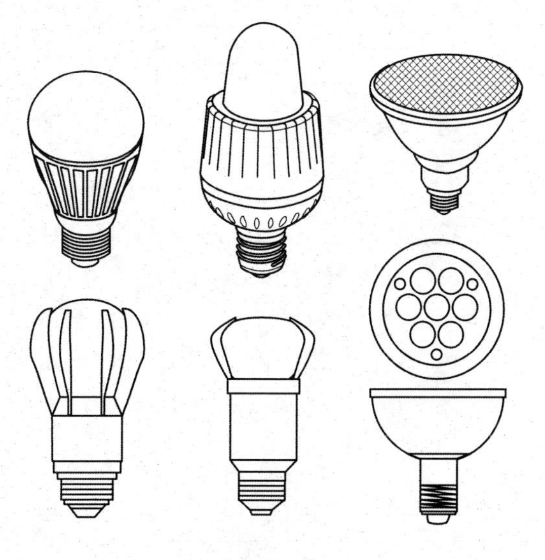

Figure 11-9 Examples of LED lamps

LEDs are sensitive to temperature and electrical conditions, so LED fixtures must be designed carefully to take this into account. Luminaires that are manufactured with the LED's diodes already installed as an integral part of the device are a practical approach because thermal management is addressed and incorporated into the luminaire design thus usually ensuring a long life.

The U.S. Department of Energy estimates that the energy efficiency of LEDs already rivals the most efficient white light sources. A U.S. Department of Energy funded study found that LEDs have less negative environmental impacts than incandescent bulbs and a slight edge over CFLs. The study evaluated use, manufacturing, transport, and disposal of LED, CFL, and incandescent lamps throughout each product's lifecycle.

The ENERGY STAR Program has established LED standards for some applications and will continue to develop new standards. Lamps meeting the standards are being used by several large 'big box' retail chains and others. As these standards are implemented, they offer huge potential savings for public and commercial buildings.

Purchase prices for LEDs are currently higher than fluorescents; however, their long life can result in a lower total cost per year. Evaluate each situation to determine the best option. As LED prices continually decline, they are likely to take over a larger share of the general lighting market.

OLEDs

While Light Emitting Diodes (LEDs) and all other lamps starting in history with candlelight are point light sources, Organic Light Emitting Diodes (OLEDs) are a surface light source. OLEDs like LEDs are a semi-conductor device. LEDs are a single diode point of light whereas OLEDs combine thin layers of red, green, and blue film to emit a surface white light. OLEDs can also be laid onto a variety of flexible and rigid surfaces, such as glass, metal or plastic, to create a light source that radiates along a broader area rather than from a single point.

The OLED panel is very thin, produces little heat and can be integrated into many design mediums such as furniture, clothing, and wall coverings as a few examples. OLEDs are the next innovation for home lighting. Imagine an entire wall or ceiling that glows or windows that simulate daylight once it's dark outside. Cost for OLEDs is still high in comparison to LEDs but is expected to come down as the technology advances. A big box building supply store is already selling OLED luminaries that are fairly affordable.

Some Other Forms of Lighting

Low-Pressure Sodium

Low-pressure sodium lamps are similar to high-pressure sodium lamps, but with the sodium gas at a lower temperature. This characteristic limits the light output. Although they have the highest efficacy of any light source, their CRI of zero severely limits visibility. You've probably encountered low-pressure sodium lamps in street lighting situations with the familiar yellow monochromatic color.

Common uses for low-pressure sodium lamps include street and highway lighting, security perimeter lighting around buildings, parking lots and garages, automobile and train tunnels (driver fatigue in tunnels is reduced because the lights eliminate the stroboscopic effect of driving past high brightness lights at speed). A very important use is in areas that are sensitive to light pollution because the monochromatic light does not interfere with astronomical observation. See the "Light Pollution" section of this chapter for more information.

Low-pressure sodium lamps are relatively inexpensive. Of increasing importance, to many people, is the fact that they contain zero mercury, and can be easily disposed of as non-toxic waste without incurring extra expense at its end of life. Care should be taken when disposing as there may be a small risk of fire if a broken bulb comes into contact with water. Read the product data safety sheet for disposal instructions. Finally, another advantage is that being a low pressure discharge lamp, its striking voltage is not sensitive to temperature as is the case for other discharge lamps. Therefore, in the case of a momentary power supply interruption, the lamp will restrike as soon as the power is restored and no cooling down time is required.

Some important disadvantages include poor color rendering and a rated life shorter than other types of discharge lamps. Typically, installations have to be re-lamped every two or three years whereas the expensive maintenance schedule can be extended to three or four years with high pressure sodium. This reduced maintenance cost can offset the energy savings of low pressure sodium for certain installations.

Induction

Pioneered by Nikola Tesla, induction lighting has been around for over 100 years. Induction lamps are designed to light immediately with no warm-up period or flicker. They operate in much the same manner as fluorescent lamps with an electromagnetic field inside the glass housing causing the mercury atoms to emit ultraviolet (UV) light. When the UV light passes through the phosphor coating of the lamp, it is converted into visible light.

Sometimes referred to as the "electronic light bulb," it is emerging as one of the newest rediscovered technologies in lighting. Induction lamps operate without an electrode, so they are not affected by on-off cycling. Currently, the long life and durability of induction lamps make them promising alternatives for hard-to-reach areas and outdoor applications since output does not seem to be significantly influenced by ambient temperatures. However, currently their high cost serves as an impediment to market penetration. Models of lamps using high frequency ballasts can cause radio frequency interference (think cell phone) although most newer lamps with low frequency ballasts do not interfere and carry certification that they comply with FCC regulations. There is still work to be done to improve the induction lamp.

Sulfur (Sulfur Plasma)

The sulfur lamp is a golfball-size sphere of quartz that emits very bright and efficient light when energized by the same type of microwave energy that cooks food in microwave ovens (electromagnetic waves). Light is generated from sulfur, argon, and other gases. Laboratory results indicate that sulfur lamps can potentially produce well over 100 lumens per watt without using mercury.

Developed in the 1990s, they did not catch on commercially, but now are back on the market. They produce intense high quality white light, and very little heat. They are suitable for outdoors or large interior spaces. Efficacy is about 90-100 Lumens/watt; CRI is about 80 and their design life is about 60,000 hours. High initial cost is their major drawback.

Light pipes are used with sulfur lamps to reflect and distribute the light over a large area. For example, only three lamps with 89-foot light pipes were substituted for 94 HID lamps in the National Air and Space Museum.

The lamp's brightness and longevity make it ideal for large facilities such as aircraft hangars, large warehouses and similar buildings. While the strength of light from a conventional source can weaken over time, the output from a sulfur lamp remains constant. It produces sunlight-quality illumination, with very low heat, virtually no ultraviolet emissions, and a full color spectrum. (Color rendition is particularly important in aircraft maintenance, since parts, like wires, are color-coded.) Another use is as a horticultural grow light. Life expectancy is three times longer than conventional light sources since it does not have filaments or electrodes to burn out.

LIGHTING NEEDS

There is great opportunity for originality and ingenuity in residential lighting design. A home combines more functions and needs than most other buildings, yet energy efficient lighting can be achieved at minimal cost. Although the needs of each home must be considered individually, certain conservation measures are applicable to all home designs, including:

- Energy-efficient fixtures and lamps (see the ENERGY STAR website for lists of ENERGY STAR residential lighting products and fixtures) for areas of high continuous lighting use, such as the kitchen, sitting areas, and outside the home for safety and security.

Figure 11-10 Some examples of common residential lighting (Figure by Mahshad Kazem-Sadeh)

- Task lighting for specific activities such as working at a desk, on a kitchen counter, or in a workshop.
- Accent lighting for areas that need more light. This enables the overall level of lighting in a room to be reduced.
- Timers, motion sensors, and light-sensitive switches for exterior lighting.
- Using sunlight as the light source (daylighting) in areas normally occupied during the day. However, choose windows that have low solar heat gain.
- Solid-state dimmers and multilevel switches that allow variable lighting levels.
- Occupancy sensors that monitor human presence, turning light fixtures on the moment someone enters a room and off automatically after they leave.

Common lighting needs are (see Figure 11-10):

- *Ambient lighting* provides illumination for performing routine daily activities—such as watching television—and safety—in a hallway, for example. Low light levels are usually fine for ambient lighting.
- *Indirect lighting*, or *uplighting*, where light is directed to the ceiling and upper part of the walls, is a specific technique commonly used to provide ambient lighting or lighting where glare can be a problem. It provides a very evenly distributed light and helps prevent reflected glare from glossy surfaces, such as televisions.
- *Activity lighting* provides illumination for a specific task, such as reading or woodworking. The light needed will vary by task and by how long the task is performed. Sewing for a few minutes does not require as much light as sewing for an extended period of time, for example.
- *Accent lighting* focuses light on an object or an area in the room to emphasize it. Accent light is used to draw attention to artwork or interesting architectural details.
- *Wall washing* is similar to accent lighting because it, too, draws attention. It can also

provide ambient lighting because the light is reflected off the wall. Wall washing can make rooms feel larger. *Wall grazing* is used to accentuate textures on wall surfaces. For wall washing, luminaires are typically placed at least 12" away from wall plane; for wall grazing, luminaires are placed very close.

- *Special purpose lighting* refers to such uses as medicine cabinets and under-cabinet lighting in the kitchen.

Treatments and Luminaires

Once the amount of lumens that are needed is determined, choose a luminaire that uses the fewest watts. In designing a lighting plan, consult with knowledgeable professionals about optimum lighting levels, CCT, CRI, and different types of fixtures and lamps. In general, commonly installed luminaries include ceiling-mounted, suspended, recessed, cove, soffit, valence, wall-mounted, and furniture or cabinet-integrated.

Of these, recessed lighting can be used for almost every lighting need. However, these fixtures are often installed into the ceiling and are notorious for air leakage, which can greatly increase heating and cooling costs. Some argue that it is virtually impossible to install a recessed fixture so that it does not leak air. Make sure that the housing—or "can"—does not have perforations in it (i.e., that it is airtight). These perforations are a direct pathway for losing heated or cooled air. Look for a housing that meets the energy code air infiltration standards. (See the FBC, Energy Conservation, Section R402.4.4 Recessed lighting.)

Further, insulation cannot come into contact with many recessed fixtures. Look for housing types rated IC, "IC" meaning that it can be covered by insulation. (Refer to Chapter 5, Insulation Materials and Techniques, for more information.) Additionally, install a luminaire that uses fluorescent lamps or LEDs to further reduce energy use. Long-lasting LEDs or compact fluorescent lamps are convenient for difficult-to-reach recessed cans in cathedral ceilings; be sure they are labeled for such use.

Light Pollution

Light pollution is largely the result of bad lighting design, which allows artificial light to shine outward and upward into the sky—where it's not wanted—instead of focusing the light downward, where it's needed. Of all the pollutions we face, light pollution is perhaps the most easily remedied. Simple changes in lighting design and installation yield immediate changes in the amount of light spilled into the atmosphere and, often, immediate energy savings. Quality lighting can reduce electricity consumption and thereby reduce carbon dioxide emissions. Many cities have enacted light pollution laws to protect the quality of life in their community. Some lighting codes, in use in other cities, now include the concept of lighting zones to distinguish different types of lighting areas. For example, parks, wildlife refuges, coastal areas and areas near astronomical observatories require much lower levels of lighting than in city centers.

Shielded or "Full Cut Off" light fixtures that are properly aimed downward are a must. This will solve most light trespass and glare problems, and it will significantly reduce sky glow. Glare always reduces visibility, which reduces traffic and personal safety. For driver, cyclist and pedestrian safely, lights should be shielded and aimed so that they are not directly visible from the roads, alleys, and pathways, and so that they do not obscure traffic signs or cause confusion.

Lamp type used for night lighting is also important in terms of spectrum. Research shows that blue rich white lighting (such as LED, fluorescent and metal halide) has a dramatically greater impact – lumen-for-lumen – on sky glow than the currently most common high-pressure sodium (HPS) and especially low-pressure sodium (LPS). Outdoor lighting with a high blue light content increases glare which in turn creates potential road safety problems for motorists and pedestrians. Exposure to blue rich white light at night also affects human and wildlife health. LEDs, if used for their efficiency and long life, should be warm-white lamps (< 3000 K), in a shielded fixture and can be fitted with timers, dimmers and/or motion detectors to save even more energy. See the "International Dark-Sky Association" website listed in resources for more information.

Commercial Lighting Energy Efficiency

FBC, Energy Conservation, Section C405 Electrical Power and Lighting Systems incorporates many changes for commercial lighting controls and daylighting, tandem wiring requirements, interior and exterior power requirements and maximum power densities, including many exceptions to the rules. Read the code carefully before beginning commercial lighting design.

Lighting that is recessed in the building thermal envelope must be sealed with a gasket or caulk between the housing and interior wall or ceiling to reduce air leakage between conditioned and unconditioned areas. See **FBC, Energy Conservation, Section R402.4.5 Recessed lighting**.

For more information specific to commercial lighting, see the following tables: **FBC, Energy Conservation, Table C405.4.2(2) Interior Lighting Power Allowances: Space-By-Space Method**; **Table C405.5.1(1) Exterior Lighting Zones**; and **Table C405.5.1(2) Individual Lighting Power Allowances for Building Exteriors.**

RESOURCES

Note: Web links were current at the time of publication, but can change over time.

Appliances

Appliance Standard Awareness Project (ASAP). Products: Residential. Retrieved June 5, 2015, from http://www.appliance-standards.org/products#residential

Earth911. (n.d.). Recycling Center Search & Recycling Guides. Retrieved from http://www.earth911.com/recycling-center-search-guides/

Federal Trade Commission (FTC):

EnergyGuide Labels: Transitional label example. Retrieved May 28, 2015, from https://www.ftc.gov/tips-advice/business-center/guidance/energyguide-labels-templates-manufacturers

Consumer Information. Shopping for Home Appliances? Use the EnergyGuide Label. Retrieved May 28, 2015, from http://www.consumer.ftc.gov/articles/0072-shopping-home-appliances-use-energyguide-label

Energy Conservation Program for Consumer Products, 10 C.F.R § 430. Retrieved from http://www.ecfr.gov/cgi-bin/text-idx?SID=553b868651c7d1b7a5f1cd6dbd90f50f&mc=true&node=pt10.3.430&rgn=div5

Energy Efficiency Program for Certain Commerical and Industrial Equipment, 10 C.F.R. § Part 431. Retrieved from http://www.ecfr.gov/cgi-bin/text-idx?SID=553b868651c7d1b7a5f1cd6dbd90f50f&mc=true&node=pt10.3.431&rgn=div5

Energy and Water Use Labeling for Consumer Products Under the Energy Policy and Conservation Act ("Energy Labeling Rule"), 16 C.F.R. § 305 . Retrieved from http://www.ecfr.gov/cgi-bin/searchECFR?ob=r&idno=16&q1=Part+305&r=&SID=553b868651c7d1b7a5f1cd6dbd90f50f&mc=true

International Association of Certified Home Inspectors (InterNACHI). InterNACHI's Estimated Life Expectancy Chart for Florida Homes. Retrieved June 5, 2015, from http://www.nachi.org/florida-life-expectancy.htm

Laitner, J. A. "Skip," McDonell, M. T., & Ehrhardt-Martinez, K. (2014). *The Energy Efficiency and Productivity Benefits of Smart Appliances and ICT-Enabled Networks: An Initial Assessment* (Research Report No. F1402). American Council for an Energy-Efficient Economy. Retrieved from http://aceee.org/research-report/f1402

Mauer, J., deLaski, A., Nadel, S., Fryer, A., & Young, R. (2013). *Better Appliances: An Analysis of Performance, Features, And Price as Efficiency Has Improved* (Report No. A132). American Council for an Energy-Efficient Economy (ACEEE) and Appliance Standards Awareness Project. Retrieved from http://www.appliance-standards.org/documents/reports/better-appliances-analysis-performance-features-and-price-efficiency-has-improved

My Florida Home Energy. (n.d.). Retrieved December 20, 2017 from http://www.myfloridahomeenergy.com/

A useful resource with a wide array of information on energy and water efficiency, including The Energy Efficient Home series of fact sheets, available at http://www.myfloridahomeenergy.com/help/library

Ruppert, K. C., Cantrell, R. A., & Lee, H.-J. (2015). Dishwashers (Energy Education Library Fact Sheet). Florida Energy Systems Consortium (FESC): University of Florida. Retrieved from http://www.myfloridahomeenergy.com/help/library/appliances-electronics/dishwashers

Ruppert, K. C., Cantrell, R. A., & Lee, H.-J. (2015). Refrigerators and Freezers (Energy Education Library Fact Sheet). Retrieved from http://www.myfloridahomeenergy.com/help/library/appliances-electronics/refrigerators-freezers

Ruppert, K. C., Porter, W. A., Cantrell, R. A., & Lee, H.-J. (2015). Clothes Washers and Dryers (Energy Education Library Fact Sheet). Florida Energy Systems Consortium (FESC): University of Florida. Retrieved from http://www.myfloridahomeenergy.com/help/library/appliances-electronics/laundry

Searcy, J. K., Jerome, L., & Taylor, N. W. (2015a). Energy Efficient Appliances (Energy Education Library Fact Sheet). Florida Energy Systems Consortium (FESC): University of Florida. Retrieved from http://www.myfloridahomeenergy.com/help/library/appliances-electronics/appliances

Searcy, J. K., Jerome, L., & Taylor, N. W. (2015b). Home Energy: Basics (Energy Education Library Fact Sheet). Florida Energy Systems Consortium (FESC): University of Florida. Retrieved from http://www.myfloridahomeenergy.com/help/library/choices/home-energy-basics

North Carolina Clean Energy Technology Center at N.C. State University. Database of State Incentives for Renewables & Efficiency®. Retrieved August 3, 2015, from http://www.dsireusa.org/

"Range Hoods Affect Indoor Air Quality." (1997, October). *Energy Source Builder*, 53. Retrieved from http://oikos.com/esb/53/rangehood.html

Steel Recycling Institute (SRI). (n.d.). Recycling Resources: Recycling Locator. Retrieved July 20, 2015, from http://www.recycle-steel.org/Recycling%20Resources/Locator.aspx

U.S. Code, Title 42 Chapter 77, Subchapter III - Improving Energy Efficiency:

Part A: Energy Conservation Program for Consumer Products Other Than Automobiles. Retrieved November 22, 2017, from https://www.gpo.gov/fdsys/pkg/USCODE-2009-title42/pdf/USCODE-2009-title42-chap77-subchapIII-partA.pdf

Part A-1: Certain Industrial Equipment. Retrieved from http://www.gpo.gov/fdsys/pkg/USCODE-2009-title42/pdf/USCODE-2009-title42-chap77-subchapIII-partA-1.pdf

U.S. Department of Energy (DOE) Energy Saver:

(2014a). Energy Saver: Tips on Saving Money & Energy at Home. U.S. Department of Energy, Office of Energy Efficiency and Renewable Energy. Retrieved from http://energy.gov/sites/prod/files/2014/09/f18/61628_BK_EERE-EnergySavers_w150.pdf

(2014b). Resolve to Save Energy This Year. Retrieved November 22, 2017, from https://energy.gov/articles/resolve-save-energy-year

(2014c). Tips: Home Office and Electronics. Retrieved June 4, 2015, from http://energy.gov/energysaver/articles/tips-home-office-and-electronics

(2015). Estimating Appliance and Home Electronic Energy Use. Retrieved May 28, 2015, from http://energy.gov/energysaver/articles/estimating-appliance-and-home-electronic-energy-use

(n.d.-a). Energy Saver Guide: Tips on Saving Money and Energy at Home. Retrieved August 12, 2015, from http://energy.gov/energysaver/articles/energy-saver-guide-tips-saving-money-and-energy-home

Online version of the *Energy Saver Guide:*

(n.d.-b). *Energy Saver Guide: Tips on Saving Money & Energy at Home*. Retrieved August 14, 2015, from http://energy.gov/energysaver/downloads/energy-saver-guide

Downloadable versions of the *Energy Saver Guide* in PDF format and for a variety of mobile devices.

U.S. Department of Energy (DOE), Office of Energy Efficiency and Renewable Energy (EERE).

(n.d.-a). Appliance and Equipment Standards Program. Retrieved July 20, 2015, from http://energy.gov/eere/buildings/appliance-and-equipment-standards-program

(n.d.-b). Energy- and Water-Efficient Products. Retrieved August 17, 2015, from http://energy.gov/eere/femp/energy-and-water-efficient-products

(n.d.-c). Statutory Rules and Authorities. Retrieved July 15, 2015, from http://energy.gov/eere/buildings/statutory-rules-and-authorities

U.S. Environmental Protection Agency (EPA). (n.d.). Responsible Appliance Disposal (RAD) Partners and Affiliates [Collections and Lists]. Retrieved June 3, 2015, from http://www2.epa.gov/rad/rad-partners-and-affiliates

U.S. Environmental Protection Agency (EPA) ENERGY STAR. (n.d.). Retrieved August 18, 2015, from http://www.energystar.gov/

(n.d.-a). 2014 ENERGY STAR Certified Homes Market Share map (graphic). Retrieved August 18, 2015, from http://public.tableau.com/views/ENERGYSTARMarketShare2014/ENERGYSTARMarketShare2014?:embed=y&:showVizHome=no&:host_url=https://public.tableau.com/&:tabs=no&:toolbar=yes&:animate_transition=yes&:display_static_image=yes&:display_spinner=yes&:display_overlay=yes&:display_count=yes&:showVizHome=no&:showTabs=y&:loadOrderID=0

(n.d.-b). Clothes Washers for Consumers. Retrieved July 20, 2015, from http://www.energystar.gov/products/certified-products/detail/clothes-washers

(n.d.-c). Commercial Food Service Equipment for Businesses and Operators. Retrieved July 20, 2015, from http://www.energystar.gov/products/certified-products/detail/commercial-food-service-equipment

(n.d.-d). ENERGY STAR Most Efficient 2015 Criteria. Retrieved June 3, 2015, from http://www.energystar.gov/index.cfm?c=partners.most_efficient_criteria

(n.d.-e). ENERGY STAR Qualified Products. Retrieved July 20, 2015, from http://www.energystar.gov/products/certified-products

(n.d.-f). Energy Use of Standard Household Appliances (pie chart). Retrieved July 20, 2015, from https://www.energystar.gov/sites/default/files/assets/images/ES_Appliance_Energy_Use_pie_graph_8%2022%2014-large.jpg

(n.d.-g). Make a Cool Change: Recycle Your Old Fridge (or Freezer) Campaign : ENERGY STAR. Retrieved July 20, 2015, from https://www.energystar.gov/index.cfm?c=promotions.fridge_recycle_cmpgn

(n.d.-h). Product Specifications & Partner Commitments Search. Retrieved August 18, 2015, from https://www.energystar.gov/products/spec

(n.d.-i). Refrigerator Retirement Savings Calculator. Retrieved July 20, 2015, from http://www.energystar.gov/index.cfm?fuseaction=refrig.calculator

(n.d.-k). Save Energy at Home. Retrieved May 26, 2015, from http://www.energystar.gov/index.cfm?c=products.pr_save_energy_at_home

U.S. Environmental Protection Agency (EPA) ENERGY STAR, & U.S. Department of Energy (DOE):

(n.d.-a). Special Offers and Rebates from ENERGY STAR Partners. Retrieved May 29, 2015, from http://www.energystar.gov/rebate-finder

(n.d.-b). Ventilation Fans for Consumers. Retrieved July 15, 2015, from http://www.energystar.gov/products/certified-products/detail/fans-ventilating

Some Specific Organizations/Groups Supporting Energy and/or Water Efficiency

Alliance to Save Energy. Retrieved August 18, 2015, from http://www.ase.org/

A coalition that promotes the efficient and clean use of energy worldwide.

American Council for an Energy-Efficient Economy (ACEEE)

A nonprofit that acts as a catalyst to advance energy efficiency policies, programs, technologies, investments, and behaviors. Projects are carried out by ACEEE staff and collaborators from government, the private sector, research institutions, and other nonprofit organizations. Their publication *Consumer Guide to Home Energy Savings* is available is two formats:

Online guide: Consumer Guide to Home Energy Savings Online. Retrieved June 10, 2015, from http://aceee.org/consumer-guide-home-energy-savings-online

Co-publisher with New Society Publishers of the hard copy version:
Amann, J. T., Wilson, A., & Ackerly, K. (2012). *Consumer Guide to Home Energy Savings* (Tenth edition). Gabriola Island, BC: New Society Publishers.

Appliance Standard Awareness Project (ASAP). Retrieved August 18, 2015, from http://www.appliance-standards.org/

Seeks to develop awareness of appliance standards and their benefits.

Association of Home Appliance Manufacturers (AHAM). (n.d.). Consumer Information. Retrieved August 18, 2015, from http://www.aham.org/consumer/

Find advice on the use, maintenance, and repair of major appliances.

California Energy Commission. (n.d.). Appliance Efficiency Program. Retrieved August 18, 2015, from http://www.energy.ca.gov/appliances/

Directory of appliances by brand and model number that exceed California and NAECA standards by at least 10 percent.

Consortium for Energy Efficiency (CEE). (n.d.). CEE Program Resources Residential. Retrieved August 17, 2015, from http://www.cee1.org/content/cee-program-resources

A U.S. and Canadian consortium of gas and electric efficiency program administrators working together to accelerate the development and availability of energy efficient products and services for lasting public benefit.

Consumer Federation of America (CFA). Retrieved August 18, 2015, from http://www.consumerfed.org/

Threefold mission: to assist state and local organizations, to provide information to the public on consumer issues, and to conduct research projects.

Consumer Reports. (n.d.). Consumer Reports Online. Retrieved August 18, 2015, from http://www.consumerreports.org/cro/homepage1/index.htm

Unbiased evaluations of thousands of products based on Consumer Reports' expert testing.

Federal Energy Management Program (FEMP). (n.d.). Energy- and Water-Efficient Products. Retrieved August 18, 2015, from http://energy.gov/eere/femp/energy-and-water-efficient-products

Provides resources and tools to help agencies purchase energy- and water-efficient products.

Federal Trade Commission (FTC). (n.d.). Consumer Information. Retrieved August 18, 2015, from http://www.consumer.ftc.gov/

Consumer information on a wide variety of topics and products.

Green Seal. (n.d.). Green Seal. Retrieved August 18, 2015, from http://www.greenseal.org/

An independent, nonprofit organization that certifies different products that meet energy efficient guidelines, as well as other environmental criteria.

Multihousing Laundry Association (MLA). Retrieved August 18, 2015, from http://www.mla-online.com/

Issues relating to multi-housing laundry operations including laundry room guide recommendations.

National Association of State Energy Officials (NASEO). Retrieved August 18, 2015, from http://www.naseo.org/

Affiliated with the National Governors' Association, this nonprofit association improves the effectiveness and quality of state energy programs and policies, and serves as a collector and repository of energy-related information.

Lighting

Building Green. (n.d.). GreenSpec: Electrical (including lighting). Retrieved November 22, 2017, from www.buildinggreen.com/lighting-design

Ecova. (2012). *Best-in-Class LED Reflector Lamps Summary Report* (IEE Whitepaper). Washington, DC: Institute for Electric Efficiency. Retrieved from http://www.edisonfoundation.net/iee/documents/iee_ecova_led.pdf

Edison Tech Center. August 18, 2015, from http://www.edisontechcenter.org/

(n.d.-a). Induction Lamps. Retrieved August 18, 2015, from http://www.edisontechcenter.org/InductionLamps.html

(n.d.-b). The Sodium Lamp - How it works and history. Retrieved August 18, 2015, from http://www.edisontechcenter.org/SodiumLamps.html

Energy Information Administration (EIA). (2013). Household Energy Use in Florida, A closer look at residential energy consumption (Residential Energy Consumption Survey (RECS) 2009). Retrieved from http://www.eia.gov/consumption/residential/reports/2009/state_briefs/pdf/fl.pdf

Federal Trade Commission (FTC). Energy Conservation Program for Consumer Products. 10 C.F.R. § 430. Retrieved from http://www.ecfr.gov/cgi-bin/text-idx?SID=553b868651c7d1b7a5f1cd6dbd90f50f&mc=true&node=pt10.3.430&rgn=div5

Illuminating Engineering Society (IES). Retrieved August 18, 2015, from http://www.ies.org/

International Dark-Sky Association. Retrieved August 18, 2015, from http://darksky.org/

Lighting Research Center (LRC), Rensselaer Polytechnic Institute. (n.d.). National Lighting Product Information Program (NLPIP). Retrieved August 18, 2015, from http://www.lrc.rpi.edu/nlpip/

Lighting Understanding for a More Efficient Nation (LUMEN). (n.d.). LUMEN Coalition. Retrieved August 18, 2015, from http://lumennow.org/

Livingston, J. 2014. *Designing with Light: The Art, Science, and Practice of Architectural Lighting Design.* 1st edition. Hoboken, NJ: Wiley.

National Lighting Product Information Program. (2000). "Electronic Ballasts: Non-dimming electronic ballasts for 4-foot and 8-foot fluorescent lamps." *Specifier Reports*, 8(1). Retrieved from http://www.lrc.rpi.edu/programs/NLPIP/PDF/VIEW/SREB2.pdf

The ENERGY STAR Choose A Light Guide (an interactive guide). (n.d.). Retrieved August 18, 2015, from http://www.drmediaserver.com/CFLGuide/index.html

University of Florida. (n.d.). Energy Education Library | My Florida Home Energy. Retrieved from http://www.myfloridahomeenergy.com/help/library

Helpful support from the My Florida Home Energy website: the Energy Efficient Home series of fact sheets, covering a wide ranges of topics on energy and water efficiency

U.S. Department of Energy (DOE):

(n.d.-a). Building Technologies Office. Retrieved August 18, 2015, from http://energy.gov/eere/efficiency/buildings

(n.d.-b). Lighting Basics. Retrieved August 18, 2015, from http://energy.gov/eere/energy-basics/articles/lighting-basics

(n.d.-c). Solid-State Lighting. Retrieved August 18, 2015, from http://energy.gov/eere/ssl/solid-state-lighting

U.S. Department of Energy (DOE) Energy Efficiency and Renewable Energy, Building Technologies Program. (2013). *Energy Efficiency of LEDs* (Solid-State Lighting Technology Fact Sheet No. PNNL-SA-94206). Retrieved November 22, 2017, from http://www.hi-led.eu/wp-content/themes/hiled/pdf/led_energy_efficiency.pdf

U.S. Department of Energy (DOE) Energy Efficiency & Renewable Energy, Building Technologies Office:

(2013). *Light Bulbs* (CALiPER Snapshot Report No. PNNL-SA-99597). Retrieved from http://apps1.eere.energy.gov/buildings/publications/pdfs/ssl/snapshot2013_a-lamp.pdf

(2014). *Indoor LED Luminaires* (CALiPER Snapshot Report No. PNNL-SA-102823). Retrieved from http://apps1.eere.energy.gov/buildings/publications/pdfs/ssl/snapshot2014_indoor-luminaires.pdf

U.S. Department of Energy (DOE) Energy Saver:

(n.d.-a). Energy Saver Guide. Retrieved August 18, 2015, from http://energy.gov/energysaver/downloads/energy-saver-guide

(n.d.-b). Lighting. Retrieved August 18, 2015, from http://energy.gov/public-services/homes/saving-electricity/lighting

U.S. Department of Energy (DOE), Office of Energy Efficiency and Renewable Energy (EERE):

(2013). Study: Environmental Benefits of LEDs Greater Than CFLs. Retrieved August 18, 2015, from http://energy.gov/eere/articles/study-environmental-benefits-leds-greater-cfls

(n.d.). Appliance and Equipment Standards Program. Retrieved July 20, 2015, from http://energy.gov/eere/buildings/appliance-and-equipment-standards-program

U.S. Environmental Protection Agency (EPA). (2011). *Energy Independence and Security Act of 2007 (EISA) Frequently Asked Questions* (U.S. EPA Backgrounder). U.S. Environmental Protection Agency (EPA). Retrieved from http://www.energystar.gov/ia/products/lighting/cfls/downloads/EISA_Backgrounder_FINAL_4-11_EPA.pdf

U.S. Environmental Protection Agency (EPA) ENERGY STAR. Retrieved August 18, 2015, from https://www.energystar.gov/

(n.d.-a). 2014 ENERGY STAR Certified Homes Market Share. Retrieved August 18, 2015, from http://public.tableau.com/views/ENERGYSTARMarketShare2014/ENERGYSTARMarketShare2014?:embed=y&:showVizHome=no&:host_url=https://public.tableau.com/&:tabs=no&:toolbar=yes&:animate_transition=yes&:display_static_image=yes&:display_spinner=yes&:display_overlay=yes&:display_count=yes&:-showVizHome=no&:showTabs=y&:loadOrderID=0

(n.d.-b). Learn About CFLs. Retrieved August 18, 2015, from http://www.energystar.gov/index.cfm?c=cfls.pr_cfls_about#how_work

(n.d.-c). Learn About LED Bulbs. Retrieved August 18, 2015, from http://www.energystar.gov/index.cfm?c=lighting.pr_what_are#what_are

(n.d.-d). Light Bulbs for Consumers. Retrieved August 18, 2015, from http://www.energystar.gov/products/certified-products/detail/light-bulbs

(n.d.-e). Lighting Made Easy (Fact Sheet). Retrieved from http://www.energystar.gov/ia/products/fap/purchasing_checklist_revised.pdf

(n.d.-f). Who's Who: Labeling and Certification for Lighting Products. Retrieved from https://www.energystar.gov/ia/partners/manuf_res/Whos_Who_Labeling_and_Certification_for_Lighting_Products.pdf

Appendix I: Mortgage Rate Tables

The following tables show the monthly payment for principal and interest for a $1,000 loan at various interest rates and amortization periods. For example, a $50,000 loan at 13% with a 25-year amortization period will have monthly payments of $11.28 × 50 = $564.00. This table is useful in comparing different methods of financing construction loans and permanent mortgages and their effect on the economics of energy efficient construction techniques.

Interest Rate

Years of Amortization	2.00	2.25	2.50	2.75	3.00	3.25	3.50	3.75	4.00	4.25	4.50	4.75
1	84.24	84.35	84.47	84.58	84.69	84.81	84.92	85.04	85.15	85.26	85.38	85.49
2	42.54	42.65	42.76	42.87	42.98	43.09	43.20	43.31	43.42	43.54	43.65	43.76
3	28.64	28.75	28.86	28.87	29.08	29.19	29.30	29.41	29.52	29.64	29.75	29.86
4	21.70	21.80	21.91	22.02	22.13	22.24	22.36	22.47	22.58	22.69	22.80	22.92
5	17.53	17.64	17.75	17.86	17.97	18.08	18.19w	18.30	18.42	18.53	18.64	18.76
6	14.75	14.68	14.97	15.08	15.19	15.31	15.42	1w5.53	15.65	15.76	15.87	15.99
7	12.77	12.88	12.99	13.10	13.21	13.33	13.44	13.55	13.67	13.78	13.90	14.02
8	11.28	11.39	11.50	11.62	11.73	11.84	11.96	12.07	12.19	12.31	12.42	12.54
9	10.13	10.24	10.35	10.46	10.58	10.69	10.81	10.92	11.04	11.16	11.28	11.40
10	9.20	9.31	9.43	9.54	9.66	9.77	9.89	10.01	10.12	10.24	10.36	10.48
11	8.45	8.56	8.67	8.79	8.90	9.02	9.14	9.26	9.38	9.50	9.62	9.74
12	7.82	7.93	8.05	8.16	8.28	8.40	8.51	8.63	8.76	8.88	9.00	9.12
13	7.28	7.40	7.51	7.63	7.75	7.87	7.99	8.11	8.23	8.35	8.48	8.60
14	6.83	6.94	7.06	7.18	7.30	7.42	7.54	7.66	7.78	7.91	8.03	8.16
15	6.44	6.55	6.67	6.79	6.91	7.03	7.15	7.27	7.40	7.52	7.65	7.78
17	5.79	5.90	6.02	6.14	6.26	6.39	6.51	6.64	6.76	6.89	7.02	7.15
20	5.06	5.18	5.30	5.42	5.55	5.67	5.80	5.93	6.06	6.19	6.33	6.46
25	4.24	4.36	4.49	4.61	4.74	4.87	5.01	5.14	5.28	5.42	5.56	5.70
30	3.70	3.82	3.95	4.08	4.22	4.35	4.49	4.63	4.77	4.92	5.07	5.22

Interest Rate

Years of Amortization	5.00	5.25	5.50	5.75	6.00	6.25	6.50	6.75	7.00	7.25	7.50	7.75
1	85.61	85.72	85.84	85.95	86.07	86.18	86.30	86.41	86.53	86.64	86.76	86.87
2	43.87	43.98	44.10	44.21	44.32	44.43	44.55	44.66	44.77	44.89	45.00	45.11
3	29.97	30.08	30.20	30.31	30.42	30.54	30.65	30.76	30.88	30.99	31.11	31.22
4	23.03	23.14	23.26	23.37	23.49	23.60	23.71	23.83	23.95	24.06	24.18	24.30
5	18.87	18.99	19.10	19.22	19.33	19.45	19.57	19.68	19.80	19.92	20.04	20.16
6	16.10	16.22	16.34	16.46	16.57	16.69	16.81	16.93	17.05	17.17	17.29	17.41
7	14.13	14.25	14.37	14.49	14.61	14.73	14.85	14.97	15.09	15.22	15.34	15.46
8	12.66	12.78	12.90	13.02	13.14	13.26	13.39	13.51	13.63	13.76	13.88	14.01
9	11.52	11.64	11.76	11.88	12.01	12.13	12.25	12.38	12.51	12.63	12.76	12.89
10	10.61	10.73	10.85	10.98	11.10	11.23	11.35	11.48	11.61	11.74	11.87	12.00
11	9.86	9.99	10.11	10.24	10.37	10.49	10.62	10.75	10.88	11.02	11.15	11.28
12	9.25	9.37	9.50	9.63	9.76	9.89	10.02	10.15	10.28	10.42	10.55	10.69
13	8.73	8.86	8.99	9.12	9.25	9.38	9.51	9.65	9.78	9.92	10.05	10.19
14	8.29	8.42	8.55	8.68	8.81	8.95	9.08	9.22	9.35	9.49	9.63	9.77
15	7.91	8.04	8.17	8.30	8.44	8.57	8.71	8.85	8.99	9.13	9.27	9.41
17	7.29	7.42	7.56	7.69	7.83	7.97	8.11	8.25	8.40	8.54	8.69	8.83
20	6.60	6.74	6.88	7.02	7.16	7.31	7.46	7.60	7.75	7.90	8.06	8.21
25	5.85	5.99	6.14	6.29	6.44	6.60	6.75	6.91	7.07	7.23	7.39	7.55
30	5.37	5.52	5.68	5.84	6.00	6.16	6.32	6.49	6.65	6.82	6.99	7.16

Interest Rate

Years of Amortization	8.00	8.25	8.50	8.75	9.00	9.25	9.50	9.75	10.00	10.25	10.50	10.75
1	86.99	87.10	87.22	87.34	87.45	87.57	87.68	87.80	87.92	88.03	88.15	88.27
2	45.23	45.34	45.46	45.57	45.68	45.80	45.91	46.03	46.14	46.26	46.38	46.49
3	31.34	31.45	31.57	31.68	31.80	31.92	32.03	32.15	32.27	32.38	32.50	32.62
4	24.4	24.53	24.65	24.77	24.89	25.00	25.12	25.24	25.36	25.48	25.60	25.72
5	20.28	20.40	20.52	20.64	20.76	20.88	21.00	21.12	21.25	21.37	21.49	21.62
6	17.53	17.66	17.78	17.90	18.03	18.15	18.27	18.40	18.53	18.65	18.78	18.91
7	15.59	15.71	15.84	15.96	16.09	16.22	16.34	16.47	16.60	16.73	16.86	16.99
8	14.14	14.26	14.39	14.52	14.65	14.78	14.91	15.04	15.17	15.31	15.44	15.57
9	13.02	13.15	13.28	13.41	13.54	13.68	13.81	13.94	14.08	14.21	14.35	14.49
10	12.13	12.27	12.40	12.53	12.67	12.80	12.94	13.08	13.22	13.35	13.49	13.63
11	11.42	11.55	11.69	11.82	11.96	12.10	12.24	12.38	12.52	12.66	12.80	12.95
12	10.82	10.96	11.10	11.24	11.38	11.52	11.66	11.81	11.95	12.10	12.24	12.39
13	10.33	10.47	10.61	10.75	10.90	11.04	11.19	11.33	11.48	11.63	11.78	11.92
14	9.91	10.06	10.20	10.34	10.49	10.64	10.78	10.93	11.08	11.23	11.38	11.54
15	9.56	9.70	9.85	9.99	10.14	10.29	10.44	10.59	10.75	10.90	11.05	11.21
17	8.98	9.13	9.28	9.43	9.59	9.74	9.90	10.05	10.21	10.37	10.53	10.69
20	8.36	8.52	8.68	8.84	9.00	9.16	9.32	9.49	9.65	9.82	9.98	10.15
25	7.72	7.88	8.05	8.22	8.39	8.56	8.74	8.91	9.09	9.26	9.44	9.62
30	7.34	7.51	7.69	7.87	8.05	8.23	8.41	8.59	8.78	8.96	9.15	9.33

Interest Rate

Years of Amortization	11.00	11.25	11.50	11.75	12.00	12.25	12.50	12.75	13.00	13.25	13.50	13.75
1	88.38	88.50	88.62	88.73	88.85	88.97	89.08	89.20	89.32	89.43	89.55	89.67
2	46.61	46.72	46.84	46.96	47.07	47.19	47.31	47.42	47.54	47.66	47.78	47.89
3	32.74	32.86	32.98	33.10	33.21	33.33	33.45	33.57	33.69	33.81	33.94	34.06
4	25.85	25.97	26.09	26.21	26.33	26.46	26.58	26.70	26.83	26.95	27.08	27.20
5	21.74	21.87	21.99	22.12	22.24	22.37	22.50	22.63	22.75	22.88	23.01	23.14
6	19.03	19.16	19.29	19.42	19.55	19.68	19.81	19.94	20.07	20.21	20.34	20.47
7	17.12	17.25	17.39	17.52	17.65	17.79	17.92	18.06	18.19	18.33	18.46	18.60
8	15.71	15.84	15.98	16.12	16.25	16.39	16.53	16.67	16.81	16.95	17.09	17.23
9	14.63	14.76	14.90	15.04	15.18	15.33	15.47	15.61	15.75	15.90	16.04	16.19
10	13.78	13.92	14.06	14.20	14.35	14.49	14.64	14.78	14.93	15.08	15.23	15.38
11	13.09	13.24	13.38	13.53	13.68	13.83	13.98	14.13	14.28	14.43	14.58	14.73
12	12.54	12.68	12.83	12.98	13.13	13.29	13.44	13.59	13.75	13.90	14.06	14.21
13	12.08	12.23	12.38	12.53	12.69	12.84	13.00	13.15	13.31	13.47	13.63	13.79
14	11.69	11.85	12.00	12.16	12.31	12.47	12.63	12.79	12.95	13.11	13.28	13.44
15	11.37	11.52	11.68	11.84	12.00	12.16	12.33	12.49	12.65	12.82	12.98	13.15
17	10.85	11.02	11.18	11.35	11.51	11.68	11.85	12.02	12.19	12.36	12.53	12.70
20	10.32	10.49	10.66	10.84	11.01	11.19	11.36	11.54	11.72	11.89	12.07	12.25
25	9.80	9.98	10.16	10.35	10.53	10.72	10.90	11.09	11.28	11.47	11.66	11.85
30	9.52	9.71	9.90	10.09	10.29	10.48	10.67	10.87	11.06	11.26	11.45	11.65

Interest Rate

Years of Amortization	14.00	14.25	14.50	14.75	15.00	15.25	15.50	15.75	16.00	16.25	16.50	16.75
1	89.79	89.90	90.02	90.14	90.26	90.38	90.49	90.61	90.73	90.85	90.97	91.09
2	48.01	48.13	48.25	48.37	48.49	48.61	48.72	48.84	48.96	49.08	49.20	49.32
3	34.18	34.30	34.42	34.54	34.67	34.79	34.91	35.03	35.16	35.28	35.40	35.53
4	27.33	27.45	27.58	27.70	27.83	27.96	28.08	28.21	28.34	28.47	28.60	28.73
5	23.27	23.40	23.53	23.66	23.79	23.92	24.05	24.19	24.32	24.45	24.58	24.72
6	20.61	20.74	20.87	21.01	21.15	21.28	21.42	21.55	21.69	21.83	21.97	22.11
7	18.74	18.88	19.02	19.16	19.30	19.44	19.58	19.72	19.86	20.00	20.15	20.29
8	17.37	17.51	17.66	17.80	17.95	18.09	18.24	18.38	18.53	18.68	18.82	18.97
9	16.33	16.48	16.63	16.78	16.92	17.07	17.22	17.37	17.53	17.68	17.83	17.98
10	15.53	15.68	15.83	15.98	16.13	16.29	16.44	16.60	16.75	16.91	17.06	17.22
11	14.89	15.04	15.20	15.35	15.51	15.67	15.82	15.98	16.14	16.30	16.46	16.63
12	14.37	14.53	14.69	14.85	15.01	15.17	15.33	15.49	15.66	15.82	15.99	16.15
13	13.95	14.11	14.28	14.44	14.60	14.77	14.93	15.10	15.27	15.43	15.60	15.77
14	13.60	13.77	13.94	14.10	14.27	14.44	14.61	14.78	14.95	15.12	15.29	15.46
15	13.32	13.49	13.66	13.83	14.00	14.17	14.34	14.51	14.69	14.86	15.04	15.21
17	12.87	13.05	13.22	13.40	13.58	13.75	13.93	14.11	14.29	14.47	14.65	14.84
20	12.44	12.62	12.80	12.98	13.17	13.35	13.54	13.73	13.91	14.10	14.29	14.48
25	12.04	12.23	12.42	12.61	12.81	13.00	13.20	13.39	13.59	13.79	13.98	14.18
30	11.85	12.05	12.25	12.44	12.64	12.84	13.05	13.25	13.45	13.65	13.85	14.05

Interest Rate

Years of Amortization	17.00	17.25	17.50	17.75	18.00	18.25	18.50	18.75	19.00	19.25	19.50	19.75
1	91.20	91.32	91.44	91.56	91.68	91.80	91.92	92.04	92.16	92.28	92.40	92.51
2	49.44	49.56	49.68	49.80	49.92	50.04	50.17	50.29	50.41	50.53	50.65	50.77
3	35.65	35.78	35.90	36.03	36.15	36.28	36.40	36.53	36.66	36.78	36.91	37.04
4	28.86	28.98	29.11	29.24	29.37	29.51	29.64	29.77	29.90	30.03	30.16	30.30
5	24.85	24.99	25.12	25.26	25.39	25.53	25.67	25.80	25.94	26.08	26.22	26.35
6	22.25	22.39	22.53	22.67	22.81	22.95	23.09	23.23	23.38	23.52	23.66	23.81
7	20.44	20.58	20.73	20.87	21.02	21.16	21.31	21.46	21.61	21.76	21.91	22.06
8	19.12	19.27	19.42	19.57	19.72	19.88	20.03	20.18	20.33	20.49	20.64	20.80
9	18.14	18.29	18.45	18.60	18.76	18.91	19.07	19.23	19.39	19.55	19.71	19.87
10	17.38	17.54	17.70	17.86	18.02	18.18	18.34	18.50	18.67	18.83	19.00	19.16
11	16.79	16.95	17.11	17.28	17.44	17.61	17.78	17.94	18.11	18.28	18.45	18.62
12	16.32	16.49	16.65	16.82	16.99	17.16	17.33	17.50	17.67	17.85	18.02	18.19
13	15.94	16.11	16.29	16.46	16.63	16.80	16.98	17.15	17.33	17.50	17.68	17.86
14	15.64	15.81	15.99	16.16	16.34	16.52	16.69	16.87	17.05	17.23	17.41	17.59
15	15.39	15.57	15.75	15.92	16.10	16.28	16.47	16.65	16.83	17.01	17.19	17.38
17	15.02	15.20	15.39	15.57	15.76	15.94	16.13	16.32	16.50	16.69	16.88	17.07
20	14.67	14.86	15.05	15.24	15.43	15.63	15.82	16.01	16.21	16.40	16.60	16.79
25	14.38	14.58	14.78	14.97	15.17	15.37	15.57	15.78	15.98	16.18	16.38	16.58
30	14.26	14.46	14.66	14.86	15.07	15.28	15.48	15.68	15.89	16.09	16.30	16.50

Appendix II: Fingertip Facts

This section contains statistical energy information—conversion factors, R-values, fuel prices, and energy efficiency recommendations. It serves as a reference guide for those seeking a quick answer to an energy question.

Abbreviations

Btu British Thermal Unit, the amount of heat needed to increase the temperature of one pound of water one degree Fahrenheit (about the amount of heat released when a kitchen match burns)

1° F	one degree Fahrenheit	**cf**	cubic foot
MMBtu	million Btu	**cfm**	cubic foot per minute
kWh	kilowatt-hour	**bbl**	barrel
kW	kilowatt	**gal**	gallon

Energy and Fuel Data

Energy Units

1 kWh	=	3,412 Btu
1 MMBtu	=	293 kWh
1 Btu	=	252 calories
1 Btu	=	1,055 joules

Power Units

1 watt	=	3.412 Btu/hour
1 kW	=	3,412 Btu/hour
1 horsepower	=	746 watts
1 ton of heating/cooling	=	12,000 Btu/hour

Fuel Units

1 cf of natural gas	=	1,000 Btu
1 therm	=	100,000 Btu
1 bbl fuel oil	=	42 gallons
1 bbl fuel oil	=	5.8 MMBtu
1 ton fuel oil	=	6.8 bbl
1 gallon fuel oil	=	136,000 Btu
1 gallon propane	=	91,500 Btu
1 ton bituminous (Eastern) coal	=	21–26 MMBtu
1 ton sub-bituminous (Western) coal	=	14–18 MMBtu
1 cord wood	=	128 cubic feet (4 ft × 4 ft × 8 ft)
1 cord dried oak	=	23.9 MMBtu
1 cord dried pine	=	14.2 MMBtu

HVAC Equipment Efficiencies

Annual Fuel Utilization Efficiency (AFUE)
Shows the average annual efficiency at which fuel-burning furnaces operate.

Coefficient of Performance (COP)
Measures how many units of heating or cooling are delivered for every unit of electricity used in a heat pump or air conditioner.

Heating Season Performance Factor (HSPF)
Measures the average number of Btu of heating delivered for every watt-hour of electricity used by a heat pump.

Seasonal Energy Efficiency Ratio (SEER)
Measures how readily air conditioners convert electricity into cooling—a SEER of 13 means the unit provides 13 Btu's of cooling per watt-hour of electricity.

INSULATING VALUES

The R-value is the measure of resistance to heat flow via conduction. R-values *vary* according to specific materials and installation.

Material	R-value
Insulation	**R-value per inch**
Fiber glass batts/rolls	3.2
Fiber glass loose-fill	2.2
Rock wool loose-fill	2.6
Cellulose	3.7
Vermiculite	2.1
Perlite	3.3
Polyicynene spray	3.6
Rigid Insulation Boards	
Fiberboard sheathing (noninsulating blackboard)	2.6
Expanded polystyrene (beadboard)	4.0
Extruded polystyrene	5.0
Polyisocyanurate and polyurethane	7.2
Building Materials	
Drywall	.9
Wood siding	.9 to 1.2
Common brick	.2
Lumber and siding	
Hardwood	.8 to .94
Softwood	.9 to 1.5
Plywood	1.3
Particle Board (medium density)	1.1
Asbestos-cement (entire shingle)	.21
Building Materials	**Total R-value**
Concrete block (entire block)	
Unfilled	1.3 to 1.7
Filled with perlite/vermiculite	2.3 to 3.7
Filled with cement mortar	1.6
Dead Air Spaces	**R-value of air space**
½-inch	.75
¾-inch	.77
3½-inch	.80
3½-inch, reflecting surface on one side	1.6
3½-inch, reflecting surface both sides	2.2

APPENDICES

Appendix III: Chapter 8 Notes

Fuel Cost Conversions

Different energy sources are used to operate our homes and businesses. As prices fluctuate some method must be found to compare these sources on an equal basis. The most popular and easiest method is to convert each energy source to a dollar cost per million BTUs (MBTU). Our home utility bills show the following energy sources purchased in these customary units:

- Electricity is purchased in units of dollars per kilowatt hour (kWh)
- Propane is bought by the gallon
- Natural gas is sold by the therm

How do we convert each of these fuels to a cost per million BTU basis?

Skipping the math...

1. Multiply dollars per kWh by **293** to obtain electricity costs in dollars per MBTU
2. Multiply dollars per gallon by **10.9** to obtain propane costs in dollars per MBTU
3. Multiply dollars per therm by **10** to obtain natural gas costs in dollars per MBTU

Examples (be sure to use the prices in your area as these prices may fluctuate greatly):

- I pay approximately $0.15/kWh (including taxes and other fees) for electricity. What is the cost in MBTUs?

  $0.15/kWh × 293 = $43.95 per MBTU

- I pay approximately $3.00/gallon for propane. What is the cost in MBTU's?

  $3.00/gallon × 10.9 = $32.70 per MBTU

- Let's assume natural gas can be bought for approximately $1.20/therm (outside city).

  $1.20/therm × 10 = $12.00 per MBTU

This conversion clearly shows that electrical energy is the most costly form of energy and natural gas is the least costly. *However,* this comparison does not say how much each of these energy sources costs for what we could call "delivered energy". Let's use the following assumptions*:

- Electric water heaters can deliver approximately 92% of the electrical energy that is input to the water heater.
- Natural gas or propane water heaters can be 52– 60% efficient (60% is used in the following example).
- Electric heat pumps deliver 2.2 to 2.6 units of heat for every unit consumed (2.2 is used in the example).
- Natural gas furnaces can be anywhere from 60% efficient to 96% efficient, depending on age and type of furnace (85% is used in the example).

Using these numbers, let's calculate the cost of delivered energy for the initial costs listed above:

- Water Heating

Natural Gas:	$/MBTU delivered	= $12.00/MBTU/0.60	= $20.00 per MBTU
Propane:	$/MBTU delivered	= $32.70/MBTU/0.60	= $54.50 per MBTU
Electric:	$/MBTU delivered	= $43.95/MBTU/0.92	= $47.77 per MBTU

- Space Heating

Heat pump:	$/MBTU delivered	= $43.95/MBTU/2.2	= $19.97 per MBTU
Natural Gas:	$/MBTU delivered	= $12.00/MBTU/0.85	= $14.11 per MBTU
Propane:	$/MBTU delivered	= $32.70/MBTU/0.85	= $38.47 per MBTU
Strip Heat:	$/MBTU delivered	= $43.95/MBTU	

*Note: This analysis is meant to be a general guideline only. It does not include fuel source conversion factors that are dependent on the origin of the specific fossil fuels. The efficiencies listed for the combustion devices are indicative of well maintained, fairly up-to-date devices. Older, less maintained items can be much less efficient.

Requirements for AHU's Located in Attic Spaces

FBC, Energy Conservation, Section 403.2.4 Air-handling units (Additional requirements for location in attics)

- AHU service panel is less than 6 feet from an attic access
- Condensation drain alarm is required
- Attic access sized for replacement of AHU
- Notice that the AHU is in the attic, is required to be posted on the electric service panel. As per code:

NOTICE TO HOMEOWNER

A PART OF YOUR AIR CONDITIONING SYSTEM, THE AIR HANDLER, IS LOCATED IN THE ATTIC. FOR PROPER, EFFICIENT, AND ECONOMIC OPERATION OF THE AIR CONDITIONING SYSTEM, YOU MUST ENSURE THAT REGULAR MAINTENANCE IS PERFORMED.

YOUR AIR CONDITIONING SYSTEM IS EQUIPPED WITH ONE OR MORE OF THE FOLLOWING: 1) A DEVICE THAT WILL WARN YOU WHEN THE CONDENSATION DRAIN IS NOT WORKING PROPERLY OR 2) A DEVICE THAT WILL SHUT THE SYSTEM DOWN WHEN THE CONDENSATION DRAIN IS NOT WORKING. TO LIMIT POTENTIAL DAMAGE TO YOUR HOME, AND TO AVOID DISRUPTION OF SERVICE, IT IS RECOMMENDED THAT YOU ENSURE PROPER WORKING ORDER OF THESE DEVICES BEFORE EACH SEASON OF PEAK OPERATION.

Index

A

B

C

D

E

F

H

I

L

T

U

V

W

